Advisory Board

ACS Symposium Series

Foreword

THE ACS SYMPOSIUM SERIES was first published in 1974 to provide a mechanism for publishing symposia quickly in book form. The purpose of this series is to publish comprehensive books developed from symposia, which are usually "snapshots in time" of the current research being done on a topic, plus some review material on the topic. For this reason, it is necessary that the papers be published as quickly as possible.

Before a symposium-based book is put under contract, the proposed table of contents is reviewed for appropriateness to the topic and for comprehensiveness of the collection. Some papers are excluded at this point, and others are added to round out the scope of the volume. In addition, a draft of each paper is peer-reviewed prior to final acceptance or rejection. This anonymous review process is supervised by the organizer(s) of the symposium, who become the editor(s) of the book. The authors then revise their papers according to the recommendations of both the reviewers and the editors, prepare camera-ready copy, and submit the final papers to the editors, who check that all necessary revisions have been made.

As a rule, only original research papers and original review papers are included in the volumes. Verbatim reproductions of previously published papers are not accepted.

ACS BOOKS DEPARTMENT

Contents

DIMERS, OLIGOMERS, AND CLUSTERS

EXTENDED ORGANIC SYSTEMS

with tutorial papers on the theory of magnetic exchange and the various techniques by which the magnetic properties of these materials may be investigated. Subsequent sections are devoted to finite systems (organic and metal-based dimer, oligomer, and cluster molecules), extended organic systems, and extended metal-based systems. Here, the distinction between organic and metal-based systems is made by the source of the magnetic moment in the sample. Each of these sections begins with tutorial chapters encompassing the fundamentals specific to that area followed by chapters relating recent results.

We hope that this book will be of interest to a broad audience. Scientists working with organic radical systems or paramagnetic transition or lanthanide metals will find a general overview of molecular magnetism that may suggest new directions for their studies or applications for their materials. Scientists joining established molecular magnetism groups will find that the chapters relating to exchange theory and measurement techniques provide a useful introduction to their studies. Finally, the book will provide a reference on the potential applications of magnetic materials for workers in related areas of materials science such as organic and inorganic superconductors, low-dimensional systems, and magneto-optics.

Acknowledgments

We are grateful to all those who have supported the production of this book and the symposium upon which it is based. Acknowledgment is made to the Donors of The Petroleum Research Fund, administered by the American Chemical Society, for partial support of the symposium. Financial support from Lake Shore Cryotronics, Quantum Design, and the ACS Division of Inorganic Chemistry, Inc., is also gratefully acknowledged. We also appreciate the assistance of Kunio Awaga and Hisashi Okawa in the organization of the symposium. Finally, we thank all those who have contributed their time and efforts to the work described in this book.

TOYONARI SUGIMOTO
University of Osaka Prefecture
Sakai, Osaka 593, Japan

MARK M. TURNBULL
Clark University
Worcester, MA 01610

LAURENCE K. THOMPSON
Memorial University of Newfoundland
St. John's, Newfoundland A1B 3X7, Canada

June 3, 1996

Preface

THE FIELD OF MOLECULAR MAGNETISM began in 1983 at a NATO Advanced Study Institute two-week summer school in Castigleone della Pescaia, Italy (Magneto-Structural Correlations in Exchange Coupled Systems). Organized by Olivier Kahn, Dante Gatteschi, and Roger Willett, it was devoted to the study of magnetic interactions in insulators. The school was designed to bring together chemists and physicists with common interests, but differing backgrounds, and was therefore of a tutorial nature. A second NATO conference (Organic and Inorganic Low-Dimensional Crystalline Materials, Minorca, Spain, 1987) strengthened this interdisciplinary field by bringing together the emerging "magneto-chemists" with the organic conductors community. Since that time, there has been a growing interest in the field as judged by the number of national and international conferences dedicated to the subject.

As a result of these meetings, a solid core of chemists and physicists working in the theory, design, and study of these new classes of molecule-based magnetic materials has grown over the past twelve years. The focus of their work continues to be understanding the relationship between structure at the molecular level and the resulting magnetic properties of the bulk substance, with the objective of designing new and better magnetic materials. Molecular magnetism has become so well established that "outsiders" working in fringe areas, who might otherwise be daunted by the breadth of expertise required to make a contribution, are now taking a serious interest in the field.

Molecular magnetism also lends itself to, or perhaps even requires, collaborative work largely because of the dearth of magnetometers, particularly in chemistry departments. For the field to continue its development, "membership" needs to be expanded further. Thus, a symposium titled "Molecule-Based Magnetic Materials" was sponsored by the 1995 International Chemical Congress of Pacific Basin Societies (Pacifichem '95) and the ACS Division of Inorganic Chemistry, Inc., on December 17–22, 1995, in Honolulu, Hawaii. The symposium presented a mixture of tutorial talks on the various theoretical and practical aspects of work in molecular magnetism along with lectures describing the most recent developments in the field. The goal of the symposium was to present an introduction to the field that showed both its fundamentals and its excitement.

This book has the same purpose. The first section deals primarily

ACS SYMPOSIUM SERIES **644**

Molecule-Based Magnetic Materials

Theory, Techniques, and Applications

Mark M. Turnbull, Editor
Clark University

Toyonari Sugimoto, Editor
University of Osaka Prefecture

Laurence K. Thompson, Editor
Memorial University of Newfoundland

Developed from a symposium sponsored
by the International Chemical Congress of Pacific Basin Societies
and the ACS Division of Inorganic Chemistry, Inc., at the
1995 International Chemical Congress of Pacific Basin Societies

American Chemical Society, Washington, DC

Library of Congress Cataloging-in-Publication Data

Molecule-based magnetic materials: theory, techniques, and applications /
Mark M. Turnbull, Toyonari Sugimoto, Laurence K. Thompson, editors.

p. cm.—(ACS symposium series, ISSN 0097–6156; 644)

"Developed from a symposium sponsored by the International Chemical Congress of Pacific Basin Societies and the ACS Division of Inorganic Chemistry, Inc., at the 1995 International Chemical Congress of Pacific Basin Societies, Honolulu, Hawaii, December 17–22, 1995."

Includes bibliographical references and indexes.

ISBN 0–8412–3452–3

1. Molecular crystals—Magnetic properties—Congresses.
2. Molecules—Magnetic properties—Congresses.

I. Turnbull, Mark M., 1956– . II. Sugimoto, Toyonari, 1945– .
III. Thompson, Laurence K., 1943– . IV. American Chemical Society.
Division of Inorganic Chemistry. V. International Chemical Congress of Pacific Basin Societies (1995: Honolulu, Hawaii) VI. Series.

QD940.M66 1996
548′.85—dc20 96–29235
CIP

This book is printed on acid-free, recycled paper.

PRINTED IN THE UNITED STATES OF AMERICA

Chapter 1

Molecule-Based Magnets: An Introduction

Joel S. Miller[1] and Arthur J. Epstein[2]

[1]Department of Chemistry, University of Utah, Salt Lake City, UT 84112
[2]Departments of Physics and Chemistry, Ohio State University, Columbus, OH 43210–1106

History of magnets and magnetism

The shepherd Magnes in the ancient Greek province of Magnesia in Thessaly is attributed to be the first to note that some rocks stuck to the nails of his shoes and the tip of his staff as he minded his flock of sheep. These seemingly magical rocks called *Magnes Lapis* at the time were later known as lodestone (leading or way stone) a naturally magnetic variety of magnetite (Fe_3O_4). Later in ca. 1200, in what was perhaps the first example of technology transfer, the floating compass, Figure 1, was invented in China. The compass was crucial to future explorations including those of the new world. In addition to being crucial to the evolution of society, magnetic materials were crucial to the development of science as well as technology.[1]

Figure 1. Example of an early floating compass.
Reproduced with permission of Dover Publications, Inc. Copyright 1986.

0097–6156/96/0644–0001$15.00/0

Although Galileo Gallelei is often attributed to be the father of experimental science, in 1269 Petrus Peregrinus de Maricourt preceded him with his important empirical observations aimed to prove/disprove theory. He (1) identified and named the north /south magnetic poles, (2) noted that opposite magnetic poles attracted while

Figure 2. William Gilbert
Reproduced with permission from reference 2. Copyright 1958. Dover Publications, Inc.

similar magnetic poles repelled, (3) noted that breaking a magnet was not destructive, and (4) improved the compass by mounting it on a pin.

Man's fascination with magnets continued and William Gilbert, Figure 2, attributed by Galileo as the first experimentalist, (1) created a new not naturally occurring magnet - iron that (2) upon heating lost its magnetic force which returned upon cooling and (3) identified the Earth as a magnet. Furthermore, in 1600 he wrote *De Magnete Magneticisque Corporibus et de Magno Magnete Tellure Physiologia Nova* - the first systematic Treatise based on an experimental scientific study.[2]

In the nineteen century additional key development occurred, namely (1) Hans Christian Øersted observed that electricity affected magnets (1819); (2) Michael Faraday invented the electromagnet (1823); and (3) the use of magnets enabled the production of low-cost electricity at an ac generating station built by Westinghouse in Buffalo, NY (1886). In a major triumph for the then-fledging theory of quantum mechanics, Niels Bohr in 1913 identified the underlying physics from which magnetism results, *i. e.*, the minute spin associated with an unpaired electron.[1]

Throughout this time magnets were limited to metallurgically processed inorganic materials with little, if any, thought given to organic materials.

History of molecule-based magnets

Use of molecules (or molecular ions) instead of atom (or ions)presents an exciting prospect as it is expected to enable (i) the modulation of the magnetic properties by energy efficient low-temperature organic-synthesis methodologies, (ii) the improvement of commercially useful magnetic properties, and (iii) the combination of magnetic properties with other physical properties (*e. g.*, mechanical, electrical, and/or optical). Nonetheless, despite these potential attributes the concept of molecule-based magnets, let alone its realization, was slow to evolve.

The early history of molecule-based magnets was dominated by theory. In 1963 Harden H. McConnell then at the California Institute of Technology suggested a mechanism for ferromagnetic coupling between molecules that involved the spatial arrangement of neighboring radicals.[3] In 1967 McConnell published another approach aimed at stabilizing ferromagnetic coupling between molecular ions that involved the admixing of an change transfer excited state into the ground state.[4] Neither of the models dealt with magnetic ordering - only ferromagnetic coupling between a pair of molecules - and research focused toward the testing these models did not appear until 1987[5] for the former and 1982[6] for the latter.

Also in 1967 H. Hollis Wickman *et. al.* then at Bell Laboratories reported that intermediate-spin (S = 3/2) chlorobis(diethlydithiocarbamate)iron(III), **1**, $ClFe^{III}$-$(S_2CNEt)_2$, was a ferromagnet with a critical temperature, T_c, of 2.46 K.[7] In 1970 R. L. Martin and co-workers at the University of Melbourne reported that another intermediate-spin (S = 3/2) complex, manganese phthalocyanine (MnPc), **2**, was a ferromagnet.[8] Its 8.3 K T_c as well as its detailed characterization as being a canted-ferromagnet were reported by William E. Hatfield's University of North Carolina group in 1983.[9] In 1973 Claudine Veyret and co-workers in Grenoble, France reported that bis(2,2,4,4-tetramethyl-4-piperidinol-1-oxyl), **3**, tanol suberate, was a ferromagnet with a T_c of 0.38 K.[10] In 1981, additional data from the same group led to the recharacterization of tanol suberate as being a metamagnet, *i. e.*, it had an antiferromagnetic ground state; however, above a critical applied magnetic field of 100 Oe and below a T_c of 0.38 K it had a high moment, ferromagnetic-like state.[11] In 1979 the 7,7,8,8-tetracyano-*p*-quinodimethane (TCNQ), **4**, electron-transfer salt of decamethylferrocene, $FeCp^*_2$, **5**, was reported to be a metamagnet below the T_c of 2.55 K with a critical applied magnetic field of 1600 Oe.[12]

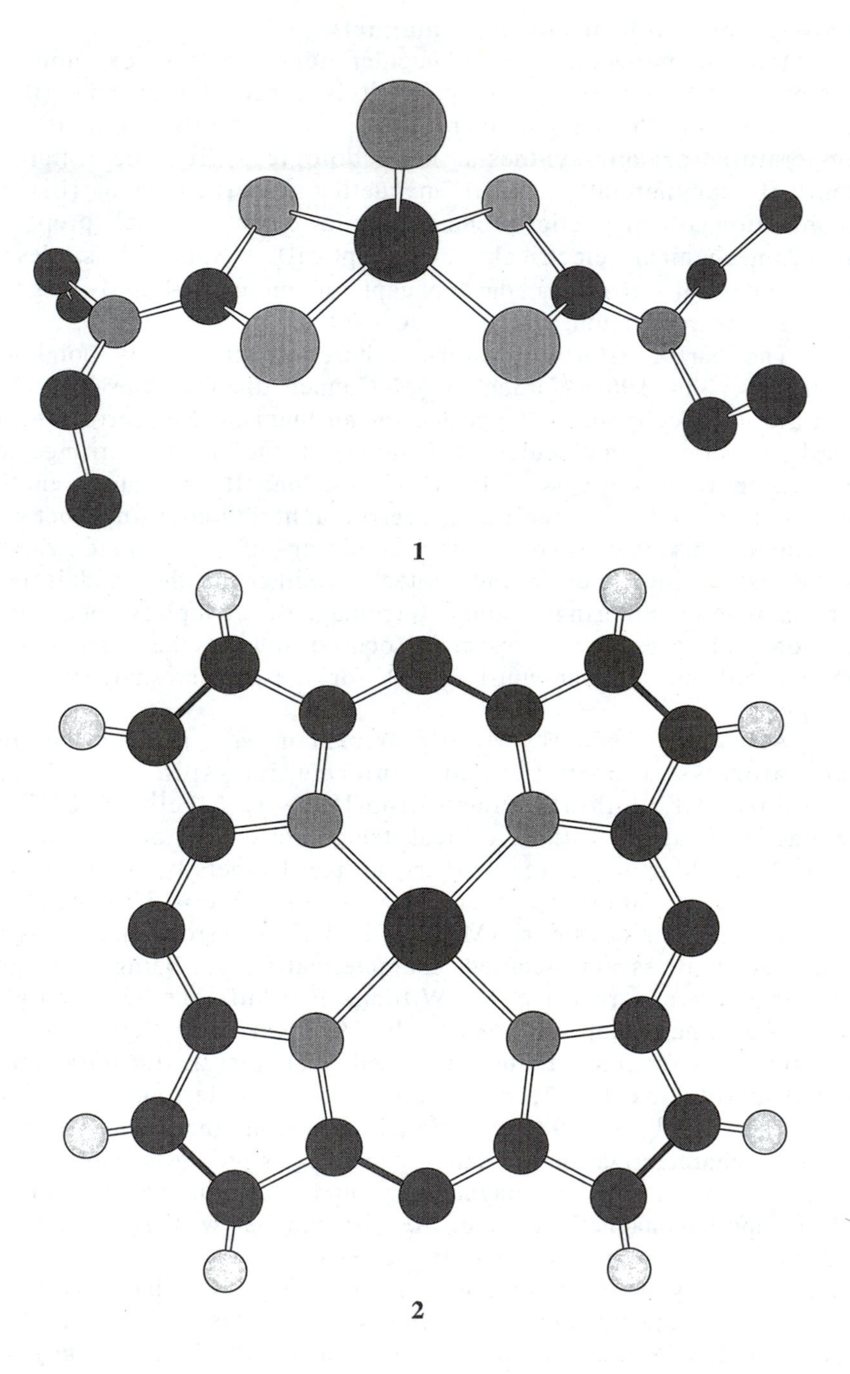
1
2

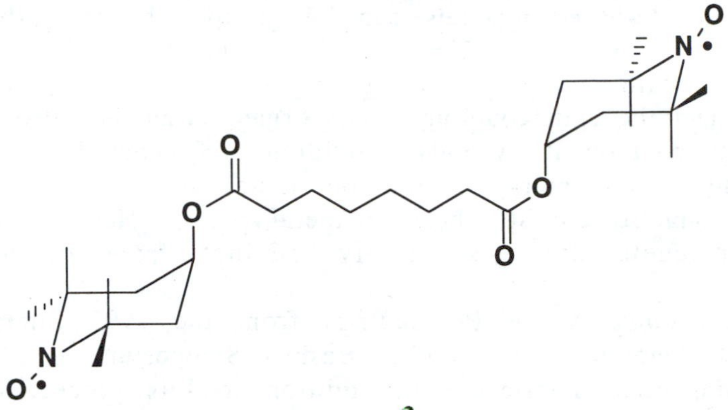

3

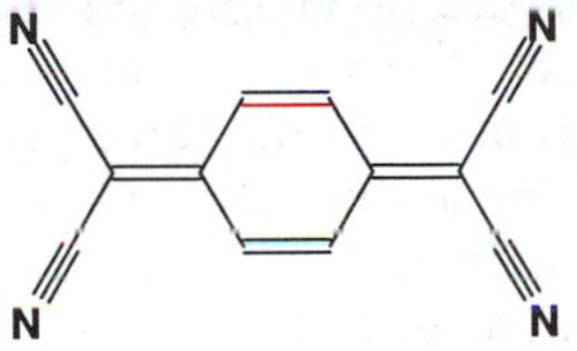

4 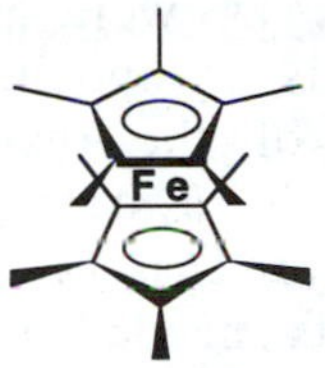 **5** 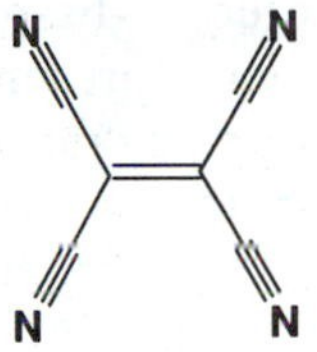 **6**

Molecular ion-based salts exhibiting magnetic ordering were first reported by the William M. Reiff and Richard L. Carlin groups. $[Cr^{III}(NH_3)_6]^{3+}[Fe^{III}Cl_6]^{3-}$ (S = 3/2 Cr^{III}; S = 5/2 Fe^{III})[13] and $[Cr^{III}(NH_3)_6]^{3+}[Cr^{III}(CN)_6]^{3-}$ (S = 3/2 Cr^{III}),[14] Figure 3, were respectively reported to be ferro- (T_c = 0.66 K) and ferrimagnets (T_c = 2.85 K). The formation of the salts were not a consequence of electron-transfer, but this pioneering work led to systems with d-orbital spins sites connected via covalent bonds and exhibiting magnetic ordering.

Unlike classical magnets each of these magnets (a) was prepared from conventional organic preparative methodologies, (b) do not have extended ionic, covalent, or metallic bonding in the solid state and as a consequence are readily soluble in conventional organic solvents, but (c) do not exhibit magnetic hysteresis characteristic of hard ferromagnets. Hysteresis for a molecule-based magnet was first reported for the tetracyanoethylene (TCNE), **6**, electron-transfer salt of $FeCp^*_2$. Due to the smaller size of $[TCNE]^{\cdot -}$ with respect to $[TCNQ]^{\cdot -}$ and expected greater spin density, $[FeCp^*_2]^{\cdot +}$-$[TCNE]^{\cdot -}$ was expected to have enhanced ferromagnetic coupling and order ferro-magnetically at low temperature. Subsequent research verified these expectations.[15]

Molecule-based magnets can be grouped by the orbitals in which the spins reside. The organic fragment may be an active component with spin sites contributing to both the high magnetic moment and the spin coupling. The organic fragment also may be a passive component by only providing a framework to position the spins which solely reside on metal ions and facilitating coupling among the spin-bearing metal ions. Hence, molecule-based magnets are qualitatively distinct from conventional magnets.

This book is the Proceedings from the 1995 International Chemical Congress of the Pacific Basin's Symposium on Molecule-based Magnetic Materials. In addition to this proceedings, the proceedings from the past four International Conferences[16] on molecule-based magnets as well as reviews are available.[17] The papers presented at this symposium covered the wide gamut of interests representative of the researcher groups in the area and papers covering magnetic systems with both active and passive organic fragments were discussed. To introduce the general reader with the broad concepts needed to understand general aspects of this research area. magnetic behavior and mechanism to enable spin coupling are introduced.

Magnetic Behavior - An Overview

Magnetism is frequently measured by a materials' response (attraction or repulsion) to a magnet. It is a consequence of the spin associated with an unpaired electron ($m_s = 1/2$ '↑' or $m_s = -1/2$ '↓') and how nearby unpaired electron spins interact with each other. Molecules are typically sufficiently large and far apart that their spin-spin coupling energy, J (as deduced from the Hamiltonian $H = -2J\mathbf{S_a} \bullet \mathbf{S_b}$) is small compared to the coupling-breaking thermal energy. Their spins do not couple, but instead form a very weak paramagnet, Figure 4a. Upon bringing the spins closer, J can become sufficiently large to enable an effective parallel (or antiparallel) alignment. This increases (or decreases) the measured susceptibility, χ. For uncoupled or weakly coupled spins the susceptibility can be modeled by the Curie-Weiss expression, $\chi \propto (T - \theta)^{-1}$. When the spins couple in a parallel manner χ is enhanced, *i. e.*, $\theta > 0$, and when spins couple in an antiparallel manner χ is surpressed, *i. e.*, $\theta < 0$. The magnetization, M, is χ/H where H is the applied magnetic field as a function of H

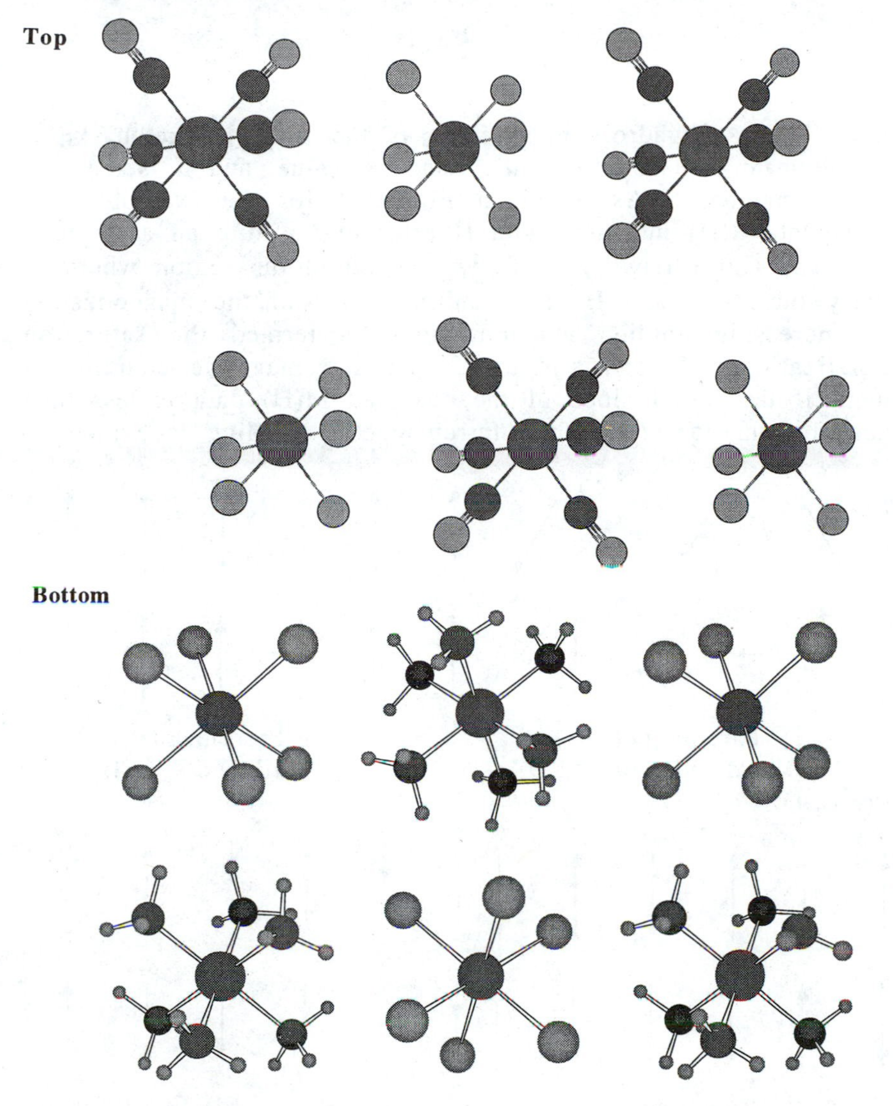

Figure 3. Segments of the salt-like structures of $[Cr^{III}(NH_3)_6]^{3+}[Cr^{III}(CN)_6]^{3-}$ a ferrimagnet with a T_c of 2.85 K (the H's are not illustrated) (top) and $[Cr^{III}(NH_3)_6]^{3+}[Fe^{III}Cl_6]^{3-}$ a ferromagnet with T_c of 0.66 K (bottom).

can be calculated from the Brillouin function, eqns. (1) for a system with noninteracting spins, *i. e.*, $\theta = 0$.

$$M = \mu_B N S g B \quad (1)$$

where
$$B = \frac{2S+1}{2S}\coth\left(\frac{2S+1}{2S}x\right) - \frac{1}{2S}\coth\left(\frac{x}{2S}\right)$$

and $$x = \frac{gS\mu_B H}{k_B T}$$

where N is Avrogadro's number, μ_B is the Bohr magneton, k_B is the Boltzmann constant, g is the Landé g value, and S is the spin quantum number. As noted in Figure 4 for the example of a paramagnet, M(H) increases with H prior to reaching an asymptotic value. The Curie Law, $\chi \propto (T)^{-1}$, is valid in the region where M linearly increases with H. The limiting value of the magnetization with increasing applied magnetic field is termed the saturation magnetization, M_S, and is $Ng\mu_B S$. Antiferromagnetic coupling is evident if the initial slope of the observed M(H) data is less than expected from eqn. (1), while ferromagnetic coupling is evident if

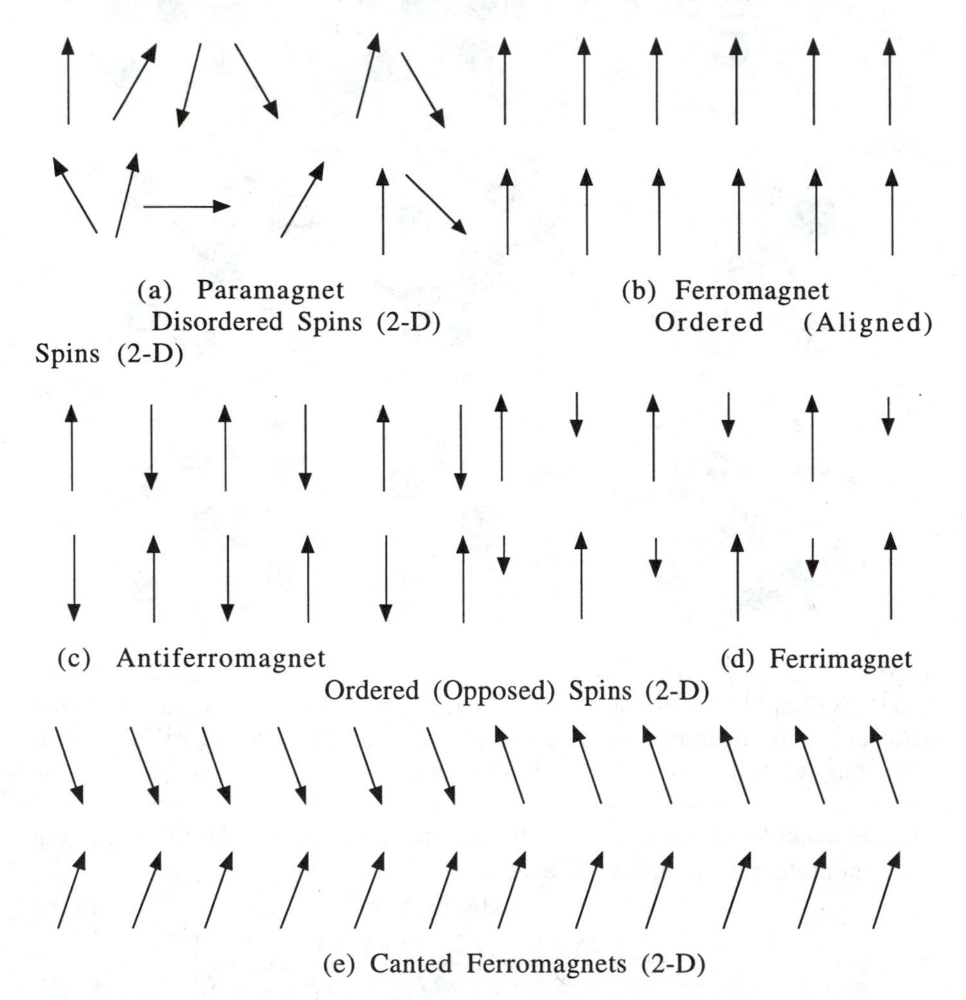

Figure 4. Schematic illustration of spin coupling behaviors.

the initial slope of the observed M(H) data exceeds the expectation from eqns. (1). A ferromagnet exhibits a spontaneous magnetization, Figure 5. Hence, the shape of the M(H) curve, Figure 5, as well as the determination of θ reveals the dominant magnetic coupling.

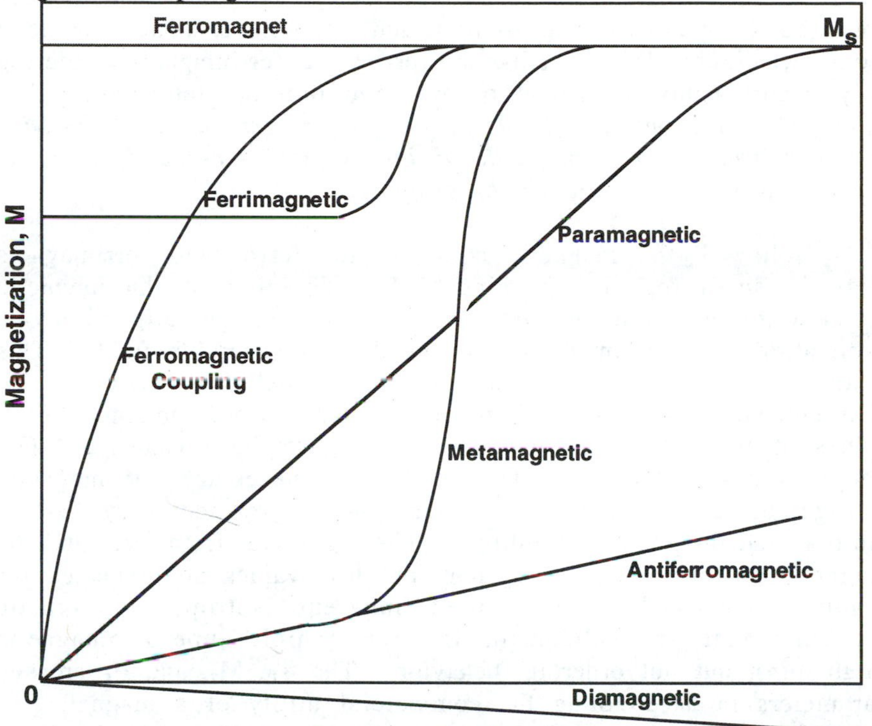

Figure 5. Schematic illustration of variation of magnetization, M, with applied magnetic field,H, for materials with different types of magnetic coupling. The magnitude of the saturation magnetization (magnetization at high applied field), M_s, depends upon the number of spins per repeat unit and the coupling among these spins (ferromagnetic or antiferromagnetic). The magnitude of the spontaneous magnetization (magnetization at zero applied field) similarly depends upon the number of spins per repeat and the coupling among these spins (ferromagnetic or antiferromagnetic) as well as the history of the sample. The applied magnetic field necessary to achieve saturation for the different types of magnets depends up the strengths of the ferromagnetic and antiferromagnetic couplings present in the materials, as well as the temperature at which the experiment is carried out.

The θ-values reflect pairwise, not long-range spin coupling. Pairwise ferromagnetic coupling ('↑↑'), albeit rare, can lead to long range ferromagnetic order, Figure 4b. Antiferromagnetic order can result from pairwise antiferromagnetic coupling ('↑↓'), Figure 4c. Ferrimagnets result from antiferromagnetic coupling which does not lead to complete cancellation and thus have a net magnetic moment, Figure 4d. Ferro-, antiferro-, or ferrimagnetic ordering only occurs below a critical or magnetic ordering temperature, T_c. *Note that the commercially useful ferro- or ferrimagnetic behavior is not a property of a molecule or ion; it, like superconductivity, is a cooperative solid state (bulk) property.*

Below T_c the magnetic moments for ferro- and ferrimagnets align in small regions (domains). The direction of the magnetic moment of adjacent domains differ, but can be aligned by a application of a minimal magnetic field, H_c (coercive field). This leads to history dependent magnetic behavior (hysteresis) characteristic of ferri- and ferromagnets. 'Hard' magnets have values of $H_c > 100$ Oe, whereas 'soft' magnets have values <10 Oe. Large values (hundreds of Oe) of H_c are necessary for magnetic storage of data, while low values (mOe) are necessary for ac motors and magnetic shielding. The coercive field reflects the magnetic anisotropy of the system and low values are expected for organic compounds due to their inherent isotropy. Magnetic materials that are subdomain in size exhibit superparamagnetic (high-spin), but not ordering, behavior. The T_c, M_s, and H_c are key parameters in ascertaining the commercial utility of a magnet.

Other magnetic ordering phenomena, *e. g.*, metamagnetism, canted ferromagnetism, and spin-glass behavior, may occur. The transformation from an antiferromagnetic state to a high moment state, *i. e.*, the spin alignment depicted in Figure 4c is transformed to that depicted in Figure 4b by an applied magnetic field, is metamagnetism, Figure 5. α-Decamethylferrocenium 7,7,8,8-tetracyano-*p*-quinodimethanide, α-$[Fe(C_5Me_5)_2][TCNQ]$, is a molecular-species-based example of a metamagnet.[12] A canted (or 'weak') ferromagnet results from the relative canting of ferromagnetically coupled spins that reduces the moment, Figure 4e. Manganese phthalocyanine is a molecule-based example of a canted ferromagnet.[8] A spin-glass occurs when local spatial correlations in the directions of neighboring spins exist, but not long range order. The spin alignment for a spin glass is that of a paramagnet, Figure 3a; however, unlike paramagnets for which the spins directions vary with time, the spins orientations for a spin-glass remain fixed or vary slowing with time. V[TCNE]$_x$·yMeCN at low temperature is an example of a correlated spin-glass.[17a,18]

All magnets require spin alignment throughout the solid. The spin coupling energy, J, describes the type [ferro- ($J > 0$) or antiferromagnetic ($J > 0$)] and magnitude of coupling between pairs of spins. Unlike the directly determinable T_c, the determination of the value of J depends of the details of the mathematical model used to fit the magnetic data and is not discussed herein. Three distinct mechanisms for spin coupling exist, Table 1.[17a,f] Additionally the antiferromagnetically coupled of adjacent sites with a differing number of spins per sites can lead to ordering as a ferrimagnet.

The primary exchange mechanism for a particular system, however, may not be clear and more that one mechanism may be operative (*e. g.*, ferromagnetic spin coupling within a chain due to orthogonal orbitals (1st mechanism) and ferro- or antiferromagnetic coupling between chains due to CI interaction (2nd mechanism). Furthermore, there are many levels of complexity of the CI model, hence, there are numerous ways to describe this mechanism. It is important to emphasize that *highly magnetic behavior is not a property of an isolated molecule; it is a cooperative solid state (bulk) property.* Thus, to achieve bulk magnetic behavior for a molecular system, intermolecular interactions must be present in all three directions. A detailed description of these models is presented in ref. 17a,f.

Key to the formation of a magnet is the presence of unpaired electron spins. The number and means of coupling of the spins dictate the magnetic properties. Molecule-based magnets affords the opportunity for organic species to provide a means to couple

Table 1. Mechanisms for Achieving Ferro- or Antiferromagnetic Spin Coupling

Mechanism	Spin Interaction	Spin Coupling[a]
1. Spins in Orthogonal Orbitals (no CI) - Hund's rule	Intramolecular	FO
2. Spin Coupling via Configuration Interactions (CI)	Intra- or intermolecular	FO or AF
3. Dipole-dipole (through-space) interactions	Intra- or intermolecular	FO or AF

[a] FO = ferromagnetic ($J > 0$); or AF = antiferromagnetic ($J < 0$)

the spins and in some cases contribute spins. Hence, as is the case with proteins, the primary, secondary, and tertiary structures are crucial for achieving the desired cooperative magnetic properties. Currently the discovery of new molecular-species-based magnets is limited by the rational design of solid state structures which remains an art. This is due to the formation of numerous polymorphs, complex and solvated compositions, as well as undesired structure types. The growth of crystals enabling the study of the single crystal structure and magnetic properties is also an important limitation. New radicals, donors, and acceptors as well as new structure types are necessary for this area of research to develop further. Finally, one needs to understand the frontier orbital overlaps as they contribute to the interchain and intrachain coupling. Given the present rapid growth in this field it is clear that major advances will be occurring over the next decade in this new multidisciplinary branch of solid state science.

Acknowledgment

The authors gratefully acknowledge the continued partial support by the Department of Energy Division of Materials Science (Grant Nos. DE-FG02-86ER45271.A000 and DE-FG03-93ER45504) as well as the National Science Foundation (Grant No. CHE9320478).

References

1 Hadfield, D., *Permanent Magnets and Magnetism*, Iliffe Books, Ltd., New York, ch. 1. Encyclopedia Britannica.
2. Gilbert, W., *De Magnete*; P. F. Mottelay, trans., Dover Publications, Inc., 1958.
3 McConnell, H. M., *J. Chem. Phys.* **1963**, *39*, 1910.
4 McConnell, H. M., *Proc. R. A. Welch Found. Chem. Res.* **1967**, *11*, 144.
5. Izoka, A.; Murata, S.; Sugawara, T.; Iwamura, H., *J. Am. Chem. Soc.* **1985**, *107*, 1786; *J. Am. Chem. Soc.* **1987**, *109*, 2631.
6. Breslow, R.; Juan, B.; Kluttz, R. Q.; Xia, C.-Z., *Tetrahedron* **1982**, *38*, 863. Breslow, R., *Pure App. Chem.* **1982**, *54*, 927.
7. Wickman, H. H.; Trozzolo, A. M.; Williams, H. J.; Hull, G. W.; Merritt, F. R., *Phys. Rev.* **1967**, *155*, 563.
8. Barraclough, C. G.; Martin, R. L.; Mitra, S.; Sherwood, R. C., *J. Chem. Phys.* **1970**, *53*, 1638.
9. Mitra, S.; Gregson, A.; Hatfield, W. E.; Weller, R. R., *Inorg. Chem.* **1983**, *22*, 1729.
10. Saint Paul, M.; Veyret, C., *Phys. Lett.* **1973**, *45A*, 362.
11. Chouteau, G.; Veyret-Jeandey, C., *J. Physique* **1981**, *42*, 1441.

12. Candela, G. A.; Swartzendruber, L. J.; Miller, J. S.; Rice, M. J., *J. Am. Chem. Soc.* **1979**, *101*, 2755.
13. Hatfield, W. E.; Helms, J. H.; Singh, P.; Reiff, W. M.; Takacs, L.; Ensling, J., *Trans. Met. Chem.* **1992**, 17, 204. Helms, J. H.; Hatfield, W. E.; Kwiecien, M. J.; Reiff, W. M., *J. Chem. Phys.* **1986**, *84*, 3993.
14. Iwata, M.; Sato, Y., *Acta Cryst.* **1973**, *B29*, 822. Burriel, R.; Casabo, J.; Pons, J.; Carnegie, Jr., D. E.; Carlin, R. L., *Physica* **1985**, *132B*, 185.
15. Miller, J. S.; Calabrese, J. C.; Dixon, D. A.; Epstein, A. J.; Bigelow, R. W.; Zhang, J. H.; Reiff, W. M., *J. Am. Chem. Soc.* **1987**, *109*, 769. Miller, J. S.; Calabrese, J. C.; Epstein, A. J.; Bigelow, R. W.; Zhang, J. H.; Reiff, W. M., *J. Chem. Soc., Chem. Commun.* **1986**, *1026*. Chittipeddi, S.; Cromack, K. R.; Miller, J. S.; Epstein, A. J., *Phys. Rev. Lett.* **1987**, *22* , 2695.
16. Conference Proceedings in the conference on *Ferromagnetic and High Spin Molecular Based Materials* (Eds.: Miller, J. S.; Dougherty, D. A.), *Mol. Cryst., Liq. Cryst.* **1989**, *176*. Conference Proceedings in the conference on *Molecular Magnetic Materials: NATO ARW Molecular Magnetic Materials* (Eds.: Kahn, O.; Gatteschi, D.; Miller, J. S.; Palacio, F.), Kluwer Acad. Pub., London, **1991**, E198. Conference Proceedings in the conference on *Chemistry and Physics of Molecular Based Magnetic Materials* (Eds.: Iwamura, H.; Miller, J. S.) *Mol. Cryst., Liq. Cryst.* **1993**, 232/233. Conference Proceedings in the conference on *Molecule-based Magnets* (Eds.: Miller, J. S.; Epstein, A. J.) *Mol. Cryst., Liq. Cryst.* **1995**, *271*.
17. (a) Miller, J. S.; Epstein, A. J., *Angew. Chem. internat. Ed.* **1994**, 106, 399; *Angew. Chem. int. Ed.* **1994**, *33*, 385. (b) Kahn, O., *Molecular Magnetism*, VCH Publishers, Inc. **1993**. (c) Gatteschi, D., *Adv. Mat.* **1994**, *6*, 635. (d) Kinoshita, M., *Jap. J. Appl. Phys.* **1994**, 33, 5718. (e) Buchachenko, A. L., *Russ. Chem. Rev.* **1990**, *59*, 307. (f) Miller, J. S.; Epstein, A. J., *Chem. & Engg. News*, **1995**,*73*#40, 30 .
18. Zhou, P.; Morin, B. G.; Miller, J. S.; Epstein, A. J., *Phys. Rev. B* **1993**, *48*, 1325.

Theory and Applications

Chapter 2

Exchange Interaction of Organic Spin Systems

Y. Teki and K. Itoh

Department of Materials Science, Faculty of Science, Osaka City University, 3–3–138 Sugimoto, Sumiyoshi-ku, Osaka 558, Japan

The exchange mechanisms in molecular-based magnetism are conveniently divided into intramolecular and intermolecular exchanges. It is shown that the ferromagnetic conditions for the intermolecular exchange in organic solids proposed with the different formalisms (McConnell (1963) and Kinoshita et al. (1987)) can be understood consistently by a Hubbard model. The intramolecular spin alignment in organic molecules through chemical bonds can also be interpreted well by an generalized Hubbard model. For conjugated π bonds, the spin polarization mechanism is predominant and topology of π electron networks determines the effective exchange, both its sign and magnitude. In addition, as a typical example for non-conjugated bonds, the intramolecular spin ordering utilizing superexchange mechanism for the ether bridge is presented and the sign and magnitude of the intramolecular effective exchange interaction is interpreted in terms of the spin densities. A variety of exchange interactions are also reviewed.

Remarkable progress has been made in the field of organic molecular-based magnetic materials, and knowledge about the magnitude and mechanism of the exchange interactions of organic spin system has been accumulated (*1*). There are a variety of the exchange interactions closely related to molecular magnetism. The exchange mechanisms in molecular-based magnetism are conveniently divided into intramolecular and intermolecular exchange. In this paper, after a brief review of the exchange interactions, we first derive the conditions for the intermolecular exchange to be ferromagnetic on the basis of a Hubbard model, and interpret consistently the different formalisms proposed by McConnell (*2*) and Kinoshita et al. (*3*). Then, the intramolecular spin alignment in organic high-spin molecules are also interpreted by a generalized Hubbard model. Finally, the sign and magnitude of the effective exchange interaction in a weakly coupled high-spin system through the ether bridge is dealt with in terms of a valence-bond (VB) formalism similar to that of McConnell for the intermolecular exchange.

Various Exchange Interactions

The direct exchange interaction which arises as a result of Pauli's exclusion principle is expressed in terms of the spin operator $\mathbf{S}_i$,

0097–6156/96/0644–0016$15.00/0

$$H_{ex} = -2J_{eff}\mathbf{S}_i \cdot \mathbf{S}_j \qquad (1),$$

$$J_{eff} = (K_{ij} - C_{ij}\gamma_{ij}^2)/(1 - \gamma_{ij}^4) \qquad (2)$$

where K_{ij}, C_{ij} and γ_{ij} are the exchange, Coulomb, and overlap integrals, respectively. Hamiltonian (1) is called Heisenberg Hamiltonian which expresses the isotropic exchange interaction between spins i and j. For the usual organic radical solids or the chemical bonds between neighboring pπ orbitals, the effective exchange integral J_{eff} is negative (antiferromagnetic). In the case of orthogonal orbitals, the sign of J_{eff} is positive, and this leads to the ferromagnetic exchange interaction. Some ferromagnetic crystals of organic radicals such as p-nitrophenyl α-nitronyl nitroxide belong to this case (*4*). The spin alignment in the pπ and σ orbitals at the divalent carbon atom of polycarbenes also belongs to this case.

It is widely known that the superexchange interaction is the dominant mechanism of magnetic oxides or transition metal fluorides. . This interaction is an indirect interaction through a closed shell orbital such as the pπ orbital of the oxygen atom. According to Anderson's theory (*5*), the effective exchange integral of the superexchange interaction is given by

$$J_{super} = -t_{ij}^2/U \qquad (3)$$

where t_{ij} and U are the hopping matrix element of electrons between neighboring orbitals and the average Coulomb repulsion between electrons in the same orbital. This type of exchange interaction is important in the molecule-based magnetic materials such as metal-oxalate complexes. The intramolecular exchange through the ether bridge can be understood by the superexchange mechanism as shown in the final section.

In some case, the antisymmetric (anisotropic) exchange called Dzyaloshinski-Moriya interaction exists. This exchange is expressed as

$$H_{DM} = -2D_{ij} \cdot [\mathbf{S}_i \times \mathbf{S}_j] \qquad (4)$$

where D_{ij} is a constant vector describing antisymmetric exchange. The origin of D_{ij} comes from the spin-orbit coupling. Thus the magnitude is proportional to the anisotropy of g value, i.e. $(g\text{-}2)/g$. When the crystal structure has an inversion symmetry, this interaction should disappear from the symmetry requirement. This anisotropic exchange interaction is responsible for the weak ferromagnetism. The weak ferromagnetism is due to an almost antiferromagnetic alignment of spins on the two sublattices which have the orientations slightly canted with each other. The canting sublattice leads to the non-zero magnetization at zero magnetic field in the antiferromagnetic ordered states. The weak ferromagnetism has recently been observed in organic spin systems such as 1,3,5-Triphenyl-6-oxoverdazyl (*6*), 1,3,5-Triphenyl-phenylverdazyl (*7*), and Li-tetrafuluoro-TCNQ (*8*) at low-temperature.

In itinerant-electron magnetic systems, the exchange interaction between conduction electrons and localized electrons causes indirect exchange coupling between the localized magnetic moments. This indirect exchange interaction is called Ruderman-Kittel-Kasuya-Yoshida (RKKY) interaction, which is occasionally the most important interaction leading to the magnetic order in

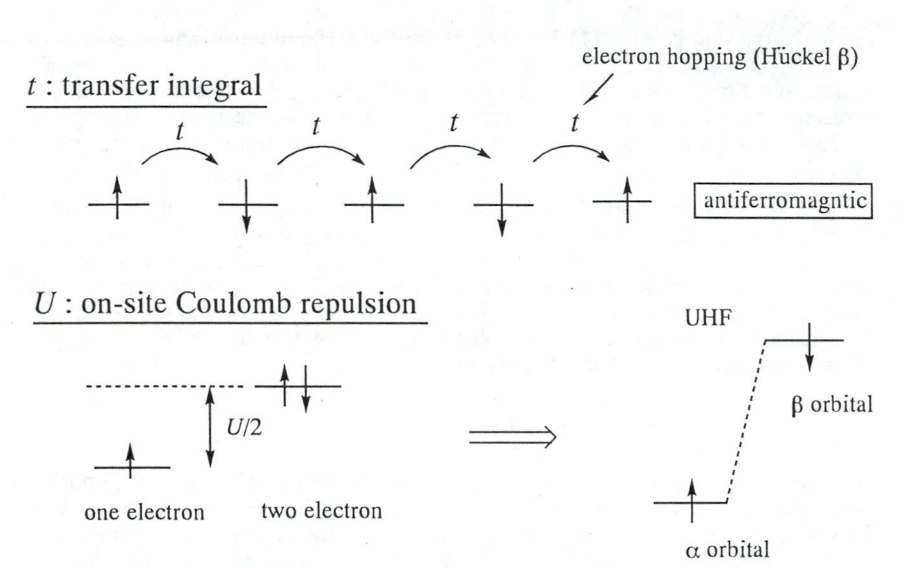

Figure 1. Physical pictures of the Hubbard model. The transfer integral and the on-site Coulomb repulsion.

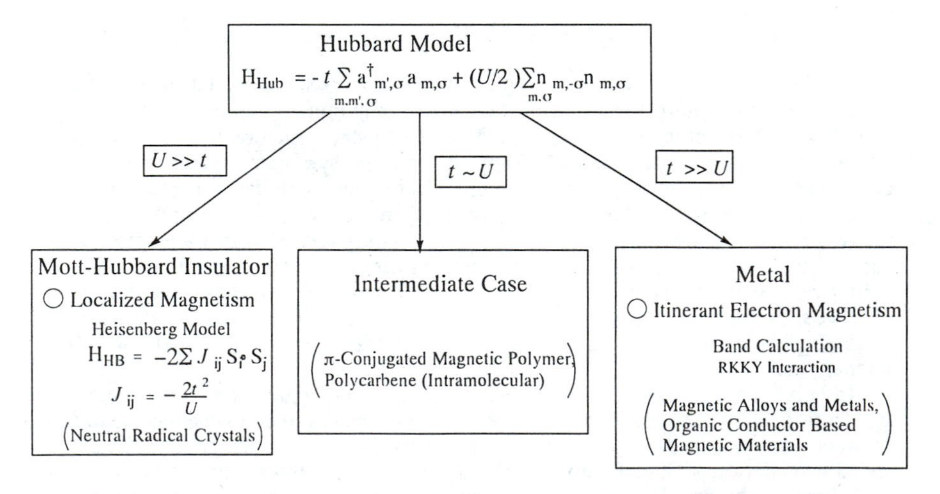

Figure 2. General features of the electron magnetism derived from the Hubbard model.

magnetic metals. The characteristic nature of this exchange interaction is the following spatial oscillation.

$$J_{RKKY} = A\{2k_fR\cos(2k_fR) - \sin(2k_fR)\}/R^4 \qquad (5)$$

where k_f is the Fermi wavenumber for the conduction-electron band. R is the spatial distance between the localized and conduction electron spins. Thus, when the localized spin is introduced into a metal, the conduction spins have an oscillating spin polarization. This oscillating exchange interaction is also responsible for the spin glass phase. Although the spin glass like behavior in molecular magnetism has recently been reported for transition-metal-complex-based materials $\{NBu_4[MFe(ox)_3]\}_x$ (M = Ni, Fe) (*9*), it is not clear yet whether this interaction is related to the cause of the spin glass behavior or not.

Although there are other types of the anisotropic exchange interactions arising from the spin-orbit interaction or the LS-coupling, they are not important for organic spin systems. In the next section, we deal in detail with the ferromagnetic conditions for the organic spin systems as well as the inter- and intra-molecular spin alignment in terms of the direct exchange mechanism on the basis of a Hubbard model.

Hubbard Model and Intermolecular Ferromagnetic Exchange Interaction

For a simple model which takes both electron transfer and spin correlation (the spin polarization) into account, the following Hubbard model Hamiltonian is widely used in the field of solid state physics (*10*)

$$H_{Hub} = -t\sum a^{\dagger}_{m',\sigma}a_{m,\sigma} + (U/2)\sum n_{m,-\sigma}n_{m,\sigma} \qquad (6)$$

where $a^{\dagger}_{m',\sigma}$ and $a_{m,\sigma}$ are the creation and annihilation operators of the electron with the spin σ, and $n_{m,\sigma} = a^{\dagger}_{m,\sigma}a_{m,\sigma}$. The parameter t stands for the electron transfer integral between adjacent sites, and U is the effective on-site Coulomb repulsion (Figure 1) representing the average Coulomb repulsive interaction between electrons on the same site. The first term in eq.(6) gives the itinerant nature of the electrons, leading to the energy band structures. The second term is responsible for the spin polarization.

General features of magnetism can be derived from this model. In the case of $U << t$, the Hubbard Hamiltonian well describes the itinerant electron magnetism. In the case of $U >> t$, the Hubbard Hamiltonian leads to the Heisenberg exchange Hamiltonian (1) (Figure 2) which is widely used to describe the localized electron magnetism. The intramolecular spin alignment of polycarbenes and π-conjugated magnetic polymers belongs to the intermediate case (Figure 2). In the case of $U >> t$, the electron can not easily transfer to the adjacent site and, therefore, the system becomes an insulator. This type of insulators arising from the effect of U are called the Mott-Hubbard insulator. In the half-filled Mott-Hubbard insulators, the electron transfer t can take place only for antiparallel spin configurations, so that exchange is antiferromagnetic in the single-band Hubbard model (Figure 1). Therefore, in order to realize ferromagnetic exchange in neutral organic radical solids, a 2-band model is necessary. The ferromagnetic conditions for organic solids which have been proposed with the different formalisms (McConnell (*2*) and Kinoshita et al. (*3*)) can be described consistently using a 2-band Hubbard model (Figure 3).

2-Band Hubbard Model

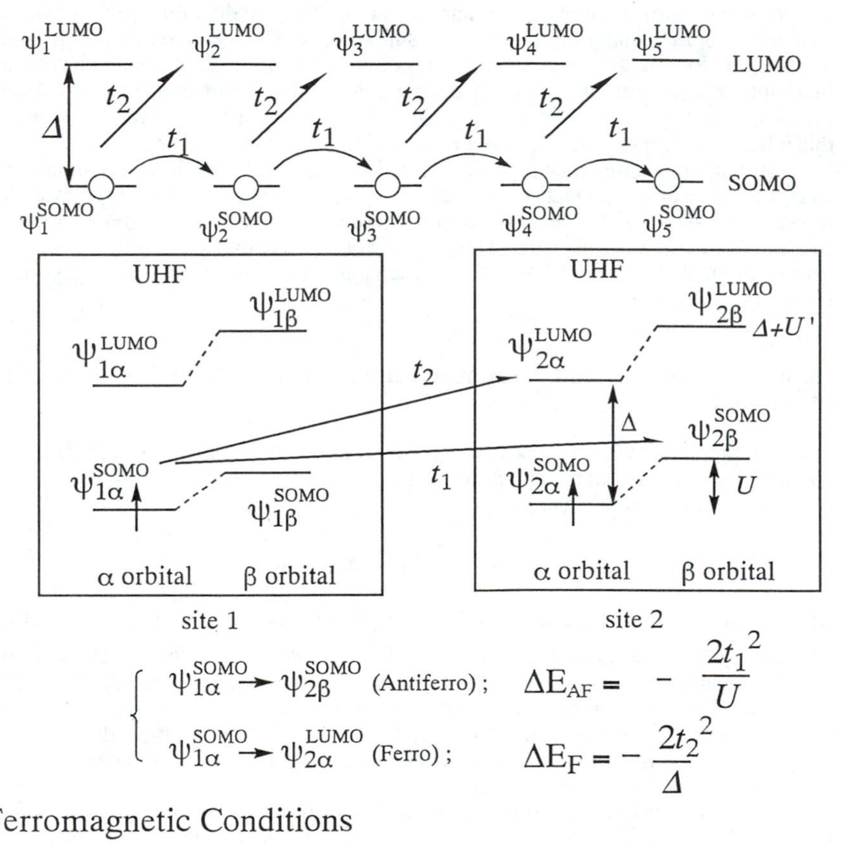

Ferromagnetic Conditions

$$\frac{t_2^2}{\Delta} > \frac{t_1^2}{U} \Longrightarrow \begin{cases} t_2 > t_1 \ (\langle \text{SOMO-}\alpha | \text{LUMO-}\alpha \rangle \neq 0 \\ \qquad \langle \text{SOMO-}\alpha | \text{SOMO-}\beta \rangle = 0 \ ) \\ \Delta < U \ \text{(Large Spin Polarization)} \end{cases}$$

Figure 3. 2-band Hubbard model and the ferromagnetic conditions in organic radical solids.

We first derive the ferromagnetic conditions based on this model. In order to simplify the argument, we treat the 2-band system made up of singly occupied molecular orbitals (SOMOs) and lowest unoccupied molecular orbitals (LUMOs) which are separated by an energy gap Δ, their on-site Coulomb repulsions (intra-orbital repulsions) being U and U', respectively. In addition, we take into account only the two important paths of the electron transfer between adjacent sites, and denote their transfer integrals as t_1 and t_2. In the case of $U >> t_1$, the electron transfer $\psi_{1\alpha}^{SOMO} \rightarrow \psi_{2\beta}^{SOMO}$ for t_1 stabilizes the antiferromagnetic (anti-parallel) spin configuration by $2t_1^2/U$ in energy. On the other hand, the electron transfer $\psi_{1\alpha}^{SOMO} \rightarrow \psi_{2\alpha}^{LUMO}$ for t_2 stabilizes the ferromagnetic (parallel) spin configuration by $2t_2^2/\Delta$ in energy. Therefore, the ferromagnetic condition is given by

$$t_2^2/\Delta > t_1^2/U \quad (7).$$

In order to realize this requirement, we need the condition $t_2 > t_1$ which leads to $\langle SOMO\text{-}\alpha | SOMO\text{-}\beta \rangle \sim 0$ and $\langle SOMO\text{-}\alpha | LUMO\text{-}\alpha \rangle \neq 0$ (the ferromagnetic conditions proposed by Kinoshita et al. (*3*)). The large value of U and the small value of Δ also favor the stabilization of the ferromagnetic spin configuration. The necessity of the large intramolecular spin polarization which has been proposed by Kinoshita et al. (*3*) corresponds to the condition of large U. Similar ferromagnetic conditions can also be derived for the electron transfers between NHOMOs and SOMOs using the corresponding 2-band Hubbard model.

In 1963, McConnell pointed out that the exchange interaction between two aromatic molecules A and B can be written in the form

$$H^{AB} = -\mathbf{S}^A \cdot \mathbf{S}^B \Sigma\, 2J_{ij}^{AB} \rho_i^A \rho_j^B \quad (8)$$

where $\mathbf{S}^A$ and $\mathbf{S}^B$ are the total spin operators, and ρ_i^A and ρ_j^B are the π-spin densities on atoms i and j of A and B, respectively (*2*). This equation can be generalized to a spin system with arbitrary spin S within the approximation of the first order perturbation using the Wigner-Eckart theorem (*11*), where the zeroth order Hamiltonian is H(A) + H(B) and the perturbing Hamiltonian is $-\Sigma\, 2J_{ij}^{AB} \mathbf{s}_i^A \cdot \mathbf{s}_j^B$. The total energy of the interacting system is given by

$$E = E^A + E^B - J_{ex}[S(S+1) - S^A(S^A+1) - S^B(S^B+1)] \quad (9)$$

where

$$J_{ex} = \Sigma\, J_{ij}^{AB} \rho_i^A \rho_j^B / (4S^A S^B). \quad (10)$$

The Hamiltonian of the effective exchange interaction between molecules A and B with the spin $\mathbf{S}^A$ and $\mathbf{S}^B$, respectively, is given by

$$H^{AB} = -\mathbf{S}^A \cdot \mathbf{S}^B \Sigma\, 2J_{ij}^{AB} \rho_i^A \rho_j^B / (4S^A S^B) \quad (11)$$

where

$$J_{ij}^{AB} = -2|t_{ij}^{AB}|^2/U + K_{ij} \quad (12).$$

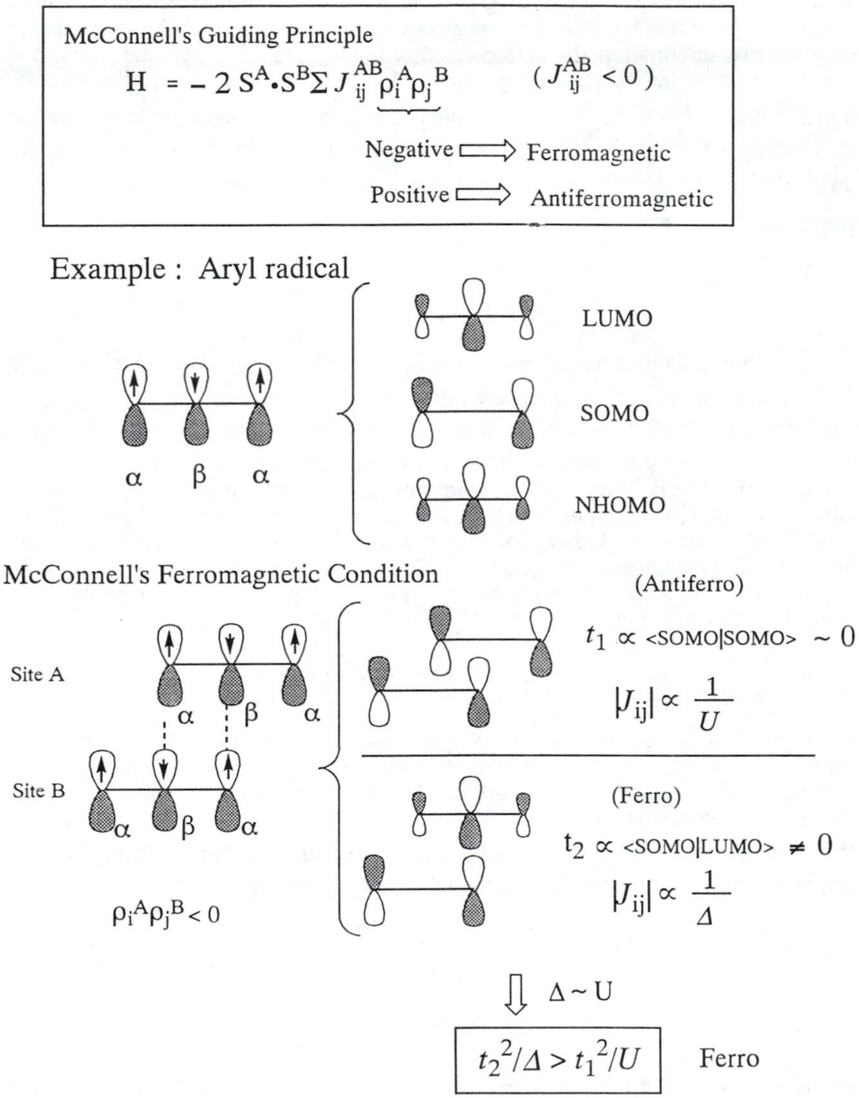

McConnell's Guiding Principle

$$H = -2\,S^A \cdot S^B \Sigma J_{ij}^{AB} \underbrace{\rho_i^A \rho_j^B} \qquad (J_{ij}^{AB} < 0)$$

Negative ⟹ Ferromagnetic

Positive ⟹ Antiferromagnetic

Example : Aryl radical

α β α

LUMO

SOMO

NHOMO

McConnell's Ferromagnetic Condition

Site A

α β α

Site B

α β α

$\rho_i^A \rho_j^B < 0$

(Antiferro)

$t_1 \propto \langle SOMO|SOMO \rangle \sim 0$

$|J_{ij}| \propto \dfrac{1}{U}$

(Ferro)

$t_2 \propto \langle SOMO|LUMO \rangle \neq 0$

$|J_{ij}| \propto \dfrac{1}{\Delta}$

$\Delta \sim U$

$t_2^2/\Delta > t_1^2/U$ Ferro

Figure 4. McConnell's guiding principle and the ferromagnetic conditions derived from the 2-band Hubbard model (Aryl radical)

t_{ij} and U have the same meaning as those for the Hubbard Hamiltonian given by eq. (6), and K_{ij} is the exchange integral between sites i and j. As shown in the last section, this equation can also apply to the intramolecular case in which two high-spin moieties couple with each other by a weak exchange interaction such as the superexchange through the oxygen bridge.

McConnell proposed the following guiding principle for the ferromagnetic intermolecular interaction in free radical solids (especially for odd-alternant hydrocarbon radicals) (*2*). From equation (8), the negative spin density product leads to the ferromagnetic exchange, since the exchange interaction J_{ij}^{AB} is normally negative in organic crystals. This negative spin density product, i.e. the ferromagnetic exchange, is expected for such a molecular stacking that the pπ orbitals with positive spin densities and those of negative spin densities are most strongly overlapping with each other.

It can be shown using the aryl radical as a simple example that this McConnell's guiding principle for the ferromagnetic exchange is equivalent to the ferromagnetic condition, equation (7), derived from the 2-band Hubbard model. According to Coulson's pairing theorem for odd alternant hydrocarbons, the bonding and the corresponding antibonding orbitals have the same absolute amplitudes of the AO coefficients in their LCAO-MOs. As shown in Figure 4, the NHOMO and the LUMO of the aryl radical have the same absolute amplitudes of the AO coefficients and the SOMO has the non-zero AO coefficients only in the α spin-density sites in the simple Hückel approximation. Therefore, the particular packing for which the positive and negative spin densities are overlapping leads to the almost zero SOMO-SOMO overlap, i.e. the small value of t_1. In contrast, this packing gives the large SOMO-NHOMO overlap. As a result of the pairing theorem, this packing also gives the large SOMO-LUMO overlap, i.e. the large value of t_2. Thus, the condition of the negative spin density product proposed by McConnell is equivalent to the present ferromagnetic condition of $t_1 > t_2$. In addition, although McConnell has not pointed out in his paper (*2*), it should be noted that J_{ij}^{AB} is inversely proportional to U for the electron transfer $\psi_{1\alpha}^{SOMO} \rightarrow \psi_{2\beta}^{SOMO}$, while to Δ for the electron transfer $\psi_{1\alpha}^{SOMO} \rightarrow \psi_{2\alpha}^{LUMO}$.

Similar relationships hold in the electron transfers between NHOMOs and SOMOs. Thus, McConnell's guiding principle is certainly equivalent to the ferromagnetic conditions, equation (7), derived in terms of the 2-band Hubbard model and , therefore, to the ferromagnetic conditions in the MO representation proposed by Kinoshita et al. (*3*). We have so far discussed the intermolecular exchange in organic solids and the ferromagnetic conditions related to it. In the following sections, we describe the intramolecular exchange associated with the spin alignment through chemical bonds.

Intramolecular Spin Alignment in Organic High-Spin Polycarbenes

As described in the single-band Hubbard model, the electron transfer t between adjacent singly-occupied pπ orbitals can take place only for antiparallel spin configurations by the Pauli's exclusion principle so that the exchange J between the adjacent pπ orbitals is antiferromagnetic. In the case of the half-filled electron band due to singly occupied pπ orbitals, this leads to the simple VB picture characterized by the alternate up-spin and down-spin distribution. As a result, the topology of the π electron networks determines their spin alignments. In this case,

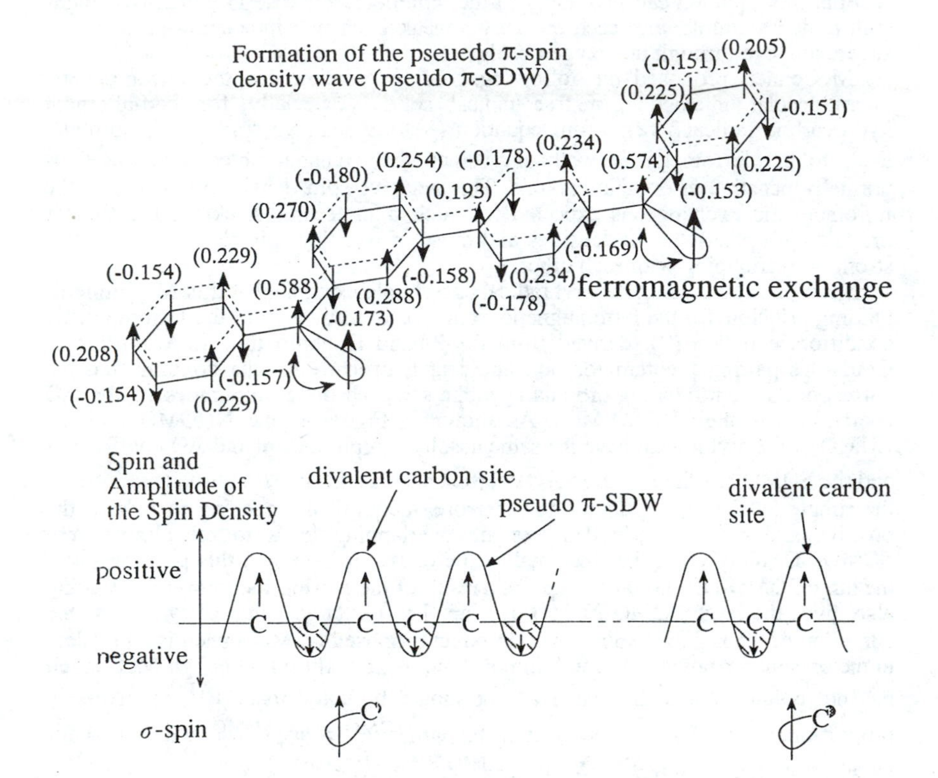

Figure 5. Mechanism of the intramolecular spin alignment in terms of the pseudo π-SDW in the high-spin polycarbenes. The π spin densities of the carbon sites obtained by the UHF calculation using a generalized Hubbard model are shown in parentheses.

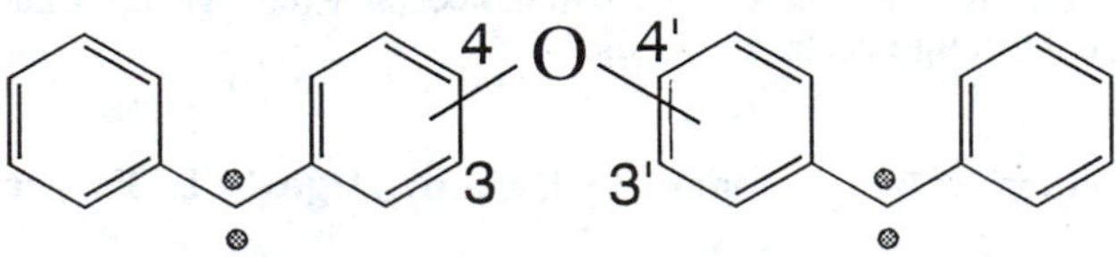

Figure 6. Model compounds for the intramolecular spin alignment (superexchange mechanism); a,a'-bis(phenylmethylene)diphenylether (a,a'-BPDE).

the Heisenberg Hamiltonian derived from the Hubbard model as a localized limit predicts well their ground and low-lying excited spin states as well as their energy gaps and spin density distributions (*12*). In order to obtain the physical picture of spin alignment as a quantitative theoretical approach, we have done a UHF calculation using the following generalized Hubbard model in which the exchange term between σ and π electrons at the divalent carbon atoms is added to the right side of equation (6),

$$H_{GHub} = -t\Sigma a^{\dagger}_{m',\sigma}a_{m,\sigma} + (U/2)\Sigma n_{m,-\sigma}n_{m,\sigma} - J\Sigma[S_k^zS_m^z+(S_k^+S_m^- + S_k^-S_m^+)/2] \qquad (13).$$

We have taken these parameters to be U/t = 2.0 and J/t = 0.25 which are the most appropriate values for a series of high-spin polycarbenes. The result of the UHF calculation on a typical high-spin (S=2) molecule, biphenyl-3,4'-bis-(phenylmethylene), is shown in Figure 5 as an example of intramolecular spin alignments. The physical picture of its spin alignment is also shown in this figure. As predicted in terms of the simple VB picture, the unpaired π electrons are distributed over the carbon skeleton with the alternate sign of spin densities from carbon to carbon. Thus, a pseudo π-spin density wave (π-SDW) is formed in the π conjugated system of this molecule. The formation of the pseudo π-SDW means that the spin polarization mechanism is predominant in the spin alignment. Both of the divalent carbon atoms have the positive sign of π-spin densities as a result of the topological nature of this molecule. In addition, the two unpaired spins in the localized σ dangling orbitals couple ferromagnetically to the π spins in the pπ orbitals at the divalent carbons. Therefore, the four parallel spin configuration is stabilized, leading to the quintet ground state.

On the other hand, in the case of biphenyl-3,3'-bis(phenylmethylene), which is a topological isomer of biphenyl-3,4'-bis(phenylmethylene), each divalent carbon atoms have opposite signs of spin densities as a result of its topological nature. The UHF calculation based on the generalized Hubbard model gives the two equivalent pseudo π-SDW states with equal energies but opposite phases of the spin densities. The ground state is expressed by the superposition of these two states of pseudo π-SDW, which gives the singlet spin state without net spin densities. The intramolecular spin alignment in other π-conjugated polycarbenes can also be interpreted well in terms of similar physical pictures.

Intramolecular Spin Alignment in Weakly Exchange-Coupled Systems

So far we have considered the intramolecular spin alignment arising from singly-occupied π-conjugated orbitals. For saturated bonds, the superexchange mechanism plays an important role in the intramolecular spin alignment. We have studied the exchange interaction via the ether bridge between two magnetic moieties in organic high-spin molecules. As the simplest model compounds, we have adopted a,a'-bis(phenylmethylene)diphenylether (a,a'-BPDE: (a,a') = (3,3'), (3,4'), and (4,4')), in which the spins in each of the triplet diphenylmethylene moieties interact via the oxygen bridge (Figure 6). Since the pπ orbital in the oxygen bridge forms a doubly occupied closed-shell, the superexchange mechanism predominates the spin alignment. We have determined the spin multiplicities of the ground state and the low-lying excited states, and estimated their energy gaps by ESR experiments (*13*). It was found that the effective exchange between the two triplet moieties is antiferromagnetic for the (3,3') and (4,4') bridges, whereas ferromagnetic for the (3,4') bridge. Thus, the effective exchange of the oxygen bridge depends on the bridging positions. This clearly shows the importance of the topology of the π electron network in the intramolecular spin alignment as already described in the case of π-conjugated systems.

We first interpret the above salient features of the spin alignment in terms of the Heisenberg model which is derived under the condition of $U >> t$ from the generalized Hubbard Hamiltonian in equation (13). In the case of a,a'-BPDE, the Heisenberg Hamiltonian (valence-bond Hamiltonian) may be divided into three parts:

$$H_{VB} = -2\Sigma J_{ij}^{A} S_i^{A} \cdot S_j^{A} - 2\Sigma J_{ij}^{B} S_i^{B} \cdot S_j^{B} - 2J_{kl}^{AB} S_k^{A} \cdot S_l^{B} \qquad (14)$$

where the first and second terms correspond to the spin coupling within moieties A and B, respectively, and the third term to the weak exchange interaction between A and B; k and l refer to the bridging positions of A and B. For conjugated alternant hydrocarbons, the sign of J_{ij}^{A} and J_{ij}^{B} is negative for the π-π exchange interaction ($J_{\pi\pi} = -J,\ J > 0$), whereas positive for the one-center σ-π exchange interaction ($J_{\sigma\pi} = J' > 0$) at the divalent carbon atoms. We have taken the ratio of J' to J to be 0.2 and the superexchange ($J_{kl}^{AB} = J'' < 0$, antiferromagnetic) to be an adjustable parameter. Using the Lanczos method, we exactly solved the Heisenberg Hamiltonian (14) by numerical diagonalization of the matrix derived from the full basis set of the spin system $|s_{1z}, s_{2z}, \cdots, s_{Nz}>$ ($s_{iz} = \pm 1/2$). We actually calculated the energies of the ground and the low-lying excited spin states for a,a'-bis(methylene)diphenylether (a,a'-BMDE), in which each of the two end phenyl groups of a,a'-BPDE was replaced by the hydrogen atom to reduce the dimensionality of the Hamiltonian matrix (Figure 7). This figure depicts the energy level diagram of a,a'-BMDE as a function of J''/J. The salient features of this diagram are consistent with our experimental results, i.e., the effective exchange via the oxygen bridge is ferromagnetic for 3,4'-BPDE and antiferromagnetic for 4,4'- and 3,3'-BPDE. The calculated energy splitting ratios among the ground and low-lying excited spin states well reproduce the observed ones, if we choose the magnitude of the superexchange parameter as indicated by the arrows.

Since the above Heisenberg Hamiltonian approach is a phenomenological treatment, we have taken into account the exchange mechanism explicitly and calculate the sign and magnitude of the effective exchange interaction. The model compounds a,a'-BPDE can be regarded as the two triplet diphenylmethylene moieties weakly interacting with each other. Therefore, equations (9) - (11) are also applicable to these weakly exchange coupled systems, although the present case corresponds to the intramolecular exchange. From the Anderson formula, the superexchange interaction through the doubly occupied pπ orbital on the oxygen atom is given by equation (3). In a,a'-BPDE, U in equation (3) is the repulsion of the two electrons on the same site and t_{ij} is the transfer integral for the electron that transfers, with preserving its spin, from the carbon site i to l, i and l being connected with each other by the oxygen bridge. The magnitude of t_{ij} is proportional to the orbital overlap between the pπ orbital on the oxygen atom and that on the adjacent carbon atom, and it can be estimated to be ca. 0.7 eV (*14*). On the other hand, the electron repulsion U is of the order of 10 eV. Therefore, we obtain ca. - 400 cm^{-1} for J_{super} in equation (3) which corresponds to the value of

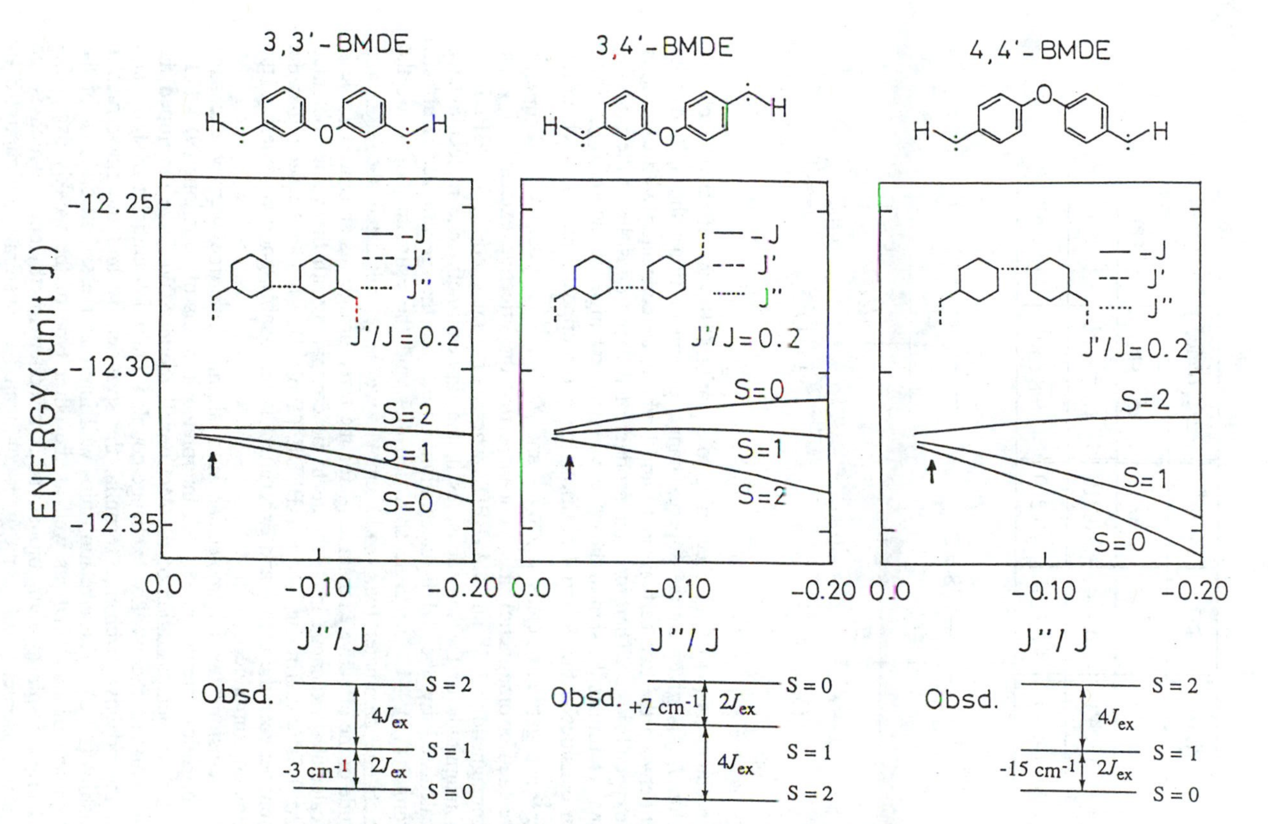

Figure 7. The ground and low-lying excited spin states of a,a'-BMDE. Their energies were calculated by the exact numerical diagonalization of the Heisenberg Hamiltonian using Lanczos method.

Table I Calculated and observed effective exchange integrals ($2J_{ex}$) of the oxygen bridge. The calculated values were obtained using equation (9) (see Text).

(i, j)	ρ_i^A	ρ_j^B	$2J_{ex}$ (calcd.)	$2J_{ex}$ (obsd.)
(3, 3')	-0.154	-0.154	-4.7 cm^{-1}	-3 cm^{-1}
(3, 4')	-0.154	+0.210	+6.5 cm^{-1}	+7 cm^{-1}
(4, 4')	+0.210	+0.210	-8.8 cm^{-1}	-15 cm^{-1}

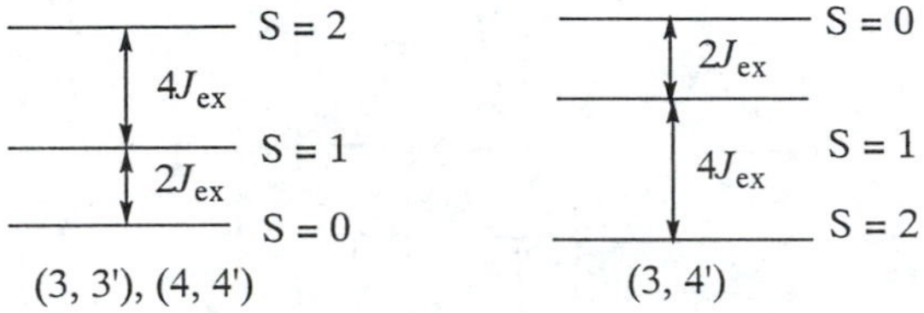

S = 2
$4J_{ex}$
S = 1
$2J_{ex}$
S = 0
(3, 3'), (4, 4')

S = 0
$2J_{ex}$
S = 1
$4J_{ex}$
S = 2
(3, 4')

J_{ij}^{AB} in equations (9) - (11). According to equation (10), the effective exchange interaction J_{ex} is given by the superexchange parameter times the spin density product between the two carbon sites i and l which are connected with each other by the oxygen bridge. Therefore, the product of the spin densities at the bridging positions determines the sign of the effective superexchange interaction J_{ex}. Since the spin densities at the 3- and 4-positions have the opposite sign to each other, $\rho_3^A\rho_{4'}^B$ is negative, whereas $\rho_3^A\rho_{3'}^B$ and $\rho_4^A\rho_{4'}^B$ are positive. As J_{super} is negative (antiferromagnetic), it is concluded that the effective exchange J_{ex} is antiferromagnetic in 3,3'- and 4,4'-BPDE whereas ferromagnetic in 3,4'-BPDE.

The magnitude of the spin densities in the diphenylmethylene molecule have been calculated by the UHF calculation based on the generalized Hubbard Hamiltonian (13). Using these spin densities and the estimated value of the superexchange parameter (ca. - 400 cm^{-1}), we have calculated the sign and magnitude of the effective exchange J_{ex} for the three bridges as shown in Table I. These calculated exchange integrals are to be compared with the values obtained from the ESR experiments for a,a'-BPDE (Table I). It can be seen that the agreement between the theory and experiment is reasonably good considering the approximation introduced here.

In conclusion, the intramolecular spin alignment in the weakly exchange coupled organic systems can be well interpreted in terms of equations (9) - (11). This result is a generalization of the McConnell model. We have developed the general theory of the weakly exchange-coupled multiplet-multiplet system, and derived the effective exchange formula, the spin-density and the fine-structure relationships (*12* c). These relationships well interpret the salient features of the spin alignment of a,a'-BPDE as well as the other type of the weakly exchange-coupled organic spin system, biphenyl-3,3'-bis(phenylmethylene), which is the topological isomer of the π-conjugated system, biphenyl-3,4'-bis(phenylmethylene).

Acknowledgements

The present work was partially supported by a Grant-in-Aid for scientific research on Priority Area "Molecular Magnetism" (Area No.228/04242105) and for General Scientific Research (No. 07454191) from the Ministry of Education, Science, Sports and Culture, Japan. The authors thank the Computer Center at the Institute for Molecular Science, Okazaki National Research Institutes.

References

1. For recent overviews on this topic, see: (a) Molecular Magnetic Materials; D. Gatteschi, O. Kahn, J. S. Miller, F. Paracio Eds., NATO ASI Series E, Kluwer Academic, Dordrecht, **1991**, Vol. 198. (b) H. Iwamura, J. S. Miller Eds., Mol. Cryst. Liq. Cryst. **1993**, 232/233, 1- 720. (c) J. S. Miller, A. J. Epstein Eds., Mol. Cryst. Liq. Cryst. **1995**, 271/272/273/274, 1- 867.
2. H. M. McConnell, J. Chem. Phys. **1963**, 39, 1910.
3. K. Awaga, T. Sugano, and M. Kinoshita, Chem. Phys. Lett. **1987**, 141, 540 - 544.
4. (a) K. Awaga and Y. Maruyama, Chem. Phys. Lett. **1989**, 158, 556 - 559. (b) M. Kinoshita, P. Turek, M. Tamura, K. Nozawa. D. Shiomi. Y. Nakazawa, M. Ishikawa, M. Takahashi, K. Awaga, T. Inaba, and Y. Maruyama, Chem. Lett. **1991**, 1225 - 1228.
5. P.W. Anderson, Phys. Rev. **1959**, 115, pp. 2 - 13.
6. R. K. Kremer, B. Kanellakopulos, P. Bele, H. Brunner, and F. A. Neugebauer, Chem. Phys. Lett. **1994**, 230, 255 -259.
7. S. Tomiyoshi, T. Yano, N. Azuma, M. Shoga, K. Yamada, and J. Yamauchi, Phys. Rev. **1994**, B49, 16031 - 16034.
8. T. Sugimoto, M. Tsujii, H. Matsuura, and N. Hosoito, Chem. Phys. Lett. **1995**, 235, 183 - 186.
9. H. Okawa, N. Matsumoto, H. Tamaki, and M. Ohba, Mol. Cryst. Liq. Cryst. **1993**, 233, 257 - 262.
10. J. Hubbard, Proc. Roy Soc. **1963**, 276A, 238 -257.
11. (a) Y. Teki, T. Takui, A. Yamashita, M. Okamoto, T. Kinoshita, and K. Itoh, Mol. Cryst. Liq. Cryst. **1993**, 232, 261 - 270. (b) M. Okamoto, Y. Teki, T. Kinoshita, T. Takui, and K. Itoh, Chem. Phys. Lett. **1990**, 173, 265 - 270. (c) Y. Teki, A. Yamashita, M. Okamoto, T. Kinoshita, T. Takui, and K. Itoh, J. Am. Chem. Soc. to be published.
12. Y. Teki, T. Takui, M. Kitano, and K. Itoh, **1987**, 142, 181- 186.
13. K. Itoh, T. Takui, Y. Teki, T. Kinoshita, Mol. Cryst. Liq. Cryst. 1**989**, 176, 49 - 66.
14. Y. Teki, and K. Itoh, unpublished work.

Chapter 3

Theoretical Approaches to Molecular Magnetism

T. Kawakami, S. Yamanaka, D. Yamaki, W. Mori, and K. Yamaguchi

Department of Chemistry, Faculty of Science, Osaka University, Toyonaka, Osaka 560, Japan

The orbital symmetry rules were derived based on the intermolecular interaction theories and were applied to elucidate effective exchange interactions between allyl radicals and related π-conjugated radicals. The SOMO-SOMO overlap integrals are zero at several conformations even in parallel interplane orientations for the radical pairs because of the NBMO-type SOMO, predicting ferromagnetic effective exchange interactions via nonzero-potential exchange integrals. Approximate spin-projected UHF Møller-Plesset and UHF coupled-cluster SD(T) computations showed that ferromagnetic intermolecular interactions are feasible at these conformations, supporting the orbital symmetry rules. The UHF natural orbital (UNO) CI and UNO complete active space (CAS) SCF computations were also carried out for key conformations with ferromagnetic interactions in order to elucidate the relative importance between direct kinetic plus potential exchange interactions and indirect spin polarization plus other higher-order spin-correlation effects. These CI calculations showed that the indirect mechanism overweighs the direct SOMO-SOMO interaction mechanism in some cases examined here.

Three decades after the discovery of Woodward-Hoffmann (WH) symmetry conservation rules for concerted reactions, an anti-WH world, *molecular magnetism* (1), has been developed considerably, and now attracts general interest in the fields of chemistry and material science. Therefore it is important and interesting to present guiding principles for modeling and design of molecular magnetic materials. In the anti-WH world, spin concept (2, 3) plays an essential role for extracting important characteristics from various experimental results available. For example, the Heisenberg (HB) spin Hamiltonian (4-6) has been used to derive the selection rules of free radical reactions (2, 3), which are regarded as one of the typical anti-WH worlds

$$\boldsymbol{H}(HB) = -2\sum_{a,b} J_{ab}\boldsymbol{S}_a \bullet \boldsymbol{S}_b \qquad (1)$$

where $\mathbf{S}_c$ (c = a, b) denotes the spin localized on a site C, and J_{ab} is the effective exchange integral between localized spins. The HB Hamiltonian is also utilized to describe the spin alignment rules in molecule-based magnetic materials (1).

0097–6156/96/0644–0030$15.00/0

As part of a continuing theoretical study of instability in chemical bonds (7), we wish to explore the levels of theories which provide the sign and magnitude of J_{ab} values in Eq. 1 at least in a semiquantitative manner so as to allow derivation of selection rules for spin alignments in organic crystals. To this end, both spin-restricted and spin-unrestricted post Hartree-Fock methods have been examined numerically in relation to the first principle computations of J_{ab} values in typical cases. In this paper, we summarize several theoretical and computational models (8-12) and explore their interrelationships : (i) intermolecular perturbation (PT) and configuration interaction (CI) models (8-10), (ii) approximate spin-projected (AP) unrestricted Hartree-Fock (UHF) model (10,11) and (iii) spin-restricted complete active space (CAS) CI and CASSCF methods (12-14). Reliable computations of J_{ab} by these methods enable us to elucidate origins of ferromagnetic interactions and to derive spin alignment rules in molecular magnetic materials (1).

Theoretical Backgrounds

Intermolecular Perturbation and Configuration Interaction Models

(A) Direct Exchange. The intermolecular perturbation (PT) and configuration interaction (CI) models (8-10) were extensively examined to elucidate the SOMO-SOMO, SOMO-LUMO and SOMO-HOMO orbital interactions between organic radicals (1). Here, let us consider a simple example as illustrated in Fig. 1. The intermolecular CI wave functions (15) for the singlet and triplet diradical (DR) configurations are given by

$$^{1,3}\Phi(DR) = \frac{1}{\sqrt{2(1 \pm S_{ab}^{2})}}(|\phi_a\bar{\phi}_b| \pm |\phi_b\bar{\phi}_a|) \qquad (2)$$

where 1 and 3 denote, respectively, the singlet (S) and triplet (T) states, and S_{ab} is the intermolecular orbital overlap integral.

$$S_{ab} = \int \phi_a \phi_b d\tau \qquad (3)$$

The total energies of the singlet and triplet DR configurations are given by

$$^{1,3}E(DR) = \frac{1}{(1 \pm S_{ab}^{2})}\{H_{aa} + H_{bb} + G_{ab} \pm 2H_{ab}S_{ab} \pm K_{ab}\} \qquad (4)$$

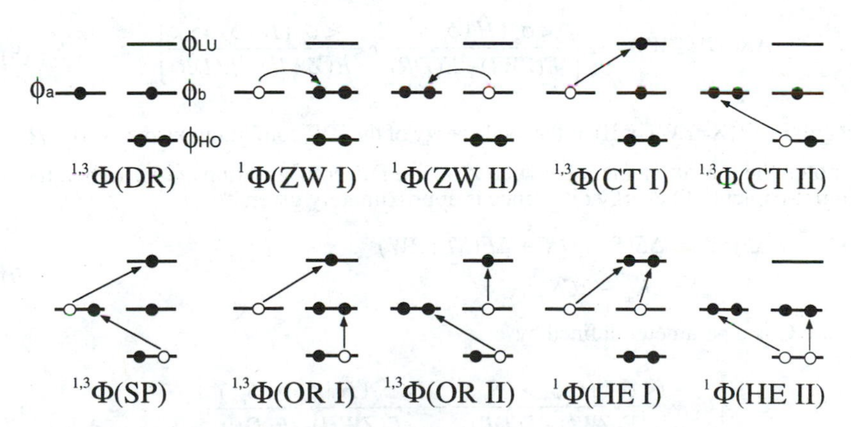

Fig.1 Intermolecular CI configurations for organic radical pair. DR, ZW, CT , SP, OR and HE denote, respectively, the diradical, zwitterionic, charge-transfer, spin polarization, orbital relaxation, and higher-order excitations.

where the Coulomb (H_{cc} (c=a,b)) and resonance (H_{ab}) integrals and the two-electron Coulombic repulsion G_{ab} and exchange integrals K_{ab} are, respectively, defined by

$$\begin{aligned} H_{ac} &= \int \phi_a \hat{H}_{core} \phi_c d\tau \\ G_{ab} &= \int \phi_a(1)\phi_a(1)\hat{G}\phi_b(2)\phi_b(2)d\tau_1 d\tau_2 \\ K_{ab} &= \int \phi_a(1)\phi_b(1)\hat{G}\phi_a(2)\phi_b(2)d\tau_1 d\tau_2 \end{aligned} \tag{5}$$

$\hat{H}_{core}$ and $\hat{G}$ denote, respectively, one and two electron parts of the total Hamiltonian $\hat{H}$ (15). Since the orbital overlap integral S_{ab} is small at the intermolecular distances determined for crystals of organic radicals, the total energies in Eq. 4 can be expanded in Taylor series of S_{ab}, and the singlet-triplet (ST) energy difference for the DR configuration is truncated at the second-order level of S_{ab} as follows:

$$\begin{aligned} \Delta E(ST:DR) &= {}^1E(DR) - {}^3E(DR) \\ &= 2\{K_{ab} - S_{ab}{}^2 C_1\} \end{aligned} \tag{6a}$$

where

$$C_1 = H_{aa} + H_{bb} + G_{ab} - 2H_{ab}/S_{ab} \tag{6b}$$

In addition to the DR configuration, the lower-lying charge transfer (CT) configurations between SOMOs are feasible for the singlet state (15) as

$${}^1\Phi(CT: a \to b) = {}^1\Phi(ZWI) = |\phi_b \bar{\phi}_b| \tag{7a}$$

$${}^1\Phi(CT: b \to a) = {}^1\Phi(ZWII) = |\phi_a \bar{\phi}_a| \tag{7b}$$

where ZWI and II denote the zwitterionic configurations a^+-b^- and a^--b^+, respectively. The stabilization energies of the singlet state by the configuration mixings between singlet DR and ZWI(II) are given by the second-order perturbation method as

$$\Delta E(ST:ZW) = -2\left\{ \frac{<\phi_a | \hat{H} | \phi_b>^2}{{}^1E(ZWI) - {}^1E(DR)} + \frac{<\phi_b | \hat{H} | \phi_a>^2}{{}^1E(ZWII) - {}^1E(DR)} \right\} \tag{8}$$

where ${}^1E(X)$ (X=ZWI or II), is the total energy of the ZW configuration and $<\phi_a | \hat{H} | \phi_b>$ denotes the matrix element between ZW and DR configurations (15). Then,the total singlet-triplet (ST) energy difference is approximately given by

$$\begin{aligned} \Delta E(ST) &= \Delta E(ST:DR) + \Delta E(ST:ZW) \\ &= 2K_{ab} - 2CS_{ab}{}^2 \end{aligned} \tag{9}$$

where C is a parameter defined by

$$C = \left\{ C_1 + \frac{(<\phi_a | \hat{H} | \phi_b> / S_{ab})^2}{{}^1E(ZWI) - {}^1E(DR)} + \frac{(<\phi_b | \hat{H} | \phi_a> / S_{ab})^2}{{}^1E(ZWII) - {}^1E(DR)} \right\} \tag{10}$$

Judging from Eq. 10, C is positive, being almost independent of the magnitude of S_{ab}.

The effective exchange integral (J_{ab}) from the CI scheme is given from Eqs. 1 and 9 as

$$J_{ab} = K_{ab} - CS_{ab}^{\ 2} \tag{11}$$

If the SOMO-SOMO overlap integral is not zero or not very small, the orbital-overlap (OO) term defined by $J_{ab}(OO)=-CS_{ab}^{\ 2}$ (8a) usually overweighs the exchange term K_{ab}, giving rise to the antiferromagnetic exchange integral ($J_{ab}<0$). So, in order to obtain the ferromagnetic interaction, we emphasized the spin alignment rule (8-10):

(Rule 1) The $J_{ab}(OO)$ term, the SOMO-SOMO overlap integral, should be suppressed in order to diminish the antiferromagnetic interaction.

Since a zero SOMO-SOMO overlap integral guarantees a zero J_{ab} value, the K_{ab} term should be non-zero in order to obtain a positive (ferromagnetic) exchange integral, leading to the following spin alignment rule (8-10) :

(Rule 2) The potential exchange (K_{ab}) term should be larger than the absolute value of $J_{ab}(OO)$.

These rules were applied to the molecular design of organic ferromagnets (8-10). For example, it was shown that nitroxide and anion radical quinone derivatives (8a) are potential blocks for the construction of organic ferromagnets from the view point of the above selection rules. In a later section of this review, we will examine several stacking modes between π-radicals, which guarantee the zero SOMO-SOMO overlap integral (Rule 1) but the non-zero potential exchange integral (Rule 2).

(B) CT Interaction Next, let us consider secondary intermolecular interactions via a charge-transfer (CT) between SOMO and LUMO (or HOMO) as illustrated in Fig. 1. This CT interaction often plays an important role in ferromagnetic interaction. The intermolecular CT configurations between these orbitals are given (15) by

$$^{1,3}\Phi(CT) = \frac{1}{\sqrt{2}}\left(\left|\phi_p\bar{\phi}_b\right| \pm \left|\phi_b\bar{\phi}_p\right|\right) \ (p = LUMO, HOMO) \tag{12}$$

where 1 and 3 denote, respectively, the singlet (S) and triplet (T) CT states. The total energies of the singlet and triplet CT configurations are approximately given by

$$^{1,3}E(CT) = \{H_{pp} + H_{bb} + G_{pb} \pm K_{pb}\} \tag{13}$$

where the Coulomb and resonance (H_{pq} (p,q=SOMO, LUMO, HOMO)) integrals and the two-electron Coulombic G_{pb} and exchange integrals K_{pb} are given by Eq 5. The stabilization energies $^{1,3}E(ST,CT)$ of the singlet and triplet diradical (DR) states from the configuration mixings with the CT configurations are given by the second-order perturbation method as

$$^{1,3}E(ST{:}CT) = -\left\{\frac{<\phi_a|\hat{H}|\phi_p>^2}{^{1,3}E(CTI)-{}^{1,3}E(DR)} + \frac{<\phi_q|\hat{H}|\phi_a>^2}{^{1,3}E(CTII)-{}^{1,3}E(DR)}\right\} \tag{14}$$

where $^{1,3}E(X)$ (X=CTI or II) is the total energy of the CT configuration and $<\phi_a | \hat{H} | \phi_c>$ (c=p,q) denotes the matrix element between the ^{1,3}CT and ^{1,3}DR configurations. The difference of the denominator between the singlet-triplet (ST) states in Eq. 14 is given by

$$\Delta = \left\{\frac{1}{^{1}E(X)-{}^{1}E(DR)} - \frac{1}{^{3}E(X)-{}^{3}E(DR)}\right\} \cong -\frac{2K_{pb}}{G_{pb}^{\ 2}} \ (p = LUMO, HOMO) \tag{15}$$

where the intramolecular Coulomb repulsion G_{pb} is assumed to be much larger than other intermolecular terms such as G_{ab} and K_{ab}. The ST energy gap given by the configuration mixing between DR and CT is approximately given by

$$\Delta E(ST{:}CT) \cong 2\left\{\frac{<\phi_a|\hat{H}|\phi_p>^2 K_{pb}}{G_{pb}^{\ 2}} + \frac{<\phi_q|\hat{H}|\phi_a>^2 K_{qb}}{G_{qb}^{\ 2}}\right\} \tag{16}$$

Judging from Eq. 16, the ST gap is positive (ferromagnetic). However, since the exchange integral K_{cb} (c=p,q) is much smaller than the Coulombic repulsion integral G_{cb}, its contribution is not so large. Therefore we obtain the following rule (8-10):

(Rule 3) The SOMO-LUMO and SOMO-HOMO orbital interactions via the CT excitations play an important role in stabilizing the high-spin state if the SOMO-SOMO orbital interaction is nearly zero or zero.

The second-order exchange terms given by Eq. 16 are incorporated in the potential exchange term J_{ab} (PE) (see Eq. 25) under the UHF approximation (4), which is utilized in the later ab initio computations of π-radical pairs. Theoretical possibilities of organic CT ferromagnets have been discussed previously (8c).

(C) Spin Polarization Effects The spin polarization (SP) effects are considered to be significant in the case of organic radical clusters. The SP configurations can be constructed by double CT excitations (8-10, 15), and as a result, the spin flip excitations from HOMO to LUMO are obtained as illustrated in Fig. 1. For example, the singlet SP configuration is given by

$$\begin{aligned} {}^1\Phi(SP) = \frac{1}{\sqrt{12}}(2|\phi_a\phi_b\bar{\phi}_p\bar{\phi}_q| + 2|\bar{\phi}_a\bar{\phi}_b\phi_p\phi_q| - |\phi_a\bar{\phi}_b\phi_p\bar{\phi}_q| \\ - |\phi_a\bar{\phi}_b\bar{\phi}_p\phi_q| - |\bar{\phi}_a\phi_b\phi_p\bar{\phi}_q| - |\bar{\phi}_a\phi_b\bar{\phi}_p\phi_q|) \\ (p = LUMO, q = HOMO) \end{aligned} \tag{17}$$

The matrix element between singlet DR and SP configurations is given

$$<{}^1\Phi(DR)|H|{}^1\Phi(SP)> = \sqrt{2/3}(<\phi_a\phi_p|\hat{G}|\phi_a\phi_q> - <\phi_b\phi_p|\hat{G}|\phi_b\phi_q>) \tag{18}$$

The ST gap via the SP effect is approximately given by

$$\Delta E(ST{:}SP)) = -4\frac{<\phi_a\phi_p|\hat{G}|\phi_a\phi_q><\phi_b\phi_p|\hat{G}|\phi_b\phi_q>}{(E(SP) - E(DR))} \tag{19}$$

Judging from Eq. 19, the ST gap becomes positive or negative, depending on the relative magnitude of the exchange integrals. The SP contribution is pictorially expressed by the spin density product (SDP) term by McConnell (5) in the parallel stacking mode of radical pair (8a). However, it is noteworthy that the SP term is different from the McConnell term in the case of the perpendicular conformation between π-radicals (8d,9b). For alternant hydrocarbon systems, the SP effect given by Eq. 19 can be rewritten by a simple rule based on spin density (2c, 8-10) :

(Rule 4) The alternation of the up and down spin densities on the neighboring carbon sites appears in the ground spin state.

The SP rule is easily applied to elucidate the spin alignments for many alternant hydrocarbon systems such as the pentadienyl radical and many π-conjugated nitroxides. Spin density alternations on carbon atoms (16) from the SP effect have been discussed in ref. 16. Here, they will be shown in the case of allyl and pentadienyl radicals (see Figs. 5 and 6).

Spin Unrestricted Hartree-Fock (UHF) and Kohn-Sham (UKS) Models

(A) SOMO-SOMO interactions The preceding intermolecular CI (ICI) approach becomes difficult because of the nonorthogonality of SOMOs required to include the higher-order interactions. Alternately, these interactions can be easily included by the UHF-based CI and perturbation methods (2c,4,13). Therefore, their relation to the preceding ICI model is briefly described. In order to avoid the nonorthogonality problem, the localized natural orbitals (LNO) η_a and η_b for the UHF solution are defined as the corresponding orbitals at the strong correlation limit (7a) in conformity with Anderson's theory of magnetism (4)

$$\eta_a = (\phi_S + \phi_A)/\sqrt{2} \cong \phi_a - \frac{1}{2}S_{ab}\phi_b \;\; (= \phi_a \;\; at \;\; R = \infty)$$
$$\eta_b = (\phi_S - \phi_A)/\sqrt{2} \cong \phi_b - \frac{1}{2}S_{ab}\phi_a \;\; (= \phi_b \;\; at \;\; R = \infty) \qquad (20)$$

where ϕ_S and ϕ_A denote, respectively, the bonding and antibonding delocalized (symmetry-adapted) natural orbitals given by the diagonalization of the first-order density matrix of the UHF solution (13). The orthogonal LNOs ($S=\langle\eta_a|\eta_b\rangle=0$) have tails at the partner sites, although they are, respectively, equivalent to SOMOs of fragment radicals at the dissociation limit. The UHF MOs for singly occupied MOs (SOMOs) with the up(+)- and down(-)-spins in DR species are given by LNOs (13)

$$\psi^+ = \cos\omega \;\; \eta_a + \sin\omega \;\; \eta_b \qquad (21a)$$

$$\psi^- = \cos\omega \;\; \eta_b + \sin\omega \;\; \eta_a \qquad (21b)$$

where ω is the orbital mixing parameter, and it is determined by the extent of the contribution of the zwitterionic (ZW) configurations. The singlet UHF solution is formally expanded by the CI scheme as

$$^1\Psi(^1UHF) = |\Psi^+\overline{\Psi}^-|$$
$$= \frac{1}{\sqrt{2}}{}^1\Phi(DR) + \frac{1}{2}\sin 2\omega(^1\Phi(ZWI) + {}^1\Phi(ZWII)) + \frac{1}{\sqrt{2}}\cos 2\omega\, {}^3\Phi(DR) \qquad (22)$$

where ϕ_a and ϕ_b in Eqs. 2 and 7 are replaced by η_a and η_b, respectively. The last term in Eq. 22 is a triplet configuration, which is the so-called spin contamination in the singlet UHF or unrestricted Kohn-Sham (UKS) model (16,17). Then, the singlet projected UHF wavefunction (11) is given by the configuration mixing of the DR and ZW configurations defined by LNOs. It becomes equivalent to the singlet intermolecular CI wavefunction at the dissociation limit, where $\phi_a = \eta_a$ and $\phi_b = \eta_b$.

(B) SOMO-HOMO and SOMO-LUMO Interactions. The SOMO localized essentially on a radical site in the singlet or triplet UHF solution often has small factions of HOMO and LUMO on the other site, which are responsible for the symmetry-allowed orbital interactions via the CT mechanism between SOMO and LUMO (or HOMO).

$$\eta_a \cong \phi_a + d_p\phi_p + d_q\phi_q \quad (= \phi_a \ \ at \ \ R = \infty)$$
$$\eta_b = \phi_b \tag{23}$$

where d_r(r=p,q) are the orbital-mixing coefficients and are given by the intermolecular perturbation theory (15) as

$$d_p = \frac{\sqrt{2} < \phi_a | \hat{H} | \phi_p >}{E(CTI) - E(DR)}, \quad d_q = \frac{\sqrt{2} < \phi_q | \hat{H} | \phi_a >}{E(CTII) - E(DR)} \tag{24}$$

Therefore, the UHF solution is formally expanded by using Eqs. 23 and 24 into the intermolecular CI wavefunctions. The potential exchange integral between UHF SOMOs is given by

$$\begin{aligned} J_{ab}(PE) &= \int \eta_a(1)\eta_b(1)\hat{G}\eta_a(2)\eta_b(2)d\tau_1 d\tau_2 \\ &= K_{ab} + d_p{}^2 K_{pb} + d_q{}^2 K_{qb} + (other \ \ terms) \end{aligned} \tag{25}$$

Thus the potential exchange (PE) term in the projected UHF solution involves the second-order exchange interaction arising from the DR-CT mixing in Eq. 16 (4, 8b).

(C) Spin Polarization Spin polarization (SP) effects are also included in the splitting of HOMOs with the up(+)- and down(-)-spins in diradical (DR) species under the UHF (UKS) approximation (4, 7-11)

$$\psi^+ = \cos\theta \ \ \phi_q + \sin\theta \ \ \phi_p \tag{26a}$$

$$\psi^- = \cos\theta \ \ \phi_q - \sin\theta \ \ \phi_p \tag{26b}$$

where θ is the orbital mixing parameter, and is determined by the SCF procedure. The approximate spin-projected (AP) UHF solution involves the SP configuration in addition to the CTI(II) and other higher-order excited configurations.

$$\begin{aligned} \Psi(APUHF) &= \mathbf{AP} | \eta_a \overline{\eta}_b \Psi^+ \overline{\Psi}^- | \ \ (\mathbf{AP} = projection \ \ operator) \\ &= \Phi(DR) + d_p\Phi(CTI) + d_q\Phi(CTII) + d_{pq}\Phi(SP) + d_{pqr}\Phi(SC) \end{aligned} \tag{27}$$

where the last term includes all other higher-order configurations generated by the mixings of CT and SP configurations, which are referred to as the spin-correlated (SC) terms in our terminology. The effective exchange integrals given by APUHF (8g) are given by

$$J_{ab}(APUHF) = J_{ab}(KE) + J_{ab}(PE) + J_{ab}(SP) + J_{ab}(SC) \tag{28}$$

The kinetic (KE) and potential (PE) exchange terms are approximately given by the SOMO-SOMO overlap integral S_{ab} and the intermolecular exchange integral K_{ab}, respectively, as shown in Eqs. 11 and 25. The J_{ab}(SP+SC)-term is approximately given by the product of the spin densities ($\rho_{a(b)}$) induced by the SP+SC effect (5, 8b, 8g)

$$J_{ab}(KE) = J_{ab}(OO) = -CS_{ab}^2 \ \ (C{:}constant) < 0 \tag{29a}$$

$$J_{ab}(PE) = J_{ab}(SDP \ \ I) = \sum A\rho_a\rho_b \ \ (A : constant) \tag{29b}$$

$$J_{ab}(SP + SC) = J_{ab}(SDP \ \ II) = -\sum B\rho_a\rho_b \ \ (B : constant) \tag{29c}$$

where the parameters A-C can be determined nonempirically by ab initio computations(8). It is noteworthy that the first (I) and second (II) SDP terms exhibit the reverse sign with respect to the dependency on the sign of spin density product (SDP)(8b). This difference is particularly important for understanding different spin alignment rules for parallel and perpendicular stacking modes between organic radicals as shown in Fig. 3 in ref. 9b and in many examples (8-10).

In addition to the KE, PE, SP and SC terms given by projected UHF, higher-order intermolecular interactions such as van-der-Waals interactions may contribute to the effective exchange interactions.

$$J_{ab}(APUHF\ \ X) = J_{ab}(APUHF) + J_{ab}(higher-order\ \ term) \tag{30}$$

where X denotes the UHF Møller-Plesset (MP) and coupled-cluster (CC) single, double (triple), SD(T), method. APUMP4 and APUCC SD(T) were carried out for the low- and high-spin states of conjugated radical pairs in order to elucidate important roles of the correlation corrections for J_{ab} values (8-11). Here, the APUMP2(4) calculations will be discussed for π-radical pairs for illustrative purpose.

Spin Restricted UNO CI Approach

The above UHF-based approach (4) often suffers the so-called spin contamination error, showing the necessity of the spin projection in the low-spin states (11) as shown in Eq. 22. As a spin-restricted approach, the CAS CI method from the use of the UHF natural orbitals (UNO) (13) and the UNO CASSCF method were utilized for computations of J_{ab} values for π-conjugated radicals with and without positive holes introduced by doping (12). Figure 2 illustrates the UNOs for dimers of free radical species. In our spin-restricted approach, four different CI calculations were carried out to elucidate relative contributions of the direct and indirect mechanisms for effective exchange interactions:

[1] The CAS CI by the use of the two-orbital two-electron {2,2} part includes the direct kinetic and potential exchange (PE) interactions between SOMOs. The CAS CI{2,2} involves the DR and two ZW configurations constructed SOMOs and CT configurations in the intermolecular CI scheme as illustrated in Fig. 1. Therefore, it involves the second-order potential exchange (see Eq. 25) via the CTI(II) configurations.

[2] The first-order (FO) CI based CAS CI {2,2} includes the FO corrections via all the configurations where (i) one of the electrons in the doubly occupied HOMOs is transferred into a virtual orbital and (ii) one electron or one hole is introduced into CAS . Therefore these cover the spin polarization (SP) and all other single excitations relating to CAS {2,2}. The latter involves all the orbital relaxation terms, some of which are illustrated in Fig. 1. The stabilization of the DR configuration by these excitations is referred to as the orbital relaxation (OR) term. Therefore FO CI mainly involves both OR and SP contributions.

[3] The CASCI {m,m} (m>2) includes all the higher-order excitations related to spin correlation (SC) effects and nondynamical electron correlation effects, together with the

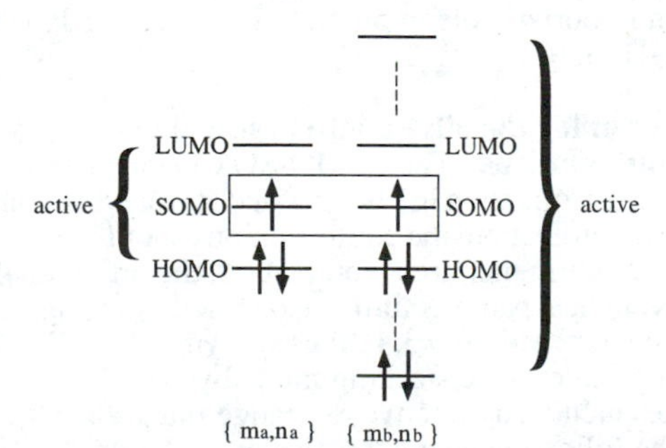

Fig.2 Classifications of UHF natural orbitals (UNO) for CAS CI and CASSCF calculations.

contributions by FO CI. The active orbitals m are selected on the basis of the occupation numbers of UNO in our scheme, and are therefore limited to those of nondynamical correlation effects.
[4] The second-order perturbation (CASPT2) calculations (13,14) based on the CASSCF {2,2} were carried out in order to investigate the dynamical correlation corrections arising from all other orbitals except for the active orbitals in the spin-restricted scheme.

Ab initio Computations of Methyl-Allyl Radical Pair

(A) Sliding conformation In order to illustrate the orbital symmetry rules described in the preceding section, the face-to-face stacking mode between allyl and methyl radicals was examined by changing the sliding distance (R_1) at a fixed interplane distance (R_2=3.4 Å) as shown in Figure 3A. Figure 4A shows variations of the J_{ab}-values with the sliding distance R_1 calculated by the APUMP2(4)/4-31G method.

The following conclusions can be obtained from Figure 4A:
[1] The J_{ab}-values become positive (ferromagnetic) at the bridge conformation **1** with R_1=1.2 Å (Fig. 3) because of the zero SOMO-SOMO overlap integral (S_{ab}) whereas they are negative in other conformations with the nonzero S_{ab}.
[2] The magnitude of the ferromagnetic interaction from APUMP4/4-31G is 76 cm^{-1} at **1**. The positive J_{ab}-values via the PE mechanism are quite large in spite of the van der Waals contact between allyl and methyl radical pair.

The potential exchange term K_{ab} overweighs the kinetic exchange term at conformation **1**, indicating the importance of the no-overlap and orientation principles (8) for the ferromagnetic interaction between NBMO and p-type SOMO. The conclusions [1] and [2] are compatible with the spin alignment rules 1 and 2.

Previously (8b), we have emphasized an important role of NBMO for ferromagnetic intermolecular interactions. Since the NBMO-type SOMO has a node at the center of the CCC bond, the pπ* orbital overlap integral $S_{p*\pi*}$ should disappear at the T-shape conformation as illustrated in Fig. 5

$$S_{p*\pi*} = S_{p1} - S_{p3} = 0 \tag{31}$$

where $S_{p1(3)}$ denotes the orbital-overlap between the atomic p-AO and the atomic site 1 (3) of the allyl radical. On the other hand, the PE term is given by

$$K_{ab} = \langle p_a(1)\pi_A^*(1) | p_a(2)\pi_A^*(2) \rangle = S_{p3}^2 U/2 > 0 \tag{32}$$

where the Mulliken approximation is employed for the atomic orbital overlap and U is the on-site Coulombic repulsion;U= U(N) = U(O). The parallel orientation is rather important for the nonzero overlap between atomic sites in Eq. 32. Therefore the effective exchange integral J_{ab} in Eq. 1 should become positive (ferromagnetic) at the structure **1** because of the nonzero potential exchange integral. Generally, the symmetry (or node) of the SOMO plays an important role in predicting the sign of J_{ab} (8g,9c) in conformity with the selection rules 1 and 2.

(B)T-shape conformation for the allyl-methyl radical pair

(1) UHF-based computations The SOMO-SOMO interaction is always symmetry-forbidden at the T-shape conformation in Figs. 5A and B, for which the potential exchange term K_{ab} is not zero. Therefore, both the conformations should exhibit the ferromagnetic interaction. In order to confirm this simple prediction, the orthogonal (T-shape) stacking mode of the methyl-allyl radical pair was further examined by changing the sliding distance (R_1) at a fixed interplane distance (R_2 =3.4 Å) as shown in Fig. 3B. The J_{ab}-values were calculated for the parallel interplane stacking mode by APUMP2(4)/4-31G. Figure 4B shows variations of the calculated effective exchange integrals with R_1.

From Fig. 4B, the following characteristics were drawn:
[1] The J_{ab}-value is negative at the bridge structure (Fig. 5A), whereas it is positive at the

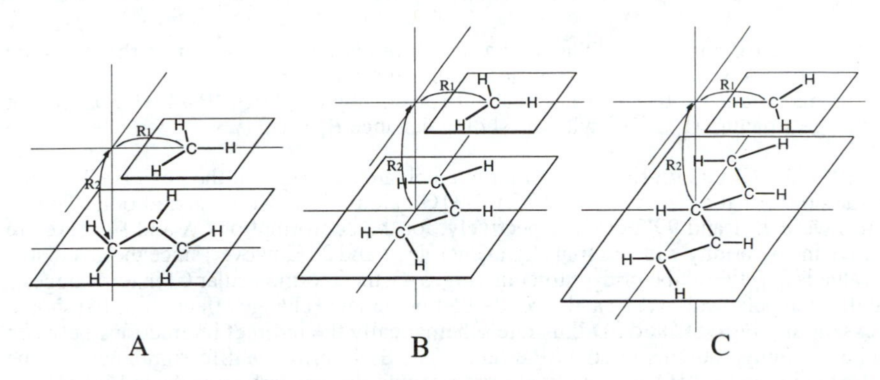

Fig.3 Schematic illustrations of the sliding (A) and T-shape (B) conformations between methyl and ally radicals. R1 and R2 denote, respectively, the sliding distance and interplane distance.

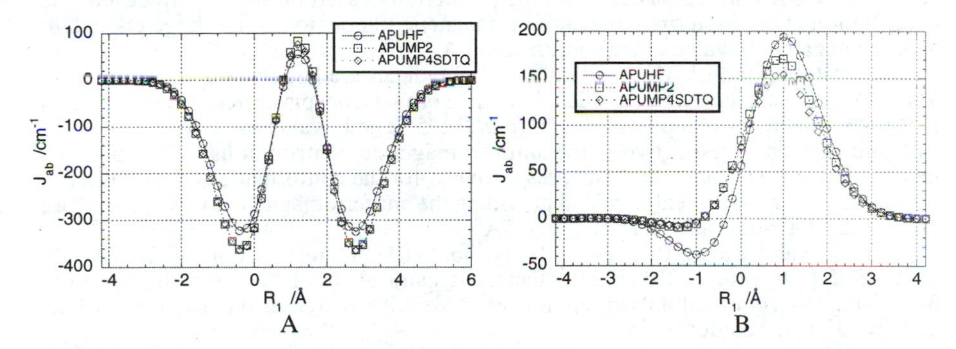

Fig.4 Variations of effective exchange integrals (J_{ab}) calculated for the parallel face-to-face stacking modes in A and B of Fig. 3 by the APUHF and APUMP4(2)/4-31G methods.

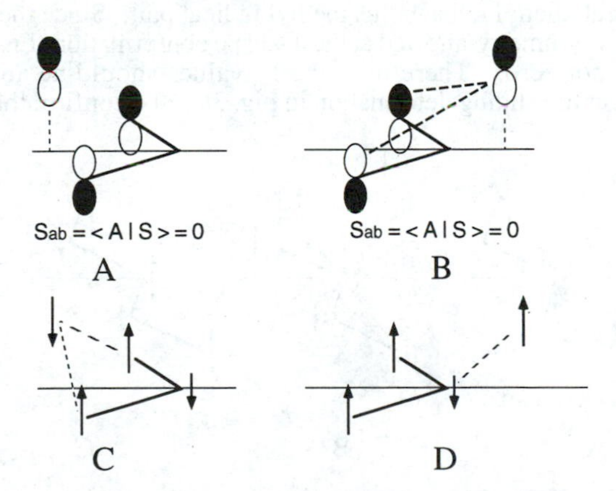

Fig.5 Schematic illustrations of the orbital-overlap (OO) and spin density product (SDP) terms for the allyl-methyl radical pair. A and B denote, respectively, the OO interactions at the T-shape conformations. C and D show, respectively, the corresponding SDP terms.

conformation B in Fig. 5. The J_{ab}-value becomes negative even at the T-shape conformation A, suggesting the significant contribution of the higher-order effects.
[2] The maximum positive J_{ab}-value is about 150 cm^{-1} by the APUMP4/4-31G method at the bridge structure (Fig. 5B) with the sliding distance R_1 = 1.0 Å.

(2) CAS CI calculations In order to elucidate the origin of the negative J_{ab} value for the conformation A, CAS CI {2,2}/4-31G calculations were carried out. The J_{ab} values were 25.1 and 99.2 cm^{-1}, respectively, for the conformations A and B: these are positive in conformity with the spin alignment rules 1 and 2. However, since the calculated J_{ab}-value is negative at the bridge structure (Fig. 5A), the intermolecular CI in turn suggests that the spin polarization (SP) and second-order potential exchange effects (Eq. 16) should be essential. Figs. 5C and 5D illustrate schematically the indirect interactions between allyl and methyl radicals via the SP effect. The SDP term is antiferromagnetic at the conformation in Fig. 5C, whereas it is ferromagnetic in the conformation in Fig. 5D. In order to examine these pictorial explanations, the first-order (FO) CI based on CAS {2,2} was carried out. The calculated J_{ab} values were 23.2 and 105 cm^{-1}, respectively, for these conformations: the corrections of the J_{ab} values by FO CI are -1.9 and 16 cm^{-1}. The calculated results are consistent with the predictions based on the SP effect, but the contribution of the SP term to the antiferromagnetic interaction is not large enough to give the negative J_{ab} value at the conformation A.

The HOMO and LUMO π-orbitals of allyl radicals are largely spin polarized (8b). The four π-orbitals should be considered as active orbitals for spin correlation (SC) effects in the allyl-methyl radical pair as illustrated in Fig. 5. The J_{ab} values by CAS CI {4,4} are -6.1 and 85.6 cm^{-1}, respectively: the antiferromagnetic contributions via SP plus SC mechanism are -31.2 and -13.6 cm^{-1}, respectively, for the conformations A and B. The SP+SC term plays an essential role in providing the antiferromagnetic exchange integral even in the T-shape conformation in Fig. 5A.

The J_{ab} value from APUMP4/4-31G is about -10 cm^{-1}, reproducing approximately the CAS CI{6,6} result. This implies that the dynamical correlation correction (see Eq. 30) arising from other orbitals except for active six-MOs increases the magnitude of the negative J_{ab} at the conformation A.

Ab initio Computations of the Pentadienyl-Methyl Radical Pair

CAS CI calculations The other orthogonal (T-shape) conformation in Fig. 3C was examined for the pentadienyl radical plus methyl radical pair. Since the SOMO-SOMO interaction is always symmetry-allowed at the T-shape conformation (Fig. 6A), the orbital overlap integral is not zero. Therefore, the J_{ab} values should be antiferromagnetic (negative) throughout the sliding deformation in Fig. 3C. To confirm this prediction, the

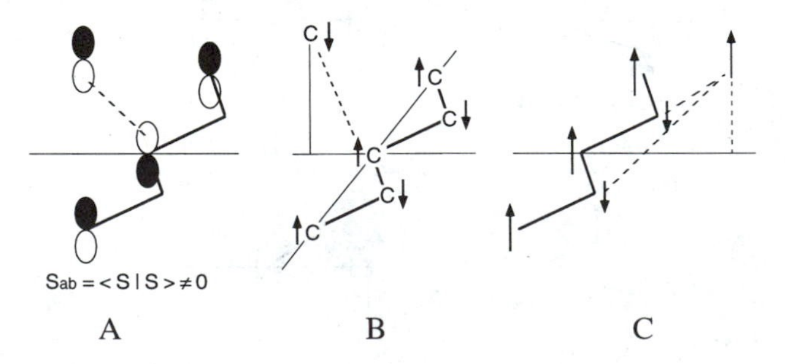

Fig.6 Schematic illustrations of the orbital-overlap (OO) and spin density product (SDP) terms for the pentadienyl-methyl radical pair. A denotes the OO interactions at the T-shape conformation. B and C show, respectively, the SDP terms.

CAS CI {2,2}/4-31G calculations were carried out. The J_{ab} values were -118 and -1.53 cm^{-1}, respectively, for the conformations A(B) and C in Fig. 6, supporting the symmetry arguments derived from the preceding intermolecular perturbation theory.

However, the spin polarization (SP) and higher-order spin-correlation effects may play an important role to provide ferromagnetic interactions between this radical pair. For example, Fig. 6B and 6C illustrate schematically the indirect interactions between pentadienyl and methyl radicals via the spin polarization (SP) effect. The SDP term is antiferromagnetic at the conformation B, whereas it is ferromagnetic in the conformation C. In order to check these explanations, the first-order (FO) CI based on CAS {2,2} was carried out, showing that the J_{ab} values were -126 and 1.21 cm^{-1}, respectively, for these conformations: the corrections of the J_{ab} values by FO CI are -8 and 2.74 cm^{-1}. These results are consistent with the predictions based on the SP effect. The SP term plays a critical role in providing the ferromagnetic interaction at the latter conformation C.

The π-orbitals of pentadienyl radicals are largely spin polarized. The six π-orbitals should be considered as active orbitals for spin correlation (SC) effects in the pentadienyl-methyl radical pair as can be recognized from Fig. 2. The J_{ab} values by CAS CI {6,6} are -132 and 6.02 cm^{-1}, respectively: the antiferromagnetic and ferromagnetic contributions via SP and SC are -14 and 7.55 cm^{-1}, respectively, for B and C in Fig. 6. Thus the SP+SC term plays an essential role in providing the ferromagnetic exchange at the T-shape conformation C.

The J_{ab} value by APUMP2/4-31G is about 14 cm^{-1} for C, reproducting approximately the CAS CI{6,6} result. The dynamical correlation correction (Eq. 30) arising from other MOs except for the active six-MOs enhances the positive J_{ab} at this conformation.

Discussions and Concluding Remarks

The preceding ab initio results clarified that the sign and magnitude of the effective exchange integrals (J_{ab}) between π-radicals are determined by subtle balances of different contributions elucidated by the intermolecular perturbation and CI methods, together with the natural orbital CI expansions of the UHF solutions. In fact, ferromagnetic effective exchange interactions are feasible in the parallel interplane orientations for π radical pairs since the potential exchange (PE) interaction overweighs the kinetic exchange (KE) interaction at some conformations. However, the spin polarization (SP) and higher-order spin correlations (SC) play important roles in many subtle cases (18-23). The SDP term arising from SP plus SC terms is useful for pictorial understanding of the spin alignment rules in these complex cases. The computational procedures summarized in this article were successfully applied to determine the sign and magnitude of J_{ab} values for clusters extracted from the X-ray structures of organic ferromagnets (18-23). The computational results are useful for theoretical explanation and understanding of many experimental results available for carbene and nitroxide derivatives (18-23). They were wholly compatible with the spin alignment rules 3 and 4 discussed in the first section.

Organic ferromagnetism is nothing but one of the cooperative phenomena. So, we must examine the transition temperatures (T) for the ferro- and ferri-magnetic phase transitions. The Curie (T_C) and Neel (T_N) temperatures for β and γ-phases of para-nitrophenyl nitronyl nitroxide, p-NPNN, (18) and related species were calculated by using the calculated J_{ab} values (24) in combination with the Langevin-Weiss-Neel mean field model (25) and extended Bethe model (26). The T_C for β-phase was reproduced by the mean field approximation (24a), whereas the Bethe-type model was essential for explanation of the T_N observed for the γ-phase because of its low-dimensionality (24b).

Here, we could not touch new opportunities in molecule-based magnetic materials; (i) organic helimagnet (6a) (ii) organic magnetic metals (12c,27a), (iii) organic Kondo and dense Kondo systems (27b,c), (iv) organic high-Tc superconductors via the spin fluctuation mechanism (27c), (v) organic Haldane system (27d), (vi) photo-induced CT magnets (27e,f) etc. Theoretical approaches to examine these new possibilities (27) were already initiated. Probably, such new fields in molecular magnetism will be realized in the future (28).

In conclusion, no SOMO-SOMO overlap and orientation principles proposed previously (8-10) are applicable to π–conjugated radical crystals (18-23) which exhibit ferromagnetic effective exchange interactions. Here, it should be emphasized that the NBMO-nature of SOMO's for alternant hydrocarbon radicals and π*-nature of SOMO's for nitroxides play important roles in reducing the orbital-overlap (OO) antiferromagnetic term even in parallel interplane orientations. Thus, controls of spatial orientations of π-conjugated radicals and nitroxides are important for ferromagnetic intermolecular interactions between these species. Present and previous computations conclude that the no overlap and orientation principle (8a) followed by the CT and SP rules (*Rules 1-4*) is a useful guide for molecular design of organic ferromagnets.

Acknowledgement

This work was supported by a Grant-in-Aid for Scientific Research on Priority Areas of Molecular Magnetism (No. 04242101). We are also grateful for the financial support of the Ministry of Education, Science and Calture of Japan (Specially Promoted Research No. 06101004).

References

1. (a) J. S. Miller and D. A. Dougherty Eds., Mol. Cryst. Liq. Cryst. Vol. 176 (Gordon and Breach, London, 1989); (b) D. Gatteschi, O. Kahn, J. S. Miller and F. Palacio Eds., Magnetic Molecular Materials (NATO ASI Series No. 198, Kluwer Academic Pub. 1991); (c) H. Iwamura and J. S. Miller Eds., Mol. Cryst. Liq. Cryst. Vols. 232 and 233 (Gordon and Breach, London, 1992); (d) J. S. Miller Eds., Mol. Cryst. Liq. Cryst. Vol 271 (Gordon and Breach, London, 1995).
2. (a) K. Yamaguchi, Chem. Phys. Lett. 28, 93 (1974); (b) idem., ibid. 34, 434 (1975);(c) idem., ibid. 35, 230 (1975).
3. L. Salem, "Electrons in Chemical Reactions: First Principles" (John Wiely & Sons, New York, 1982) chapter 7.
4. P. W. Anderson, Solid State Phys. 14, 61 (1963).
5. H. M. McConnell, J. Chem. Phys. 39 (1963) 1910.
6. (a) K. Yamaguchi, Chem. Phys. Lett. 30, 288 (1975); (b) K. Yamaguchi, Y. Yoshioka and T. Fueno, Chem. Phys. 20 (1977) 171; (c) Y. Yoshioka, K. Yamaguchi and T. Fueno, Theoret. Chim. Acta 45, 1 (1978).
7. (a) K. Yamaguchi, "Self-Consistent-Field: Theory and Applications" (R. Carbo and M. Klobukowski, Eds., Elsevier, Amsterdam, 1990) p727; (b) K. Yamaguchi, M. Okumura, K. Takada and S. Yamanaka, Int. J. Quant. Chem. S27, 501(1993).
8. (a) K. Yamaguchi, T. Fueno, K. Nakasuji and I. Murata, Chem. Lett., 629 (1986);(b) K. Yamaguchi and T. Fueno, Chem. Phys. Lett. ,159, 465 (1989); (c) K. Yamaguchi, H. Namimoto, T. Fueno, T. Nogami and Y. Shirota, idem. 166, 408 (1990); (d) K. Yamaguchi, M. Okumura and M. Nakano, idem. 191, 237 (1992); (e) T. Kawakami, S. Yamanaka, W. Mori, K. Yamaguchi, A. Kajiwara and M. Kamachi, Chem. Phys. Lett. 235, 414 (1995).
9. (a) K. Yamaguchi, H. Namimoto and T. Fueno, Mol. Cryst. Liq. Cryst. 176, 151 (1989); (b) M. Okumura, W. Mori and K. Yamaguchi, idem. 232, 35 (1993); (c) T. Kawakami, S. Yamanaka, H. Nagao, W. Mori, M. Kamachi and K. Yamaguchi, idem., 272, 117 (1995);
10. (a) K. Yamaguchi, Y. Toyoda and T. Fueno, Kagaku 41, 585 (in Japanese) (1986); (b) K. Yamaguchi, "Quant. Chem. for Molecular Design (in Japanese) " (K. Nishimoto and A. Imarura Eds. Kodansha Sci. Tokyo, 1989) part II chapter 3; (c) K. Yamaguchi, T. Kawakami and D. Yamaki, "Molecular Magnetism (in Japanese)" (K. Itoh Ed., Gakkai Pub., 1996), chapter 2.
11. (a) K. Yamaguchi, Y. Takahara and T. Fueno, in " Applied Quant. Chemistry" (V.

H. Smith et al. Eds., Reidel, Boston, 1986) p155; (b) K. Yamaguchi, F. Jensen, A. Dorigo and K. N. Houk, Chem. Phys. Lett. 149, 537 (1988); (c) K. Yamaguchi, M. Okumura, J. Maki, T. Noro, H. Namimoto, M. Nakano, T. Fueno and K. Nakasuji, ibid. 190, 353 (1992); (d) K. Yamaguchi, M. Okumura, W. Mori, J. Maki, K. Takada, T. Noro and K. Tanaka, ibid. 210, 201 (1993).

12. (a) S. Yamanaka, M. Okumura, K. Yamaguchi and K. Hirao, Chem. Phys. Lett. 225, 213 (1994); (b) S. Yamanaka, M. Okumura, H. Nagao and K. Yamaguchi, ibid. 233, 88 (1995) (c) S. Yamanaka, T. Kawakami, M. Okumura and K. Yamaguchi, ibid. 233, 257 (1995).
13. (a) K. Yamaguchi, K. Ohta, S. Yabushita and T. Fueno Chem. Phys. Lett. 49, 555 (1977); (b) K. Yamaguchi and T. Fueno, Chem. Phys. 23, 375 (1977); (c) K. Yamaguchi, S. Yabushita, T. Fueno, S. Kato, K. Morokuma and S. Iwata, Chem. Phys. Lett. 71, 563 (1980); (d) K. Yamaguchi, Int. J. Quant. Chem. S14, 269 (1980).
14. K. Andersson, P. Malqvist, B. O. Roos, A. J. Sadlej and K. Wolinski, J. Phys. Chem. 94, 5483 (1990).
15. (a) T. Fueno, S. Nagase, K. Tatsumi and K. Yamaguchi, Theoret. Chim. Acta 26, 43 (1972); (b) K. Yamaguchi, Thesis (Osaka University, 1972).
16. (a) S. Yamanaka, T. Kawakami, S. Yamada, H. Nagao, M. Nakano and K. Yamaguchi, Chem. Phys. Lett. 240, 268 (1995); (b) S. Yamanaka, T. Kawakami, H. Nagao and K. Yamaguchi, Mol. Cryst. Liq. Cryst. 271, 19 (1995).
17. S. Yamanaka, T. Kawakami, H. Nagao and K. Yamaguchi, Chem. Phys. Lett. 231, 25 (1994).
18. T. Sugawara, S. Murata, K. Kimura, H. Iwamura, Y. Sugawara and H. Iwasaki, J. Am. Chem. Soc. 106, 6449 (1984).
19. (a) K. Awaga and Y. Maruyama, Chem. Phys. Lett. 158, 556 (1989); (b) idem., J. Chem. Phys. 91, 2743 (1989),
20. (a) M. Kinoshita, P. Turek, M. Tamura, K. Nozawa, D. Shiomi, Y. Nakazawa, M. Ishikawa, M. Takahashi, K. Awaga, T. Inabe and Y. Maruyama, Chem. Lett. 1225 (1991); (b) M. Tamura, Y. Nakazawa, D. Shiomi, K. Nozawa, Y. Hosokoshi, M. Ishikawa, M. Takahashi and M. Kinoshita, Chem. Phys. Lett. 186, 401 (1991); (c) Y. Nakazawa, M. Tamura, N. Shirakawa, D. Shiomi, M. Takahashi, M. Kinoshita and M. Ishikawa, Phys. Rev. B46 (1992) 8906.
21. R. Chiarelli, A. Rassat and P. Rey, J. Chem. Commun. (1992) 1081.
22. (a) T. Ishida, K. Tomioka, T. Nogami, H. Iwamura, K. Yamaguchi, W. Mori and Y. Shirota, Mol. Cryst. Liq. Cryst. 232, 99 (1993); (b) T. Nogami, K. Tomioka, T. Ishida, H. Yoshikawa, M. Yasui, F. Iwasaki, H. Iwamura, T. Takeda and M. Ishikawa, Chem. Lett. (1994) 1723.
23. M. Kamachi, H. Sugimoto, A. Kajiwara, A. Harada, Y. Morishima, W. Mori, N. Ohmae, N. Nakano, M. Sorai, T. Kobayashi and K. Amaya, Mol. Cryst. Liq. Cryst. 232, 53 (1993).
24. (a) M. Okumura, K. Yamaguchi, M. Nakano and W. Mori, Chem. Phys. Lett. 207, 1 (1993); (b) M. Okumura, W. Mori and K. Yamaguchi, ibid. 219, 36 (1994).
25. D. C. Mattis, " Theory of Magnetism II" (Springer, Berlin, 1985).
26. (a) H. A. Bethe, Z. Physik 71, 205 (1931); (b) M. Takahashi, M. Kinoshita and M. Ishikawa, J. Phys. Soc. Jpn. 61, 3745 (1992).
27. (a) K. Yamaguchi, Y. Toyoda and T. Fueno, Synthetic Metals 19, 81 (1987); (b) K. Yamaguchi, M. Okumura, T. Fueno and K. Nakasuji, ibid. 41-43, 3631 (1991); (c) K. Yamaguchi, Int. J. Quant. Chem. 37, 167 (1990); (d) M. Okumura, S. Yamanaka, W. Mori and K. Yamaguchi, J. Mol. Structure (Theochem) 310, 177 (1994). (e) M. Okumura, W. Mori and K. Yamaguchi, Comp. Aided Innov. of New Materials II (M. Doyama et al. Eds. Elsevier, 1993) p1785. (f) G. Maruta, D. Yamaki, W. Mori, K. Yamaguchi and H. Nishide, Mol. Cryst. Liq. Cryst. 279, 19 (1996).
28. K. Yamaguchi and D. Gatteschi, NATO ASI series (1996).

Chapter 4

Magnetic-Susceptibility Measurement Techniques

Charles J. O'Connor

Department of Chemistry, University of New Orleans, Lakefront Campus, New Orleans, LA 70148

Different techniques for the measurement of magnetic susceptibility and magnetization are described. A comparison of instrumental capabilities, sensitivities, and methods of measurement is presented. Experiments involving temperature- and field-dependent measurements, ac- and dc-magnetic susceptibility, magnetic remanence, magnetic hysteresis measurements, and time dependent measurements are described. Examples of magnetic systems include superparamagnetic nanophasic particles, solid state phosphates, and other materials that exhibit magnetic phase transitions.

1. INTRODUCTION

This chapter is designed to provide an overview of instrumental and experimental techniques for the measurement of magnetic susceptibility. Much of the material that is presented here is summarized from more in depth treatments in reviews and books on magnetochemistry.[1-3] Several experimental examples taken from recent research observations are also given to help to explain the magnetochemical concepts. The chapter is divided into two parts - a brief summary of the instrumental techniques to measure magnetic susceptibility is discussed first followed by a description of various experimental measurement techniques to extract magnetic information.

2. INSTRUMENTAL TECHNIQUES

2.1. Force Methods

When an isotropic paramagnetic material is placed in an inhomogeneous magnetic field, a displacement force is exerted on the sample drawing it into a region of higher field. Since the displacement force depends on both magnetization and field gradient, measurement of the force gives direct information on the magnetic susceptibility of a material. Instruments that measure this force are collectively called force magnetometers.

0097–6156/96/0644–0044$15.75/0

2.1.1 *Gouy Method*

The simplest technique used by inorganic chemists to measure magnetic susceptibility is the Gouy method. The materials needed to assemble a Gouy balance are simple and may be found in many chemistry laboratories. With an analytical balance, a stable magnet, and some care in the fabrication of the apparatus, a sensitive instrument capable of giving a high degree of precision may be obtained.

The principle of operation is as follows. The sample is placed in a long cylindrical tube which is suspended from an analytical balance. The sample tube is positioned between the poles of the magnet such that one end of the tube is in the region of homogeneous field and the other end is in the region of zero field (see Figure 1). The force exerted on the sample is a function of the volume occupied by the sample in the region of the field gradient. This force, expressed in equation 1, may be written in scalar form as a function of the isotropic volume susceptibility,

$$f = M\frac{dH}{dz} = \chi_v vH\frac{dH}{dz} \qquad (1)$$

Where χ_v is the susceptibility per unit volume, v is the volume of the sample, and dH/dz is the field gradient of the magnet over the sample. If we integrate equation (1) over the entire length of the sample, from H = 0 to H, we obtain the equation

$$f = \frac{1}{2}\chi_v vH^2 A \qquad (2)$$

where A is the cross-sectional area of the sample. When the sample is positioned in the magnetic field as in Figure 1a, a paramagnetic material experiences an increase in weight while a diamagnetic material experiences a decrease in weight as a result of the displacement force exerted on the sample (eq. 2). It is rare that magnetic measurements are made in a vacuum and, for this reason, one must correct for magnetic buoyancy due to the gases displaced by the sample in the sample chamber. This correction is accomplished by using equation (3)

$$f = \frac{1}{2}(\chi_v - \chi_v^{(0)})H^2 A \qquad (3)$$

where $\chi_v^{(0)}$ is the magnetic susceptibility of the displaced gas. Values of $\chi_v^{(0)}$ for various gases at different temperatures have been tabulated by other authors.[4] If the temperature in the sample chamber is varied, measurement of the temperature dependent magnetic susceptibility will provide information on the electron structure of the paramagnetic species.

The Gouy experiment is capable of very precise measurements and reaches an absolute sensitivity approaching 10^{-6} to 10^{-10}emu at fields of 10kOe. However, some routine difficulties arise in the Gouy experiment that may render the technique unsuitable for many experiments. A large amount of sample is needed to fill the sample tube, usually several hundred milligrams. In addition, the susceptibility is dependent on the density of the sample in the tube. The sample packing must be highly uniform and reproducible over the entire length of sample exposed to the field gradient. Sample uniformity may be accomplished by grinding; however, this often results in the loss of solvents of crystallization. Since the sample is exposed to a continuously varying magnetic field, there must not be a field dependence to the magnetic response of the sample. As a result of these idiosyncrasies, magnetic susceptibility measurements using the Gouy method can usually be given an uncertainty of 2 to 5%.

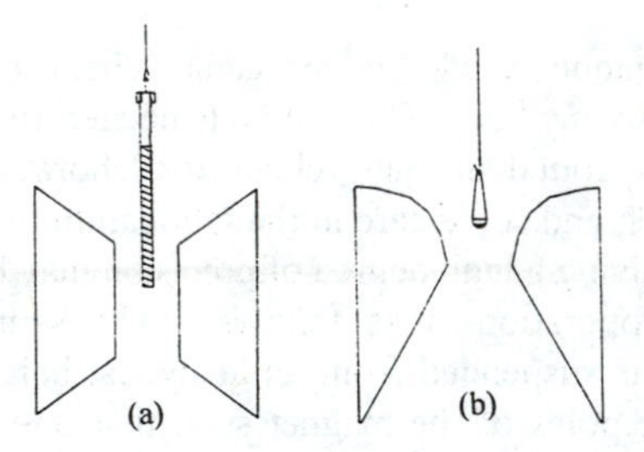

Figure 1. Sample configuration for force measurement of magnetic susceptibility using the (a) Gouy method and (b) Faraday method.

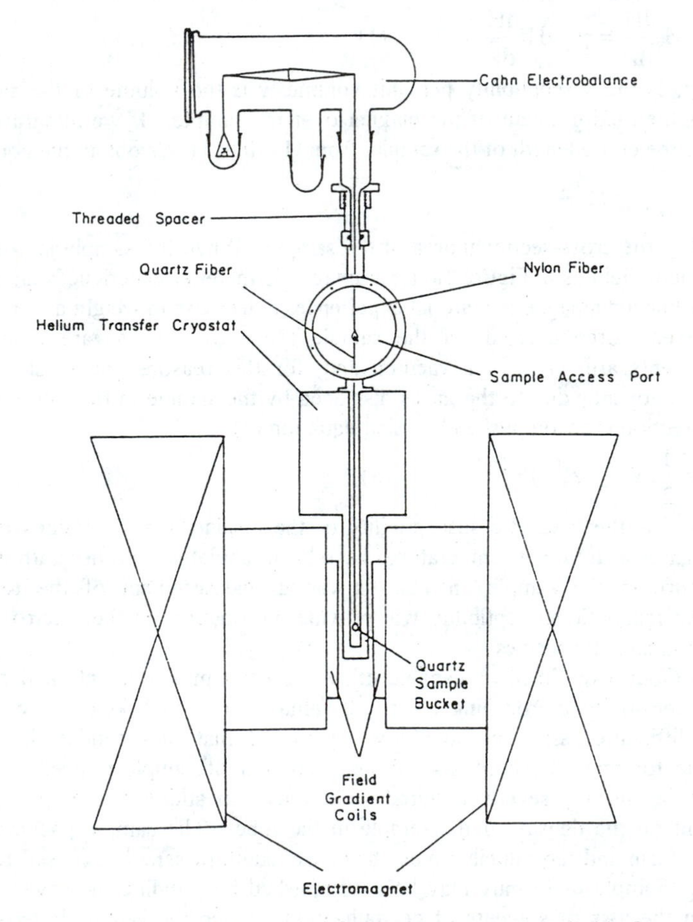

Figure 2. Schematic diagram of alternating force magnetometer suitable for measurements in the 4 to 300K range.

2.1.2. *Faraday Method*

If an isotropic sample is placed in a region of the field gradient where H(dH/dz) is constant over the volume of the sample, then the force exerted on the sample is independent of the density of the material and depends only on the total mass of the sample present. Equation (2) may now be rewritten as

$$f = M\frac{dH}{dz} = \chi_m mH\frac{dH}{dz} \qquad (4)$$

where χ_m is the magnetic susceptibility per unit mass and *m* is the mass of the sample.

In the Faraday experiment, a small sample is placed in a region of the field gradient where H(dH/dz) is constant. It is desirable to use specially designed pole faces that concentrate the field gradient and extend the region of constant H(dH/dz) over several millimeters. The experimental presentation of the sample is shown in Figure 1b.

The Faraday method has several advantages over the Gouy method. Only a small amount of sample (a few milligrams) is needed for a Faraday measurement compared to the amount (several hundred milligrams) needed for the Gouy measurement. In addition, since sample density uniformity is not required in the Faraday experiment, no sample packing error is introduced. The absolute sensitivity of the Faraday method is about the same as that for the Gouy method, a limit of about 10^{-8} emu. However, since the Faraday method measures mass susceptibility directly, a higher degree of reproducibility is usually attained. The uncertainty limit with a Faraday balance can be made less than 1%.

Single crystal measurements are possible with a Faraday balance and such measurements have been reported in the literature.[5] However, a rigid sample suspension device must be used to prevent magnetically induced torque from reorienting a sample composed of anisotropic material. If single-crystal measurements are desired, the torsion balance[6] or induction methods are also available for these types of measurements.

2.1.3. *Alternating Force Magnetometer (AFM)*

The field gradient required to induce a displacement force for the Faraday method may be produced by induction coils rather than by specially shaped pole caps. If induction coils are used, the field gradient may be varied independently of the static magnetic field.[7] Moreover, reversing the direction of current flow reverses the direction of the force on the sample. This reversal is the basis of operation of the alternating force magnetometer (AFM). The method has the definite advantage over conventional Faraday measurements that the static magnetic field may be varied independently from the magnetic field gradient, thus giving the experimentalist better control of experimental conditions. In addition, the static field may be held constant as the direction of the field gradient is periodically alternated. The force detection may then be passed through a phase-sensitive detector.[8] The result is reduced noise levels and better reproducibility of the susceptibility data. Figure 2 shows the force magnetometer that has been assembled in the author's laboratory. The field gradient is generated by Lewis gradient coils driven by a bipolar power supply.[9] Low-temperature capabilities may be added to the system through the use of a Helium-transfer Dewar. The remainder of the AFM apparatus follows a standard force magnetometer design.

2.2. Induction Methods

The change in magnetic flux density that results from placement of a material in a magnetic field may be monitored by inductive methods. If a sample is inserted into an induction detection coil, then a change in the voltage is induced in this coil concomitant with the insertion of the sample into the detection coil. The strength of the induced voltage is given by the equation

$$v = \frac{\mu N^2 A}{S}\frac{di}{dt} \qquad (5)$$

where N is the number of turns of wire, A is the cross-sectional area, S is the length of the coil, and di/dt is the frequency of current oscillation. The quantity μ is the permeability of the material within the coil and is related to the magnetic susceptibility by the equation

$$\mu = 1 + 4\pi\chi \qquad (6)$$

The general inductive response described in equation (5) is the basis of several techniques that measure magnetic susceptibility.

2.2.1. *AC Induction*

The inductive response of a specimen may be measured in the presence of a static magnetic field, an oscillating magnetic field, or both. If an oscillating magnetic field is used. the magnetic response depends on the frequency of oscillation. Static magnetic susceptibilities are measured when there is sufficient time to achieve equilibrium among the populated spin states. If an alternating field of sufficiently high frequency is applied, the magnetization tends to lag behind the oscillating magnetic field. There are some pronounced differences that may arise between the ac and dc magnetic experiments and ac measurements are often necessary to adequately characterize a material. Excellent ac-magnetic susceptibility measurement instruments are commercially available.

If we define the applied magnetic field as consisting of a static component H_0 and an oscillating component H_1, then we may write the magnetic field at any time t as

$$H(t) = H_0 + H_1 \cos(\omega t) \qquad (7)$$

Where ω is the period of oscillation.

The resulting magnetization of a sample in the oscillating magnetic field may be written as

$$M(t) = M_0 + M_1 \cos(\omega t - \varphi) \qquad (8)$$

where φ is the phase angle by which the magnetization lags the oscillating component of the magnetic field. We may then write

$$M(t) = \chi_0 H_0 + \chi' H_1 \cos(\omega t) + \chi'' H_1 \sin(\omega t) \qquad (9)$$

where

$$\chi_0 = \frac{M_0}{H_0} \qquad \chi' = \frac{M_1 \cos(\varphi)}{H_1} \qquad \chi'' = \frac{M_1 \sin(\varphi)}{H_1} \qquad (10)$$

and χ' and χ'' depend on the frequency and magnitude of the oscillating field; χ' represents the high-frequency or in-phase component of the magnetic susceptibility, and χ'' is the paramagnetic dispersion or out-of-phase component of the magnetic susceptibility. At low frequencies, there is no lag of M_1 behind H_1 ($\varphi = 0$) and therefore in the low frequency limit $\chi' = \chi_0$ and $\chi'' = 0$. At high frequencies, the two components of the susceptibility may easily be sorted out by phase-sensitive detection. With the use of a low-frequency inductive bridge, the field-dependent and near-zero field static susceptibility may be measured. When a high-frequency inductive bridge is used, information may be obtained on relaxation times of the electron spins.

A design by Carlin and coworkers[10] for an ac mutual inductance susceptometer is shown in Figure 3. The detection coil (secondary) has two components, identical in geometry, but wound in opposite senses. The sample is inserted in one component of the secondary coil. Since the components of the coil are in the homogeneous regions of the ac (primary) and dc (superconducting) solenoids, any voltage induced in the two components of the secondary coil by these external fields cancels. The output voltage from the detection coil is therefore dependent only on the magnetic susceptibility of the sample. As the temperature and static magnetic field are varied, the resultant change in the susceptibility of the sample is monitored. The apparatus described above has been primarily used for static low-frequency magnetic susceptibility measurements. Van Duyneveldt and coworkers[11] have described a similar device that is used for higher frequency measurements.

2.2.2. *Vibrating Sample Magnetometer (VSM)*

The principle of operation of the VSM depends on the inductive detection of the dipole field of an oscillating magnetic sample in a uniform magnetic field.[12] A diagram of the first VSM instrument fabricated by Foner[12] is shown in Figure 4. Sample movement was generated by a loudspeaker driven at a set frequency. The reference and the sample are mounted on a rigid rod, vibrate in phase, and are monitored by inductive coils oriented with their axes parallel to the vibration. The reference coils are used to phase detect the sample coils with respect to the frequency of vibration. The extent to which the coils respond to the vibration is proportional to the magnetic moment that is vibrating. Phase-sensitive detection removes spurious signals and provides a relatively sensitive susceptibility measurement. The VSM is ideally suited for single-crystal measurement. The accommodation of small sample size, fixed orientation with respect to axes, and homogeneous fields are all assets for single-crystal measurements. Since the sample is positioned in a homogeneous field, this method may also be used for field-dependent studies. Although the detection method is formally an induction method, the oscillation of the induction results from sample movement, not from a change in the applied magnetic field. The method therefore measures the dc-magnetic susceptibility rather than ac-susceptibility. The limit of VSM sensitivity with current commercial instruments is about 10^{-5} emu.[14]

2.2.3. *Superconducting Susceptometer (SQUID)*

In the early 1960s, three breakthroughs occurred in low-temperature physics that allowed the development of some very unique electrical devices. These breakthroughs include the effect of a weak link on a superconducting material,[15] the BCS theory of superconductivity,[16] and the quantization of magnetic flux in a superconducting ring.[17] As a result of these major advances in low temperature physics, the Superconducting QUantum Interference Device (SQUID) was developed.[18]

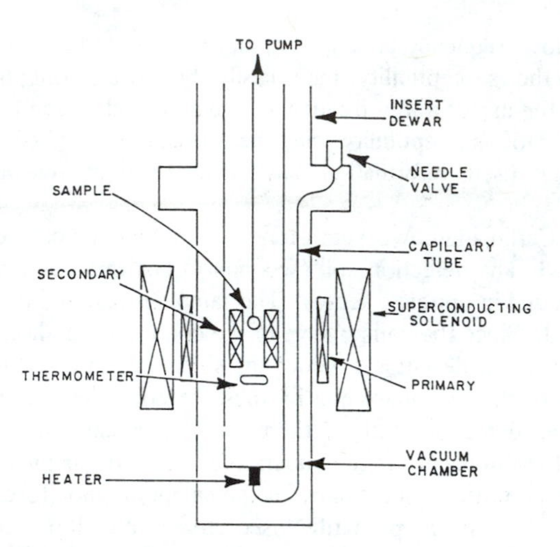

Figure 3. Experimental arrangement of insert Dewar for ,variable temperature susceptibility measurements using the mutual inductance method. (Reproduced from Ref. 10).

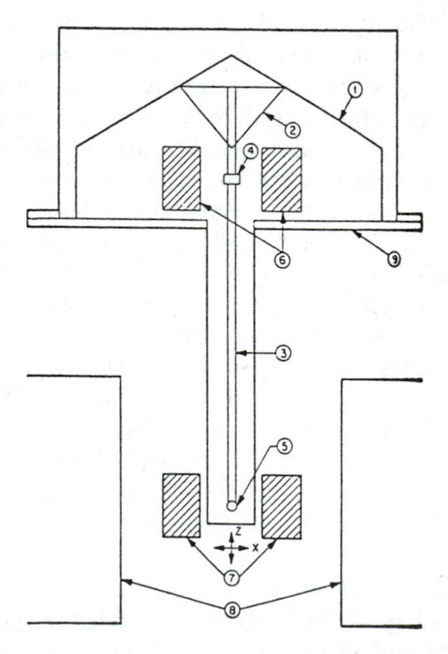

Figure 4. Vibrating sample magnetometer. (1) Loudspeaker transducer, (2) conical sample rod support, (3) sample holder rod, (4) permanent magnet reference sample, (5) sample, (6) reference coils, (7) sample coils, (8) magnet poles, (9) metal container to allow evacuation. Reproduced by permission (from Ref. 13). A more sophisticated design is described in Ref. 12

A SQUID is formed by a superconducting ring that has a weak link. The SQUID is an electrical device that can be loosely compared to a transistor. However, while the transistor is capable of amplifying small electrical signals into very large ones, the SQUID is capable of amplifying very small changes in magnetic field, even in the presence of a large static magnetic field, into a large electrical signal. When coupled to a suitable detection system, the SQUID electrical circuitry becomes the measuring probe of a superconducting magnetometer. The original design of SQUID magnetometers required the sample to be placed directly inside the SQUID loop. With current technologies, it is found that greater sensitivity is achieved by inserting the sample into a superconducting pick-up loop that is inductively coupled to a torroidal SQUID.

The sample-dependent magnetic flux detected by the SQUID may be due to either a permanent magnetic moment or the field induced magnetization of materials. When one measures the magnetic susceptibility of a material with a SQUID, the measuring magnetic field must be extremely stable to minimize noise in the SQUID detection circuitry. A superconducting susceptometer results from the addition of a superconducting solenoid to the simple SQUID magnetometer circuit.[19] To attain the ultimate sensitivity with the superconducting susceptometer, it is also necessary to isolate the SQUID itself in a superconducting enclosure. The SQUID susceptometer has an ultimate sensitivity to allow measurement of magnetic susceptibilities as small as 10^{-10} emu.[20,21]

An added attraction of the SQUID probe is its ability to resolve magnetic transients. The response time of the SQUID is limited by the radio frequency (rf) in the detection circuit. With current commercial models,[22] the SQUID susceptometer has the ability to resolve magnetic transients of less than 1ms. Attempts to increase response time and resolve even shorter-lived transients is an area of extensive research in SQUID technology.[23] The materials to be measured by the Scχ may be small single crystals, long sample tubes packed with powders, or solutions. Measurements are obtained by charging the solenoid to the desired field, operating the solenoid and shield in the persistent mode, and then passing the sample through the pickup loop. Alternatively, the sample may be held stationary in the pickup loop while an external perturbation is applied.

Though the SQUID susceptometer is the state-of-the-art-in magnetic susceptibility instrumentation, there are some problems associated with its use. For example, the SQUID susceptometer is very sensitive to mechanical vibrations and external rf noise. Another problem is the combined expense of the instrument and its cost of operation. Also, the SQUID electronics operate only in the superconducting state. Indeed, it has been observed that these instruments operate best when kept in the superconducting state for long periods of time, avoiding in the process the thermal stresses of temperature recycling. Although there is a great deal of research in the area, room temperature superconducting copper oxides have only been developed into SQUIDs for specialized applications. However, the SQUID susceptometer is still in its infancy (SQUID susceptometers have been available for *ca.* 20 years) and the instrument, because of its high sensitivity and fast response time, promises to play a crucial role in the future of magnetochemistry.

Table I

Materials That May Be Used for Calibration of Magnetic Susceptibility

Calibration material	χ_g ($x10^{-6}$ emu·g^{-1})*	Ref.
Water	-0.720	35
Benzene		36
(Air saturated)	-0.7020	
(Nitrogen Saturated)	-0.7081	
$HgCo(NCS)_4$	16.44	37
$Ni(en)_3S_2O_3$	10.82	38

*All values are for T = 20°C.

Table II

Curie-Weiss Parameters for Paramagnetic Materials That May Be Used to Verify Temperature Calibration

Compound	C (emd·mol^{-1}K^{-1})	θ (K)	Lowest temp.[d]	Ref.
$HgCo(NCS)_4$ [a,b]	2.433	-1.1	>70	39,40
$(NH_4)_2Mn(SO_4)_2 \cdot 6H_2O$ [c]	4.375	0	>1	41
$(Me_4enH_2)CuCl_4$ [b]	0.433	-0.07	>5	42

[a] $X_{dia} = -189x10^{-6}$ emu·mole^{-1}.
[b] Powder.
[c] Single crystal.
[d] Susceptibility deviates from Curie-Weiss law below these temperatures.

3. MEASUREMENT TECHNIQUES

It is often difficult and tedious to measure the absolute value of each experimental variable needed to calculate the experimental magnetic susceptibility. A much more convenient method of obtaining the experimental magnetic susceptibility is to calibrate the instrument by obtaining precise relative measurements on a calibration standard of very accurately known magnetic properties. There are several materials whose magnetic susceptibilities are known to a high degree of accuracy. Some of the more readily available calibration materials are given in Table I along with the values for their observed mass susceptibilities. When calibrating a magnetic susceptibility instrument, one should obtain calibrations from several standards. The difference in calibration constants serves as one measure of the estimated uncertainty of the instrument. It is also advisable to use a standard whose bulk susceptibility is close in magnitude to that of the sample being studied.

Temperature-dependent magnetic susceptibility measurements require special calibration procedures. In addition to calibrating the susceptibility constant of the instrument, one must also calibrate the thermometer for accurate measurement of the temperature of the sample. If a thermometer is provided with a calibration curve, several independent experiments must be performed to confirm the validity of the calibration curve. For example, published thermocouple tables often give different temperatures when applied to different thermocouple circuits. The thermometer must also be placed in a position where the recorded temperature corresponds to the actual sample temperature. When a large sample is used, as in the Gouy experiment, temperature gradients within the sample can become a significant source of error. Even when small samples are used, thermal gradients in the sample compartment must be kept to a minimum. At low temperatures, very small fluctuations in temperature can become a significant source of experimental uncertainty in magnetic susceptibility measurements. Table II provides several paramagnetic salts that may be used as checks of temperature calibrations, along with their magnetic properties and suitable temperature ranges.

Although a considerable number of magnetic susceptibility measurements are made as a function of temperature, it is worthwhile to mention that magnetic susceptibility often depends on other variables such as pressure or time.

Several experiments have been reported that investigate the dependence of magnetic susceptibility on pressure.[26-28] In general, a pressure-dependent magnetic susceptibility study is designed to monitor the shift in equilibrium toward a lower volume environment which is favored at high pressures. The pressure experiment is very difficult to execute because the massive amount of instrumental addenda required to achieve high pressures precludes a precise susceptibility measurement with most instruments. Nevertheless, there is a great deal of interest in this area and some research has been accomplished. For example, magnetic susceptibility studies have been used to investigate the pressure dependence of critical phenomena. Theory predicts a lowering of the transition temperature due to shorter superexchange pathways available at high pressures.[27,29] However, because of large uncertainties in the measurements, contradictory results have been reported.[26-28]

The quantitative measurement of magnetic response in the time domain is now becoming a powerful kinetic tool to study reaction mechanisms and transient processes. With the advent of the SQUID, magnetochemists need no longer be

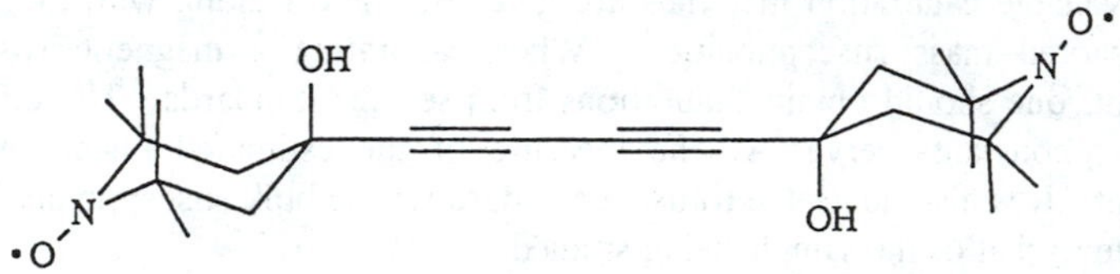

Scheme 1.

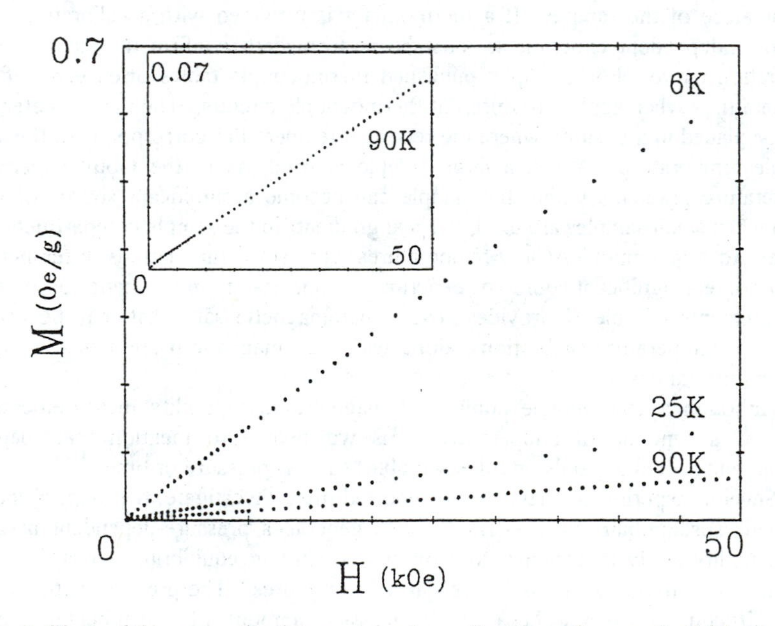

Figure 5. Plots of the magnetization measured as a function of magnetic field at various temperatures for the heat treated a-phase of (**1**) (taken from ref.45)

resigned to monitoring quasi-static processes. Superconducting susceptometers have been used by Philo[30] to monitor the kinetics of photolysls of a room temperature solution of hemoglobin-carbon monoxide. Also, the STEPS technique (susceptibility of transient excited paramagnetic states) has provided a direct measurement of the bulk magnetic properties of the phosphorescent triplet state in the aromatic hydrocarbon coronene.[31] The use of time dependent susceptibility measurements to monitor transient intermediates and reaction mechanisms shows promise of becoming a major new area of magnetochemical research.

In any discussion of the experimental aspects of magnetic susceptibility measurements, other techniques for obtaining magnetic information must be mentioned. Complementary data are often a necessity for the successful interpretation of magnetic susceptibility data. For example, magnetic exchange constants have been measured by EPR,[32] NMR,[33] and Mossbauer spectroscopy.[34] In addition, the Mossbauer experiment may be used to determine spin states when magnetic data are inconclusive.[34] Also, for the investigation of magnetic phenomena, heat capacity measurements provide a formidable and often indispensable check on the interpretation of susceptibility data.[34a]

3.1 Experimental Examples

A good illustration of techniques for the measurement and characterization of magnetic materials can be illustrated by an examination of the magnetic measurement and analysis of the organic di-radical 4,4'-butadiyne-1,4-diyl-*bis*-(,2',6,6'-tetramethyl-4-hydroxy-piperidin-1-oxyl), **(1)**. This material was reported by Ovchinnikov to exhibit a weak ferromagnetic response after it was exposed to thermal treatment.[43] Another phase of this material referred to as the beta phase was reported by Miller and coworkers.4[4] Careful measurements of the material show normal magnetic behavior expected for the diradical material.[45] Following heat treatment, the magnetic susceptibility of the materials were measured. When the data were properly corrected for contributions for the addendum in the sample area, no ferromagnetic response was detected. The magnetization data, illustrated in Figure 5 are consistent with normal Curie-Weiss magnetic behavior and can contain a maximum of 5 ppm ferromagnetic impurity, consistent with a trace amount of iron in the material. In fact, iron impurities are common in chemical materials and may arise from an innocuous source such as a spatula. Extreme care taken during sample preparation and manipulation will minimize this contamination.

Another example of an ambiguous characterization will arise in the interpretation of magnetic data when an insufficient amount of data is used to diagnose a phenomenon. Figure 6a illustrates the magnetic susceptibility of the polycrystalline vanadium compound $[NH_2(CH_2)_3NH_2](VO)_3(OH)_2(H_2O)_2(PO_4)_2$.[46] A cursory examination of the data in Figure 6a leads one to characterize this material as having an antiferromagnetic phase transition at T=5.1K. The plot of the magnetic susceptibility as a function of temperature has the shape that is characteristic of such a transition. A more complete investigation of these data begins to show inconsistencies with this initial characterization. The Curie-Weiss plot of the data illustrated in the inset of Figure 6a gives the parameters g=1.86, θ=+20K. Since θ is positive, there must be an underlying ferromagnetic interaction to the material. The drop in the data is of course the result of a spin pairing type of interaction, but the

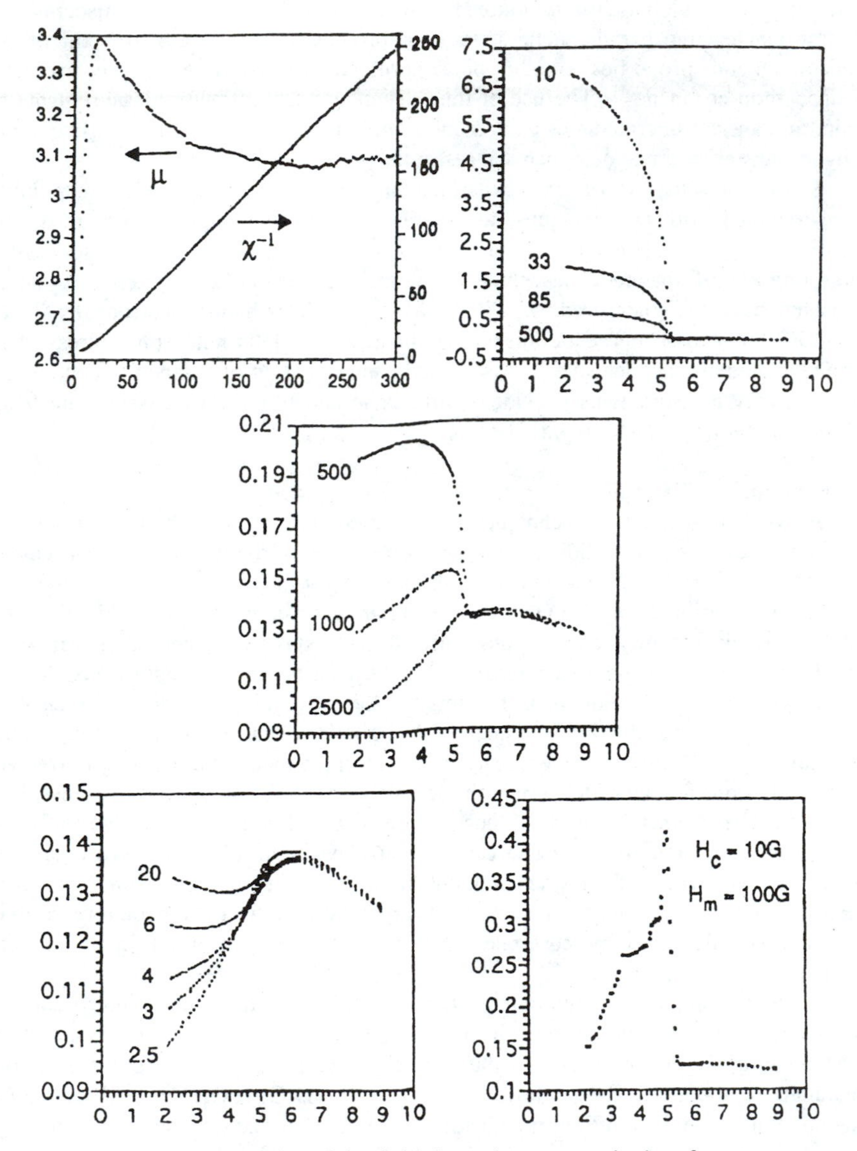

Figure 6. Plot of the field dependent magnetic data for $[NH_2(CH_2)_3NH_2](VO)_3(OH)_2(H_2O)_2(PO_4)_2$.

overall mechanism of the magnetic interaction is complicated. A ferromagnetic material generally exhibits a field-dependence and field dependent magnetic experiments demand to be done. Figure 6 illustrates the magnetic susceptibility of the material measured at several magnetic fields and shows a low temperature ferromagnetic transition at low fields that becomes antiferromagnetic at moderate fields, and then degenerated toward a paramagnetic state at still higher fields. Also shown in Figure 6e is the zero field cooled low field measurement that indicates that a very complicated phase diagram is to be expected for this material. A full characterization of this material will require single crystal anisotropy magnetic studies to adequately model the magnetic behavior. One should always be cautious about attempting to characterize a material based on data taken at only one field, especially when a three-dimensional phase transition is observed.

Most of this discussion has been limited to dc magnetic susceptibility. It is often preferred to measure the ac magnetic susceptibility for the characterization of a material since it can be done at zero applied magnetic field with an exciting field of less than one gauss. The behavior of a simple Curie-Weiss paramagnet will be the same in either ac or dc experiments. However, when complicated magnetic behavior is observed, there is often a difference in the behavior of a sample when exposed to the ac or dc experiments. The ac magnetic susceptibility of the compound $CoVO(PO)_4$ is illustrated in Figure 7 measured at several magnetic fields.[47] These data indicate a field dependence of the magnetic susceptibility of the ferromagnetic state. In the ac experiment, the magnetization vector of the electrons attempts to follow the oscillating magnetic field. The temperature and field dependence of the sample is enhanced in the ac measurement and the out-of-phase component illustrates the characteristic lag of the magnetization of a ferromagnetic material behind the oscillating component of the magnetic field to give the χ" (out-of-phase) values. It is important to note that the shapes of these curves will depend on both the frequency and the magnitude of the oscillating component of the magnetic field.

Another type of experiment that may be performed involves the characterization of samples that have history-dependent behavior. A good example of this type of behavior is found in spin glasses and superparamagnets. Both of these materials exhibit what are referred to as spin blocking temperatures where the magnetic moments loose their ability to realign themselves.

An example of an interesting new superparamagnetic material consists of nanophase ferrite particles uniformly dispersed in polymer microspheres.[48] Figure 8 illustrates the process for peppering polyphenol polymers using enzymatic catalysis in reverse micelles.[48,49] The spherical nature of the reversed micelles results in spherical sub-micron sized polymers. Nanophase ferrite particles can be synthesized and trapped in these micelles and uniformly dispersed in the polymers. There is currently a great deal of interest in nanophase magnetic particles because they can exhibit superparamagnetic behavior. Several experimental techniques can be used to fully characterize the magnetic behavior of superparamagnets.

At high temperatures, a superparamagnetic particle behaves as a classical spin with an effective spin quantum number of 10^2-10^4 and will follow Curie-Weiss Law. At lower temperatures, the large spin moments interact strongly and there is often a blocking of the orientation of the spins as the interparticle magnetic interaction begins to dominate. Figure 9 illustrates the temperature dependent magnetic susceptibility

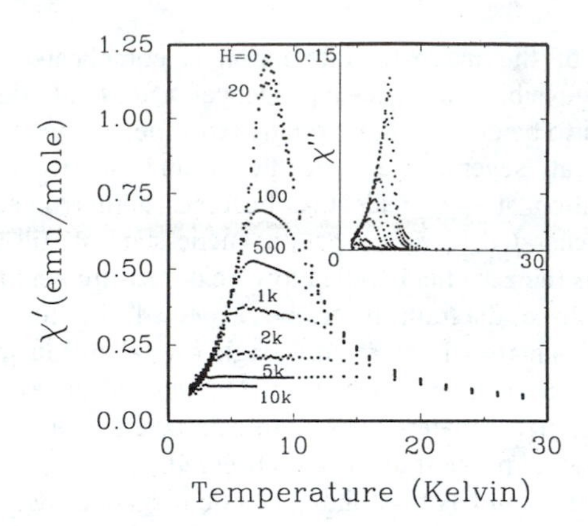

Figure 7. The ac magnetic susceptibility of the compound $CoVO(PO)_4$ is illustrated measured at several magnetic fields

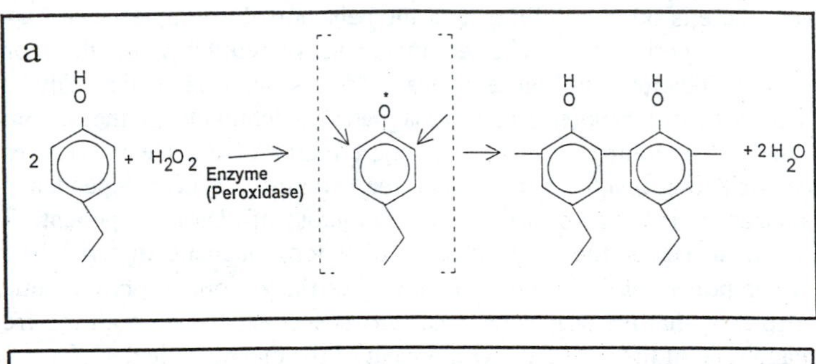

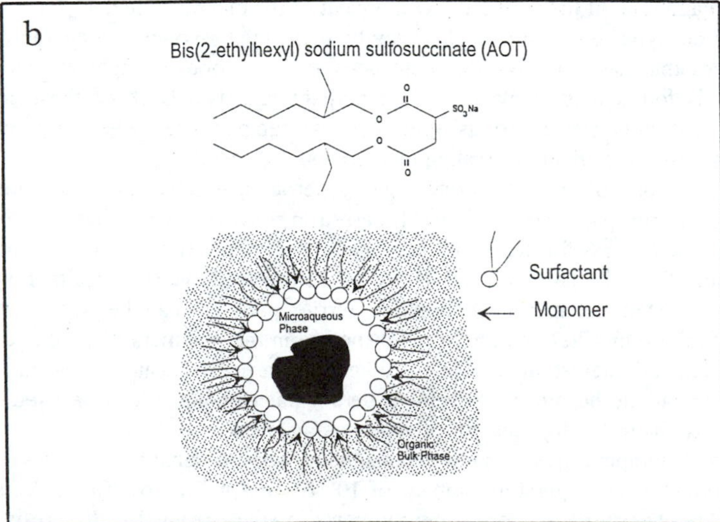

Figure 8. A schematic showing the microemulsion droplet with the various components. A simplified polymerization reaction mechanism is shown at the top. (From ref 48)

under field-cooled and zero-field cooled conditions. There is a difference in the two experiments because of the blocking of the spins at 10K. The blocking of the spins will also give a characteristic ac magnetic susceptibility that exhibits a maximum as the spins begin to loose their ability to reorient themselves.

3.1.1 *Magnetic Remanence*

Perhaps the most diagnostic experiment for the characterization of history-dependent materials such as superparamagnets and spin glasses is the analysis of the remanent magnetization. Different procedures that are followed to freeze the moments in a material can produce different results. Two common remanent-inducing procedures will be discussed.

Isothermal Remanent Magnetization (IRM): This experiment involves cooling the specimen in zero field to temperatures below the spin blocking point. The onset of the spin blocking temperature then results in a freezing of the magnetic moments in a random fashion since no external force was present for the spins to align with. The specimen is then exposed to an applied magnetic field, but the spins will tend to remain in their "frozen" random orientation. After a certain time, the magnetic field is quenched and the resultant magnetization is measured in the absence of a magnetic field. Under these conditions an ideal superparamagnet would show zero magnetization.

Thermal Remanent Magnetization (TRM): This experiment on the other hand reveals vastly different magnetic behavior. In the TRM experiment, the specimen is cooled to a temperature below the spin blocking temperature while in an applied magnetic field. The spins, which had a tendency to align with the applied magnetic field while in the paramagnetic phase, are now frozen into a position of partial alignment with the applied field. While the specimen is in the spin blocked state, the magnetic field is quenched. Since the motion of the spins is blocked, their alignment is not quenched and a remanent moment is present in the material. The TRM measurement should give a large remanent relative to the IRM.

The field-dependent IRM and TRM measurements for the spherical superparamagnetic particles are illustrated in Figure 10. There is a characteristic maximum in the TRM that occurs in both superparamagnets and spin glasses and is probably a result of the variable time dependence that is known to occur in these systems. When measuring the remanence of these materials, it is crucial that the times of each remanent measurement be exactly the same.

The remanence and history dependence characteristic of the spin-blocked state implies a memory in the material and therefore, the hysteresis measurement can provide information on the characteristics of a material. Superparamagnets show a hysteresis behavior that is useful to explain the different observations that can result from the hysteresis experiment.

3.2 Hysteresis

At high temperatures, a superparamagnetic material will follow Curie-Weiss behavior, but with a very large effective spin quantum number. As the field increases the classical spins easily follow the magnetic field but very rapidly reach saturation. Figure 11a illustrates the high temperature hysteresis loop for a superparamagnetic

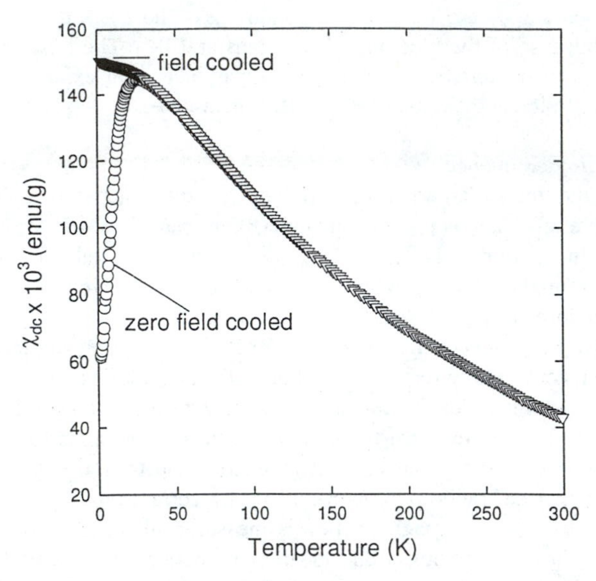

Figure 9. Field cooled (fc) and zero field cooled (zfc) magnetic susceptibility of superparamagnetic polymer microspheres plotted as a function of temperature. (From ref 48)

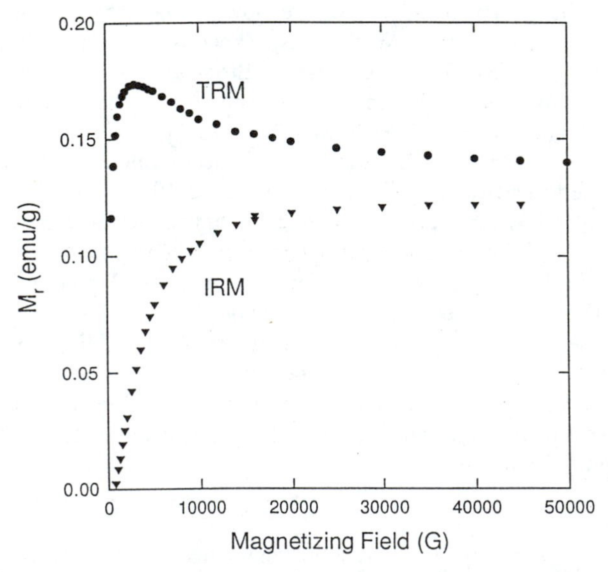

Figure 10. The IRM and TRM of superparamagnetic polymer microspheres plotted as a function of remanent inducing magnetic field. (From ref 49)

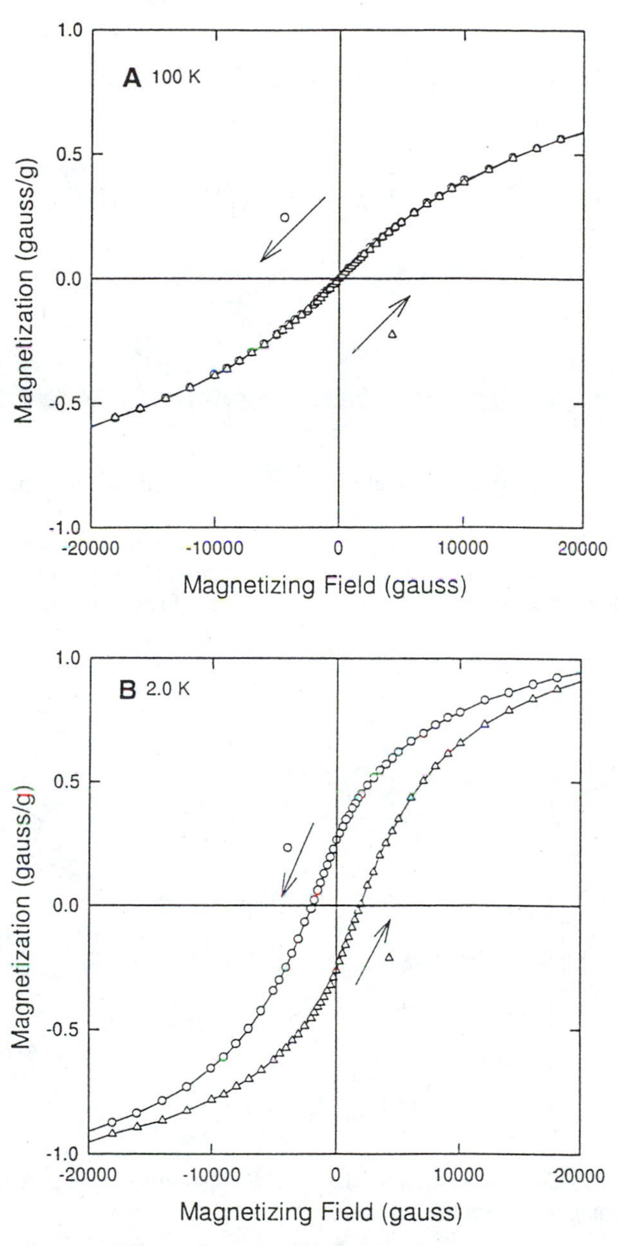

Figure 11. The magnetic hysteresis curves of superparamagnetic polymer microspheres plotted above (100K) and below (4K) the spin blocking temperature. (From ref 49)

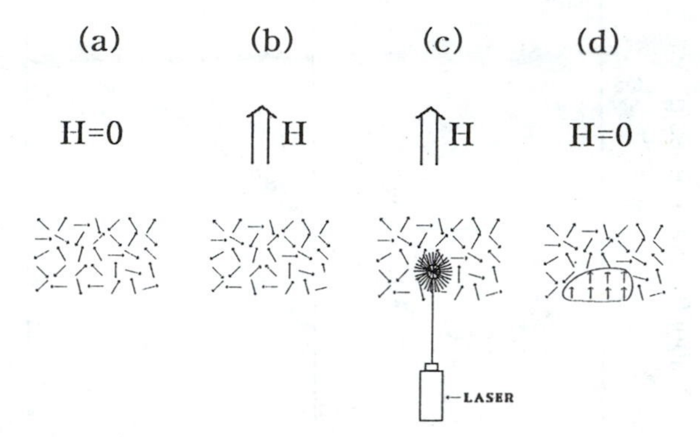

Figure 12. A schematic that illustrates photomagnetic experiment: (a) the spin-glass material is cooled to a temperature well below the spin blocking temperature in the presence of an applied magnetic field; (b) the magnetic field is then relaxed to zero field; (c) a pulse of radiation is applied to the spin blocked specimen; (d) a domain wall of demagnetization is induced in the material. (From ref. 50)

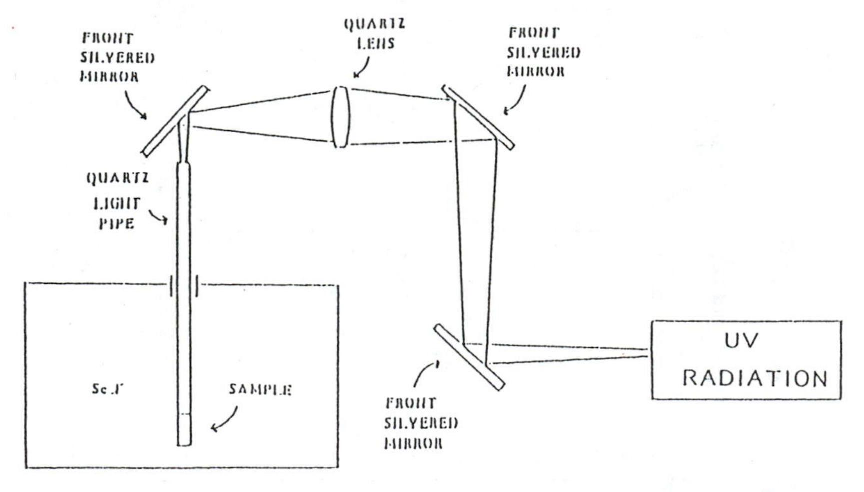

Figure 13. A schematic diagram of the STEPS apparatus for the measurement of photomagnetic response.

material. As expected for a superparamagnetic material, there is rapid saturation but no coercivity in the material.

At temperatures below the spin blocking temperature, the spins loose their ability to follow the magnetic field and there is now a coercivity in the hysteresis loop. The hysteresis experiment is shown in figure 11b recorded at T=2K, well below the spin blocking temperature. A comparison of the two hysteresis curves shows a rapidly saturating material at high temperatures and the onset of a coercive field at temperatures below the spin blocking temperature, both effects are consistent with a superparamagnetic material.

3.3 *The Photomagnetic Effect*

The unusual magnetic properties of spin-blocked materials make them excellent candidates for an experiment that uses light to generate magnetic bubbles or holes on the surface of a material.[50] These experiments use different magnetic fields for cooling and measuring. A process for generating magnetic bubbles in a spin blocked medium is illustrated schematically in Figure 12. The procedure for this experiment involves cooling the material to temperatures well below the spin-blocking temperature in the presence of an applied magnetic field [Figure 12(a)]. This results in the freezing of a large thermal remanent magnetization (TRM) in the material where the direction of the moment is coincident with the applied magnetic field direction. The magnetic field is then relaxed to zero field and the remanent will remain as a resultant TRM [Figure 12(b)]. A pulse of radiation is applied of sufficient intensity to cause local disruption of the spin state [Figure 12(c)]. The result of this pulse of radiation will destroy the remanent within the domain walls determined by the boundary of the pulse; in other words, a magnetic hole is created in the material [Figure 12(d)]

The measurement of photomagnetic effects requires the ability to deliver a pulse of radiation to the sample while it is in the sample detection area. The measuring apparatus should also have the ability to resolve time-dependent magnetic response. An experimental technique called the Susceptibility of Transient Excited Paramagnetic States (STEPS) technique has been developed and uses a SQUID susceptometer to measure the magnetic response. A schematic diagram of the STEPS apparatus is illustrated in Figure 13.

In the STEPS experiment, a sample of a spin glass ()* with a freezing temperature of T_f= 5K was frozen to a temperature of T=2.2K in a field of 1kG and the field was then relaxed to zero field in order to freeze in a resultant TRM. A pulse of 5s duration was then delivered to the sample, and a plot of the SQUID magnetometer response of the sample as a function of time, before, during, and after the sample was exposed to the radiation, is shown in Figure 14. The graph illustrates the change in remanent magnetization from the quasi-steady state value obtained before the pulse. There is a rapid drop in the remanent magnetic moment to a new steady state value that is smaller than the initial value as a direct result of the pulse of radiation. This is consistent with the process illustrated in Figure 12. Therefore, the local spin reorientation will generate a magnetic hole on the surface of the material with the boundary of the hole determined by the area that is illuminated. The STEPS technique can also be used to follow the magnetic profile of any time-dependent

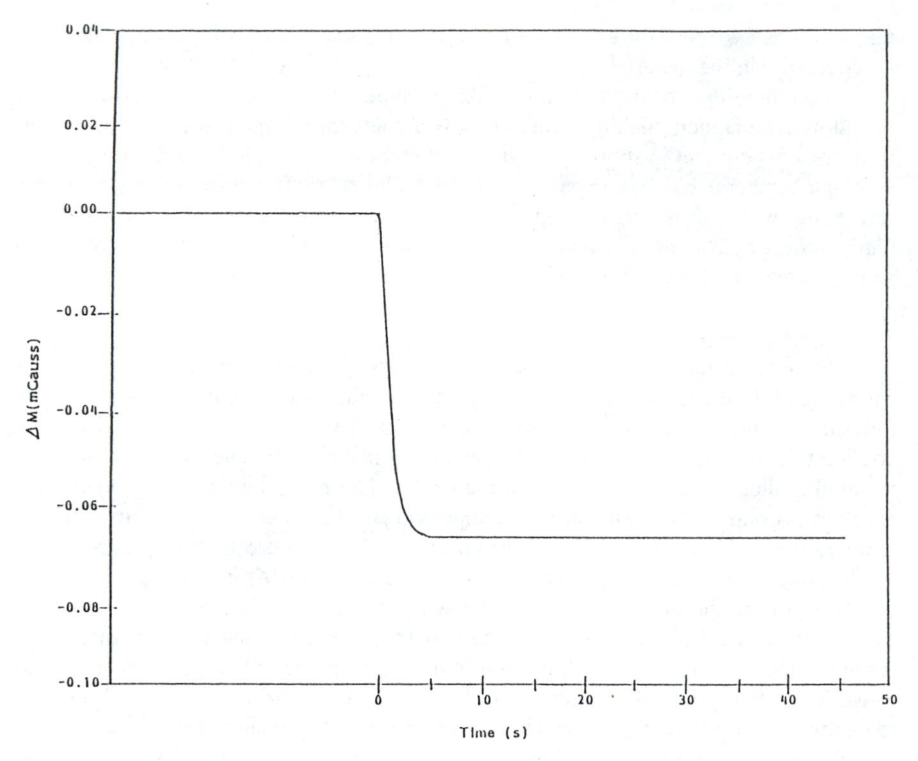

Figure 14. The magnetometer output for the photomagnetic response of the $Co_3(SbTe_3)_2$ plotted as a function of time before, during , and after a pulse of radiation about 2.0s period. (From ref. 50)

phenomenon, as long as the rate of magnetic changes are with in the response time of the magnetic detection system.

4. CONCLUSION

Several different instruments and experimental techniques have been discussed here. Development of new techniques will provide yet more information to allow characterization of novel magnetic phenomena. It is worthwhile, however, to re-emphasize some points that will help to ensure accurate magnetic measurement. Sample purity should be kept to its highest levels. This also means that extreme care must be exercised to avoid the addition of sources of magnetic response that do not arise from the specimen, and the data must be corrected for those magnetic signals that can not be avoided (e.g., sample holders, etc.).

The experimentalist must also have confidence in the calibration of the instrument. This includes a knowledge of the accuracy and precision of the temperature, magnetic field, weight, and instrument calibration as well as the calibration of the measurement of any other parameters that are introduced to perturb the sample.

When performing liquid helium measurements, it is very important to be sure that your instrument is vacuum tight. A small leak even into a pressurized sample chamber can result in spurious magnetic signals that result from gases that condense on the sample. And finally, if an interesting phenomenon is observed, it is very important to investigate what happens when one or more of the experimental conditions are changed.

5. References:

1. O'Connor, C. J., *Prog. Inorg. Chem.*, **1982**, *29*, 203.
2. Carlin, R. L., *Magnetochemistry*, Springer, Berlin, **1986**.
3. Kahn, O., *Molecular Magnets*, VCH Publishers, New York, **1993**.
4. Selwood, P. W., *Magnetochemistry*, Interscience, New York. 1956.
5. Cruse, D. A., and Gerloch, M., *J. Chem. Soc., Dalion Trans.*, **1977**, 152.
6. Krishnan, K.S. and Banerjee, S., *Philos. Trans. R. Soc. (Lond)*. **1935**,*A234*, 265.
7. Lewis, R. T.,, *Rev. Sci. Instrum.*, **1971**, *42*, 31.
8. Reeves, R., *J. Phys. E*, **1972**, *5*, 547.
9. George Associates, Berkeley, California.
10. Carlin. R. L., Joung, K. O., Paduan-Filho, A., O'Connor, C. J., and Sinn, E., *J Phys. C.*, **1979**, *12*, 293.
11. Groenendljk. H. A., van Deyneveldt, A. J., and Willett, R. D., *Physica*. **1980**, *101B*, 320.
12. Foner, S., *Rev. ScL Instr.*, **1959**, *30*, 548.
13. Foner, S., *Rev. Sci. Instr.*, **1956**, *27*, 548.
14. Princeton Applied Research (VSM). Princeton. New Jerse.,. 08580.
15. Josephson, B. D., *Phys. Lett.*, **1962**, *1*. 251.
16. Bardeen, J., Cooper. L. N., and Schrieffer, *Phys. Rev.*, **1957**, *106*, 162, **1957**, *108*. 1175.
17. Deaver, B. S. Jr., and Fairbank, W. M., *Phys. Rev,. Lett.*, **1961**, 7, 43.
18. Jacklevic, R. C., Lambe, J., Silver, A. H., and Mercereau, J. E., *Phys. Rev. Lett.*, **1964**, *12*, 159.
19. Deaver, B. S. Jr., Bucelot, T. J., and Finney, J. J., in *Future Trends in Superconducting Electronics*, AIP Conference Proceedings No. 44, Charlottesville. VA. **1978**. Deaver, B., Falco, C., Harris, J., and Wolf, S., Eds., (American Institute of Physics. New York, **1978**). p. 58.
20. Qunantum Design Corporation, San Diego, CA 92121.
21. terHarr, L. W., in *Molecular Mased Magnetic Materials*, Eds., Awaga, K., Okawa, H., Sugimoto, T., Thompson, L. K., Turnbull, M. M., ACS Symposium Series, this book.
22. Bertrand, J. A., Fujita, E. and Derveer, D. G., *Inorg. Chem.*, **1980**, *19*, 2022.
23. Long, A. P., Clark, T. D., and Prance, R. J., *Rev. Sci. Instrum.*, **1980**, *51*, 8.
24. O'Connor, C. J.,Deaver, B. S. Jr., and Sinn, E., *J. Chem. Phys.*, **1979**, *70*, 5161.
25. Bertrand, J. A., Ginsberg, A. P., Kaplan, R. I., Kirkwood, C. E., Martin, R. L., and Sherwood, R. C., *Inorg. Chem.*, **1971**, *10*, 240.
26. Barvakhtar, V. G., Galkin, A. A., and Telepa, V. T., *Phys. Status Solidi (b)*, **1977**, *80*, K37'.

27. Gorodetsk, G., Leung. R. C., Missell, F. P., and Garland, C. W., *Phys. Lett.*. **1977**, *64A. 251.*
28. Missell, F. P., Guertin, R. P., and Foner, S., *Solid State Commun.*, **1979**, *23,* 369.
29. deJongh, L. J., and Block, R., *Physica,* **1975**, *79B.* 568.
30. Philo, J. S., *Proc. Nall. Acad. Sci. U.S.A.*, **1977**, *74,* 2620.
31. O'Connor, C. J., Sinn, E., Bucelot, T. J., and Deaver. B. S. Jr., *Chem. Phys. Lett.*, **1980**, *74,* 27.
32. Takui, T., Sato, K., Shiomi, D., and Ito, K., in Molecular Mased Magnetic Materials, Eds., Awaga, K., Okawa, H., Sugimoto, T., Thompson, L. K., Turnbull, M. M., ACS Symposium Series, this book.
33. Wang, S., and Breese, S. R., in *Molecular Mased Magnetic Materials*, Eds., Awaga, K., Okawa, H., Sugimoto, T., Thompson, L. K., Turnbull, M. M., ACS Symposium Series, this book.
34. Reiff, W., in *Molecular Mased Magnetic Materials*, Eds., Awaga, K., Okawa, H., Sugimoto, T., Thompson, L. K., Turnbull, M. M., ACS Symposium Series, this book.
34a. Sorai, M, in *Molecular Mased Magnetic Materials*, Eds., Awaga, K., Okawa, H., Sugimoto, T., Thompson, L. K., Turnbull, M. M., ACS Symposium Series, this book.
35. Mulay, L. N., *Magnetic Susceptibility,* Interscience, New York, **1963**.
36. Angus, W. R., and Hill, W. K., *Trans. Faraday Soc.*, **1943**, *39,* 185.
37. Baker, G. A., Rushbrook, G. S.,and Gilbert, H. E., *Phys Rev A,* **1964**, *135*, 1272.
38. Curtis, N. F., *J. Chem. Soc.*, **1961**, 3147.
39. Bunzli. J. G., *Inorg. Chim. Acta,* **1979**, *36,* L413.
30. O'Connor, C. J., Cucauskas, E. J., Deaver, B. S. Jr., and Sinn, E., *Inorg. Chim. Acta,* **1979**, *3z* 29,.
41. Cooke, A. H., *Prog. Low. Temp. Phys.*, **1955**, *1*, 328.
42. Brown, D. B., Crawford, V. H., Hall, J. W., and Hatfield, W. H., *J. Phys. Chem.*, **1977**, *81*, 1303.
43. Korshak, Yu. V., Ovchinnikov, A. A., Shapiro, A. M., Medvedeva, T. V., Specktor, V. N., *Pisma, Zh. Eksp. Teor.Fiz.*,**1986**, *43*, 309.
44. Miller, J. S., and Epstein, A. J., *J. Amer. Chem. Soc.*, **1987**, *109*, 3850.
45. Zhang, J. H. Miller, J. S., Epstein, A. J., and O'Connor, C. J., *Molec. Cryst. Liquid Cryst.*, **1989**, *176*, 271.
46. Sohogmonian, V., Chen, Q., Haushalter, R. C., Zubieta, J., O'Connor, C. J., and Lee, Y.-S., *Chem. Mater.*, **1993**, *5*, 1690-1691
47. O'Connor, C. J., and Haushalter, R. C., manuscript in preparation.
48. Kommareddi, N. S., John, V. T., McPherson, G. L., O'Connor, C. J., Lee, Y. S., Akkara, J. A., and Kaplan, D. L., *Mater. Chem.*, **1996**, *8*, 801..
49. Kommareddi, N. S., Tata, M., Karayigitoglu, C., John, V. T., McPherson, G. L., O'Connor, C. J., Lee, Y. S., Akkara, J. A., and Kaplan, D. L., *Appl. Biochem. Biotech.* **1995**, *51*, 241.
50. O'Connor, C. J., "*The Photomagnetic and Magneto-Optic Effects*", in *"Localized and Itinerant Magnetic Materials: From Molecular Assemblies to the Devices,"* Coronado, E. Palacio, F. Gatteschi, D. and J. S. Miller, Eds., Kluwer Academic Publishers, Dordrecht, The Netherlands (**1996**), in press.

Chapter 5

Superconducting Quantum Interference Device Studies in Molecule-Based Magnetism

Pamela A. Salyer and Leonard W. ter Haar

Department of Chemistry, University of Texas, El Paso, TX 79968

Magnetometers utilizing SQUID technology are becoming increasingly popular as a tool for characterization of magnetic properties in molecular-based materials. Current SQUID magnetometer technology is described to the extent necessary for proper and practical use in fields such as materials science, chemistry, geology and physics. Sample preparation and data analysis issues associated with magnetization measurement, temperature control, and magnetic field application are reviewed. The utility of SQUID-based studies is illustrated through a specific example from our research program on materials that crystallize in anomalous morphologies such as rings, tubes or cylinders. The application of the magnetometer as a structural tool is demonstrated by the varied magnetic behavior observed for the sulfosalt mineral cylindrite ($FePb_3Sn_4Sb_2S_{14}$), an incommensurate (misfit) lattice material. Temperature and magnetic field dependent susceptibility data exhibit Curie-Weiss behavior above 70 K. Below 30 K, the data display H- and T-path dependence and two magnetic phase transitions, suggesting a magnetic model which requires the structure and magnetism of a cylindrite cylinder to vary from the inner core to the outer periphery.

Although Superconducting QUantum Interference Device (SQUID) magnetometers have been available for over a decade (*1*), their popularity has soared in recent years due to advances in the design and quality associated with the SQUID sensor, LHe cryogenics, temperature control, superconducting magnets, and computer aided data acquisition and analysis. Vibrating sample magnetometers, Faraday balances, and other magnetization measurement techniques have been around much longer (*2*) and continue to serve the scientific community, but the commercialization of the modern SQUID-based magnetometer can be readily attributed to its sensitivity and general ease of use. In particular, it is the routine characterization of new or unfamiliar materials to which they are well suited. It is from this perspective that this chapter reviews the techniques and utility of SQUID-based magnetometry.

The practical essence of any magnetometer lies in its ability to 'easily' measure, with accuracy and precision, the magnetization (M) of a sample at various temperatures (T) and applied magnetic fields (H). Experimental uncertainties in the measurement of M and in the control and measurement of T and H, must be held to a

0097–6156/96/0644–0067$15.00/0

minimum to produce high quality data. Today's SQUID magnetometers are generally capable of magnetization measurements down to 10^{-8} emu with a differential sensitivity of 10^{-9} emu. Depending on the type of magnetism exhibited by the sample and the spin concentration per formula unit, this level of sensitivity implies that very small samples (e.g., single crystals) are readily studied with a SQUID magnetometer. Ironically, the task of precisely quantifying SQUID-based magnetic data is more often limited by the inability to accurately determine the weight of a minute sample. The control, stability, spatial uniformity and measurement of cryogenic temperatures are equally important issues, and currently available commercial instruments are sufficient for the needs of most magnetic susceptibility studies (*3*). Likewise, advances in superconducting magnets have simplified the experimental aspects associated with the external application of a wide range of magnetic fields, but they too come with their own set of caveats (*4*).

The first few sections of this chapter describe the instrumentation, procedures and pitfalls associated with SQUID-based magnetometers. The latter half demonstrates how extensive studies on a single material, cylindrite, can easily become a daunting and extended series of experiments that would be impossible if magnetic susceptibility techniques other than the SQUID were the sole methods available.

SQUID-based Magnetometers

The SQUID Concept. Superconducting QUantum Interference Device (or SQUID) technologies and applications are rooted in the fundamental aspects of the Josephson effect (*5*). In its general form, this effect is a tunneling of superconducting electron pairs between two superconductors separated by a weak link, often referred to as the Josephson junction(*6*). The physical size of this junction is that of the coherence length or penetration depth of the material. Although the complex nonlinear dynamics of the general Josephson effect continue to be studied for a variety of circuit applications involving signal generation, detection, mixing, amplification and switching, one of its most common uses is as the basis for the SQUID.

The SQUID acronym refers to a superconducting electronic device conceptually based on a superconducting loop of small inductance that contains one or two Josephson junctions and can therefore be utilized in the measurement of extremely weak magnetic fields, voltages, and currents. An rf-SQUID employs a single junction, and its properties depend on changes in the applied flux as detected by an rf-voltage and feedback circuit. A dc-SQUID is obtained when two nearly identical junctions are employed. In addition to the difficulty of manufacturing two such junctions, the dc devices have historically encountered problems based on a less than complete understanding of their operation and performance (*7*). Although arrays of SQUIDs can clearly be used in digital devices (with many advantages over other logic circuits), it is their notoriety as the most sensitive detectors of magnetic flux that has led to their widespread development in the field of magnetometry.

The SQUID as a Magnetometer. Both dc- and rf-SQUIDs have been used for detecting much weaker quasistatic magnetic field changes than those that can be detected by non-SQUID techniques. It should be pointed out that the typical SQUID magnetometer is actually a flux transformer (*5,7*); it outputs a voltage that corresponds to a change in the magnetic flux-density within a volume of space defined by a superconducting loop (or series of loops), known as the SQUID detection coil(s) (Figure 1a). This extreme sensitivity in measuring differentials in magnetic flux-density is the strong point of the SQUID-based magnetometer.

Detector Geometry. The SQUID detection coils are in actuality far removed from the SQUID itself and are located within the inner core of a superconducting solenoid (Figure 1b), which is used to apply the external magnetic field. As a sample enters this cylindrical volume of space (i.e., the sample zone), the magnetic flux density associated with this zone begins to change. As the sample continues to pass through the sample zone (i.e., scanning the sample), the magnetic flux density behaves as a smoothly changing function whose magnitude and sign depend also upon the magnitude and sign of the sample's magnetic moment and position within the zone. For this reason, it is of vital importance that spatial variations in temperature and applied magnetic field be kept to a minimum over the length of the sample zone.

In order to provide the necessary magnetic stability, the SQUID itself is enclosed in a superconducting magnetic shield. If two or more detection coils are used instead of a single loop and wound with opposite sense, various gradiometer configurations can be obtained which keep interference from distant field sources and magnet fluctuations at a minimum. Gradiometer configurations are typically used in SQUID-based magnetometers because of the increased sensitivity they provide.

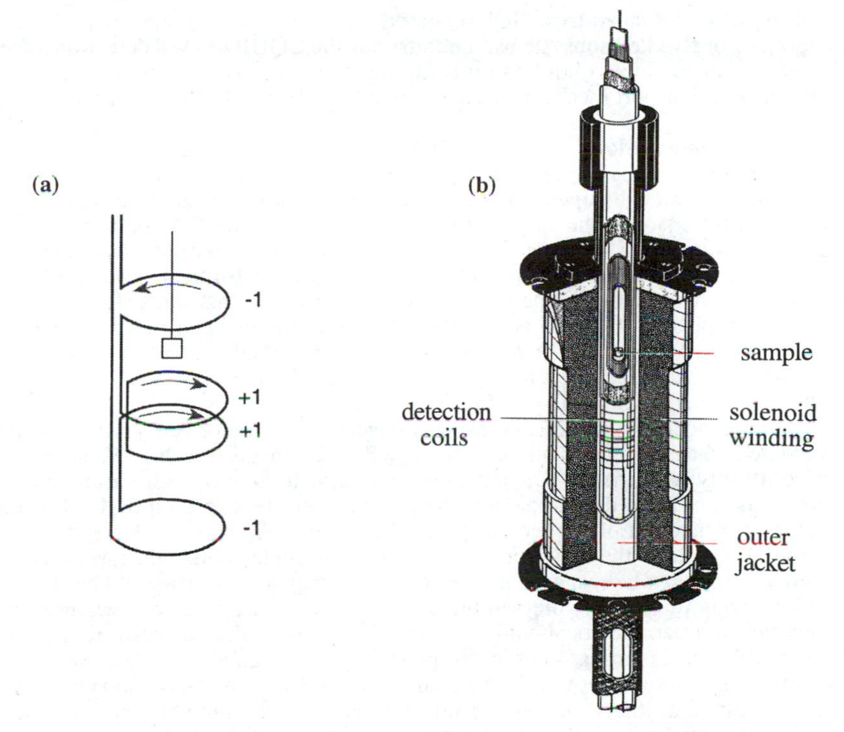

Figure 1. a) SQUID detection coil geometry; b) coil geometry within the superconducting magnet. Courtesy of *Quantum Design, San Diego, CA.*

Sample Magnetic Moment. Upon scanning the sample, the magnetic flux density changes accordingly and results in a concomitant change in the induced current in the SQUID detection coil(s). It is this induced current which is transferred by superconducting wires to the SQUID sensor, where it is inductively coupled to the thin-film SQUID. The SQUID itself is the device that finally converts the induced current to a voltage, which is strictly proportional to the changes in the magnetic flux density. The voltage therefore depends solely upon the magnetic nature of the sample and its position within the sample zone. As the sample is scanned, the SQUID's response can be recorded as a curve of voltage vs. sample position. Once calibrated with a standard (*8*), such a SQUID response curve provides a direct measurement of the sample's magnetic moment at any given temperature and field. The length of the scan is an experimental variable; repeated scans can be used to ascertain the precision of the measurement.

Since the principle of the SQUID-based measurement is the detection of a change-in-flux, calibration and measurement procedures are inherently dependent upon the sample size and shape. To minimize possible problems, small sample size (relative to the overall geometry of the detection coils) is preferred. Calibration procedures should be carried out using a sample whose size and shape are similar to that of the typical sample. It is important to note that the measured magnetic moment is that of the material of interest as well as the sample holder. Corrections for the sample holder's contribution to the measured moment must be subtracted unless its magnitude is insignificant at all temperatures and fields.

Equipment and Procedures. SQUID-based magnetometer systems are a conceptual integration of five key subsystems: circuitry for the SQUID-based detection scheme; cryogenic temperature control and measurement; a superconducting magnet; sample handling and transport mechanisms; and a computer-based data acquisition system.

Instrumentation. As discussed above, the principle of measurement is the movement of a sample through a series of superconducting detection coils contained within the core of the superconducting solenoid. Various insulating materials and vacuum jackets isolate the sample zone from the cryogenic LHe bath in which the SQUID sensor, the SQUID detection coils, and the magnet with its superconducting persistence switch must remain immersed while the instrument is in use. A key difference between SQUID magnetometers and NMRs is that the LHe in the dewar reservoir (typically 50 liters) is also used for temperature manipulation; hence the LHe consumption rate (5 liters/day) is far greater than that of NMRs and is a significant operational cost. In our own program, the magnetometer operates on a highly automated 24 hours/day, 7 days/week basis.

A number of experimental aspects concerning the five subsystems should be considered when acquiring and maintaining a SQUID magnetometer. Most important are familiarity with the SQUID detection circuitry, its tuning, calibration, and geometric relationship to the sample transport in terms of the expected SQUID response voltage as a function of sample position. Parameters which affect the sensitivity of the SQUID include the configuration of the gradiometer coils and the length over which a sample can be scanned. Temperature and magnetic field should be stable and homogeneous throughout the sample scan length. Changing temperature and/or magnetic field parameters should involve nothing more than a software command. The facility of scanning samples is generally accommodated by a sample rod mounted on a micromotor transport and a sealed antechamber entryway which prevents the admittance of oxygen into the sample scanning region. Instrument control and data collection are highly automated features that permit experiments to proceed unattended for days, if necessary. Ideally, the only aspects which require operator intervention are changing samples and refilling the LHe dewar.

Sample Selection and Preparation. Samples that can be analyzed on a SQUID magnetometer include both solids and liquids (if properly contained). Solids are acceptable in powder form when encased in a nonmagnetic container such as a gelatin capsule, or as single crystals which can be mounted directly on a quartz sample rod by aligning crystal faces with respect to the magnetic field. If absolute numbers are desired for the magnetization values, then it is important to know the weight of the sample. However, if only the relative temperature dependence of a material's magnetization is desired, a SQUID magnetometer is generally sensitive enough for very small samples (e.g., a few milligrams). Equally important for the strength (magnitude) of the signal are the volume of the sample and the material's inherent magnetization. A sample too small in volume or with weak magnetization will yield a poor signal.

The importance of sample purity (i.e., pure substance vs. mixture) is an issue that often arises in magnetic studies. In fact, the sample purity itself is not as important as knowing the composition of, or impurity levels within, the material. For example, a multi-phase mixture containing a superconductor will display the Meissner effect, but the bulk sample will not be responsible for this effect. Likewise, ferromagnetic impurities can wreak havoc in normal paramagnetic compounds, but paramagnetic or diamagnetic impurities in a ferromagnetic compound would easily go undetected. The bottom line is to be cognizant of the magnetic nature and the amount of impurities, or in the case of multi-phase mixtures, be aware that the observed behavior is a composite of all contributing components. For naturally occurring materials, such as minerals, it should be clear that the concept of impurities and multi-phase mixtures can be a significant aspect of the analysis process.

Measurement Protocol. After a powder sample has been loaded into a gelatin capsule, care should be taken not to handle the capsule with bare hands. Latex gloves are recommended during sample preparation and loading. Likewise, metal tools (i.e., spatulas, tweezers, etc.) should always be avoided in any magnetics lab. The top of the gelatin capsule should be punctured with a small hole to ensure that all oxygen is purged when the sample is outgassed inside the antechamber. After puncturing, the capsule can be loaded into the sample holder (i.e., plastic tube) which should be of uniform magnetization density and at least 3 to 4 times the scan length.

The gelatin capsule should be positioned with respect to the symmetry of the sample holder, and the desired SQUID scan length. A first "best guess" for sample alignment is to load the sample midway in the sample holder. The holder can then be attached to the end of the sample rod and the rod can be lowered into the sample chamber of the SQUID itself. The sample chamber should be purged of all oxygen, which will guarantee that the low pressure environment within the SQUID itself is preserved while the sample rod is lowered into the scanning area.

When the sample is centered properly, the second derivative curve of the SQUID response will be symmetric about the midpoint of the scan length, with a maximum voltage at the center of the scan and two minima at the location of the double gradiometer loops (the reverse is true in the case of a diamagnetic magnetization). Paramagnetic samples will exhibit a positive magnetic moment if the value of the magnetization exceeds the diamagnetism of the gelatin capsule or other sample holder. With the exception of a sizable signal from paramagnetic impurities, diamagnetic samples will always exhibit a negative magnetic moment. If data are being collected over a wide temperature range (e.g., 30 to 100 K), an allowance should be made for expansion of the sample rod during heating since the change in temperature will shift the center position of the sample and therefore the SQUID response curve. In some systems, this may be accommodated by software correction procedures, however in other systems the sample rod must be manually adjusted.

Data Collection Methodology. The initial procedure for analysis of a sample of unknown magnetic behavior is to acquire the temperature dependence of the magnetization in the range 5 to 300 K (or over a smaller range if conditions warrant). The applied field should be low, typically in the range 1 to 10 kOe. This procedure will differentiate between paramagnets and diamagnets, exhibit magnetization minima or maxima, and detect most phase transitions. For paramagnets, a Curie-Weiss analysis of the data above 5 K will generally indicate whether it is useful to spend time gathering lower temperature data. In the absence of noteworthy features above 5 K, a Weiss constant of a few degrees Kelvin is enough to warrant further data collection. A quick field dependent study should be conducted at the lowest temperature possible to check for a field dependent magnetic susceptibility (nonlinear M vs. H) and hysteresis. Likewise, if the temperature dependence of the magnetization indicates a phase transition, it is instructive to collect data while cooling to determine the order of the transition: first-order transitions will often exhibit a hysteretic path; second-order transitions will not.

Regardless of the magnetic behavior, it is important to determine the quality of the data being collected. Choosing a magnetic field in the range of 1 to 10 kOe allows for four significant figures. Temperature measurement and control should be kept at three significant figures at low temperatures and four significant figures at high temperatures. Most currently available SQUID magnetometers will deliver magnetization values with three to four significant figures, depending on the nature of the magnetization. With this type of precision, it should be possible to report parameters such as Curie-Weiss parameters with two to three significant figures.

A number of other useful features are available on the modern SQUID magnetometer which permit tailoring a data collection sequence to the behavior of a particular sample. Most are designed to enhance advanced data collection procedures for materials that display non-trivial behavior. Of these features, the most important is the ability to control over- and undershoot when changing the applied magnetic field and/or temperature of samples that display phase transitions, path dependence, or hysteretic effects. An additional useful feature is the ability to vary the scan length and the number of scans averaged to yield the magnetization value. This is particularly useful for samples with a small signal-to-noise ratio; the average of ten scans, for example, provides a significant improvement over a single scan and further homogenizes any instabilities in the temperature. Finally, an important feature which must often be considered is a pause period after changing temperature or magnetic field before measuring the magnetization. Even though the instrument electronics will indicate temperature and field stability, it is often necessary to wait an additional period of time in order to insure the sample has attained thermal and spin equilibrium. Because many of these features are programmable, the SQUID magnetometer can easily be configured to run separate, consecutive data collection schemes, unattended, over the course of several days.

Common Problems. There are a number of problems that can plague the routine usage of a SQUID magnetometer; but to list the majority of these problems and their diagnoses is beyond the scope of this chapter. In addition to the standard user and/or machine errors possible, there are three common problem areas that should be mentioned. First, samples not properly outgassed will contain adsorbed oxygen which shows up as a 'spike' in the χT vs. T data between 40 to 60 K, due to the paramagnetic/antiferromagnetic behavior of solid oxygen. Second, the instrumental logistics of data collection below 5 K are generally different from data collection above 5 K. The technique below 5 K utilizes the Joule-Thompson effect to reduce the temperature of the LHe by lowering the vapor pressure. This is accomplished by typically filling a small reservoir with LHe, which soon boils off. A limited time window therefore exists during which measurements can be conducted and often the time elapses before the data collection is complete. For samples with phase transitions and hysteretic effects, this may cause discontinuities in the data between 4 and 5 K. Finally, the interruption of a data collection scheme by a power outage can easily go unnoticed; it may be necessary to reset certain parameters (i.e., sample position and gas controls) that may have been affected due to the outage.

An Application of SQUID-based Magnetometry: Cylindrite

To appreciate the versatility of a SQUID-based magnetometer as both a magnetics and a structural characterization tool requires an example of a material with a structure that has been difficult to completely characterize with traditional structural methods. The remainder of this chapter describes how SQUID-based magnetic studies of cylindrite are being used to further develop the understanding of a very complex crystal system.

Incommensurate structures, or vernier lattice structures, are intriguing for their potential physical properties, as well as the morphologies they can assume (*9*). Anomalous morphologies such as rings and cylinders have been reported in both mineralogical and synthetic systems, for example, in the minerals cylindrite ($FePb_3Sn_4Sb_2S_{14}$) (*10*) jamesonite ($FePb_4Sb_6S_{14}$) (*11*) boulangerite ($Pb_5Sb_4S_{11}$) (*11*) and in the synthetic phase $NbSe_3$ (*12*). One aspect of our research program is to understand the morphological and compositional variances in both synthetic and natural samples. The emphasis is on exploring compositions that are expected to exhibit novel and/or interesting lattice and transport phenomena, and characterizing them using instrumentation such as the SQUID-based magnetometer.

Structure and Composition. The mineral cylindrite is a sulfosalt of nominal composition $FePb_3Sn_4Sb_2S_{14}$ found exclusively in Poopó, Bolivia, and is unique for its cylindrical crystals that can attain sizes up to several millimeters in diameter and length. The role of Fe in cylindrite is clearly a determining factor in the magnetic properties. Although the Fe content has been a matter of controversy, it is now generally accepted that Fe is integral to the cylindrite composition, which is well-defined by $Fe_{8.7}Pb_{31.3}Sn_{41.4}Sb_{18.0}S_{141.2}$ (*13*). Worth noting is that the Fe content displays a direct linear correlation to a portion of the Sb content (1:2, respectively). There is no direct Fe correlation to either the Pb or Sn contents. The formula weight of cylindrite, 1845 g/mol, indicates it to be a magnetically dilute material.

Electron microscope studies have revealed no single origin for the cylinder axis. Instead, the cylinders originate from lamellar or platy aggregates. The cores of the cylinders are generally composed of irregular sets of curved layers resulting in hemicylinders and horseshoe-shaped envelopes (Figure 2).

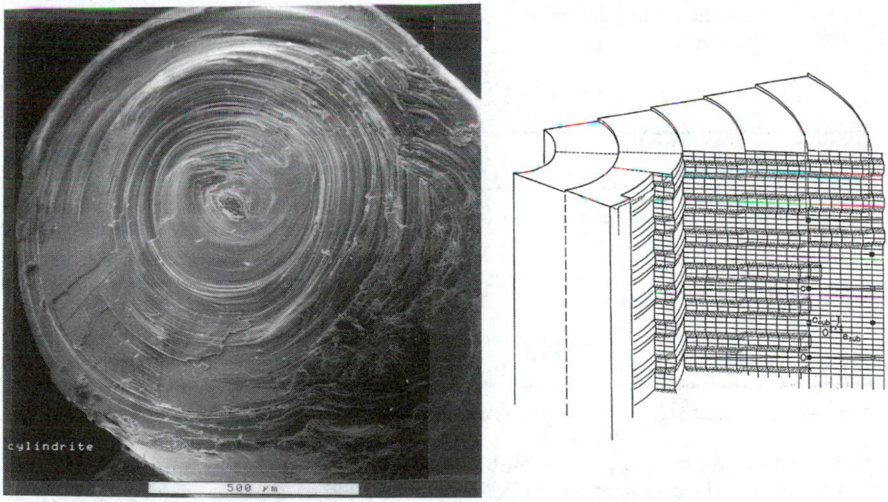

(a) (b)

Figure 2. a) ESEM micrograph of a cylindrite cylinder (approximate diameter, 1 mm); b) Graphical model of proposed growth pattern (Adapted from ref. *14*.).

Undeformed, small diameter wrappings of layers down to 2.8 μm have been observed, however arcs with radii less than 7 μm should be considered uncommon (*15*). As radii increase, layers stack with improved regularity, increased smoothness and uniformity of curvature. At the outermost layers, high magnification (25,000 x) reveals the curvature to be regular and without kinks.

The structure of cylindrite consists of cylinders formed by the incommensurate stacking of alternating pseudotetragonal (PT) and pseudohexagonal (PH) layers (*16*). The PT layers are two atomic layers thick with the approximate composition $Pb_{14.3}Sn^{2+}_{5.7}Sb_{4.4}Fe_{1.6}S_{26}$, and represent the PbS (NaCl) type structure. The layer surfaces exhibit a motif of sulfur atoms and metal atoms in square pyramidal coordination, analogous to the (100) surface of the parent compound. The PH layers are three atomic layers thick and correspond to the SnS_2 (CdI_2) type structure, with the approximate composition $Sn^{4+}_{8.2}Sb_{2.3}Fe_{1.5}S_{24}$. These layers correspond to a single layer of octahedrally coordinated Sn atoms sandwiched between two slabs of sulfur atoms. The a-axis is generally taken to be the layer stacking direction; the intralayer b-axes are incommensurable without visible modulation; and the interlayer c direction contains a semicommensurate match with a pronounced modulation. A quantitative understanding of the atomic structure of cylindrite continues to be the subject of research in terms of synthesis, single crystal x-ray diffractometry and HRTEM. In addition, understanding the magnetism of cylindrite may divulge structural information not yet revealed in the studies performed to date.

Sample Preparation and Magnetic Properties. Samples of cylindrite were acquired from four separate organizations. All samples (summarized in Table 1) contain well-defined, consistently sized cylinders resting in a dense matrix. The Harvard and Mineralogical Research samples consist of relatively large cylinders, most greater than 0.5 cm in diameter. The cylinders of the UTEP and Smithsonian samples are smaller in diameter and have a significantly larger aspect ratio. Shells were cleaved from bulk cylinders and extracted under an optical stereo microscope. Only visibly clean surfaces were accepted. Full cylinders were cleaved from the bulk material without concern for the visual purity of the outer surface. Shells and/or cylinders were slightly crushed and packed into gel capsules for analysis on a Quantum Design SQUID magnetometer.

Table 1. Cylindrite Samples

Source	Catalog #	Cylinder dia. (mm)	Aspect ratio (length/dia.)	Sample Format
Harvard Univ.	110602	5 - 20	1 - 2	pure shells
MR*	21095	3 - 15	1 - 2	1 full cylinder
UTEP	585112	< 0.5	15 - 20	several full cylinders
Smithsonian	114420	< 2	5 - 10	pure shells

*Mineralogical Research Company, E. Alta Vista Way, San Jose, CA 95127-1737

Figure 3 shows the magnetic susceptibility data for the Harvard sample in the temperature range 2.0 - 300 K at an applied magnetic field of 1.0 kOe. The data show a diverging susceptibility and an effective magnetic moment with significant deviations from Curie-Weiss behavior below 50 K. Similar data were collected for the other samples and are quantitatively similar above 50 K, however below 50 K the samples are dissimilar. All samples exhibit a linear $1/\chi$ vs. temperature above 70 K. In this region, the samples can be fit with Curie constants C = 2.83 to 3.10 emu mol^{-1} K^{-1}. These values are in agreement with the spin only magnetic moment of 5.0 μ_B expected for a high spin Fe (II) S = 2 ion. The Weiss constants are all positive (e.g., θ = 22.6 K for the Harvard sample) and are indicative of ferromagnetic behavior. This is further substantiated by the cusp in μ_{eff} which, for example, maximizes at 9.5 μ_B for the Harvard sample.

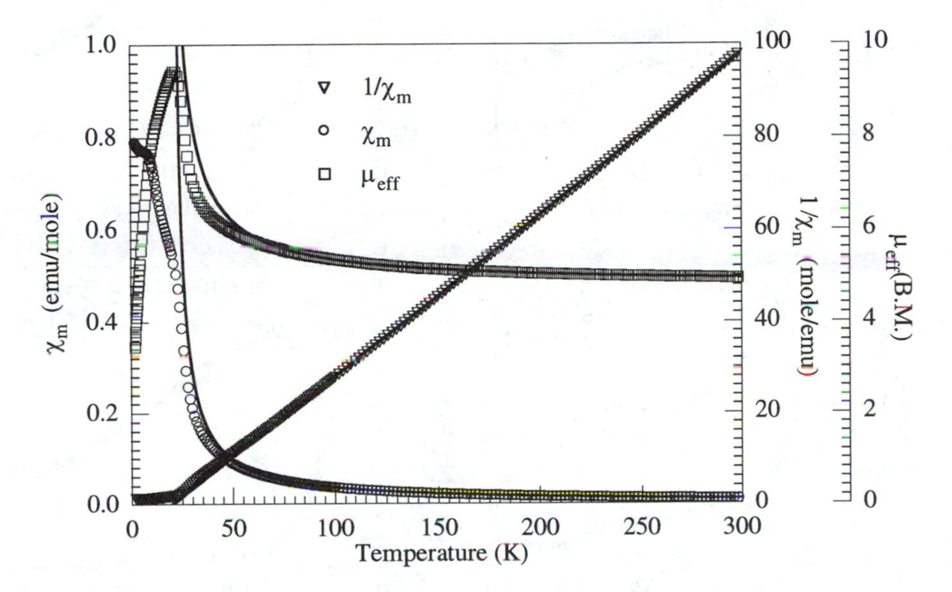

Figure 3. Magnetic data for Harvard sample #110602 using H_{app} = 1000 Oe. Lines correspond to best-fit Curie-Weiss parameters discussed in text.

Such data typically suggest a field dependence of the magnetic susceptibility. To investigate the possibility of a ferromagnetic ground state, samples were subjected to a path dependent data collection scheme at different field values. This scheme is as follows: the sample is cooled in zero field to 2.0 K, a magnetic field is applied and data are collected while the sample warms (ZFC data); sample is cooled in the field to 2.0 K, and data are collected while warming the sample (FC data); the sample is cooled, the applied field is removed and data are collected while warming the sample (REM data). The results of this data collection scheme are shown for the Harvard sample in Figures 4a-f, using various field strengths. It is evident that the magnetic susceptibility exhibits a systematic dependence on the applied field. At low field values, there are clearly two magnetic phase transitions, each of which can be associated with a remanent magnetization. As the applied field is increased, the transition at the higher temperature (23 K) visibly diminishes. The low temperature transition (8.5 K) is essentially field independent and persists to the highest field studied (10 kOe). The transition at 23 K was found to be the distinguishing feature between the cylindrite specimens. Difficulty in assigning values to the transition temperatures prompted the analysis of the data using the first derivative, dM/dT.

Data for the Harvard sample with an applied field of 0.1 kOe are repeated in Figure 5a, adjacent to the dM/dT data for the same sample (Figure 5b). The dM/dT data emphasize the transition near 8.5 K as a zero-point and the transition near 23 K as a maximum. With this method we are able to verify that the low temperature transition is essentially field independent and single valued at 8.5 K. On the other

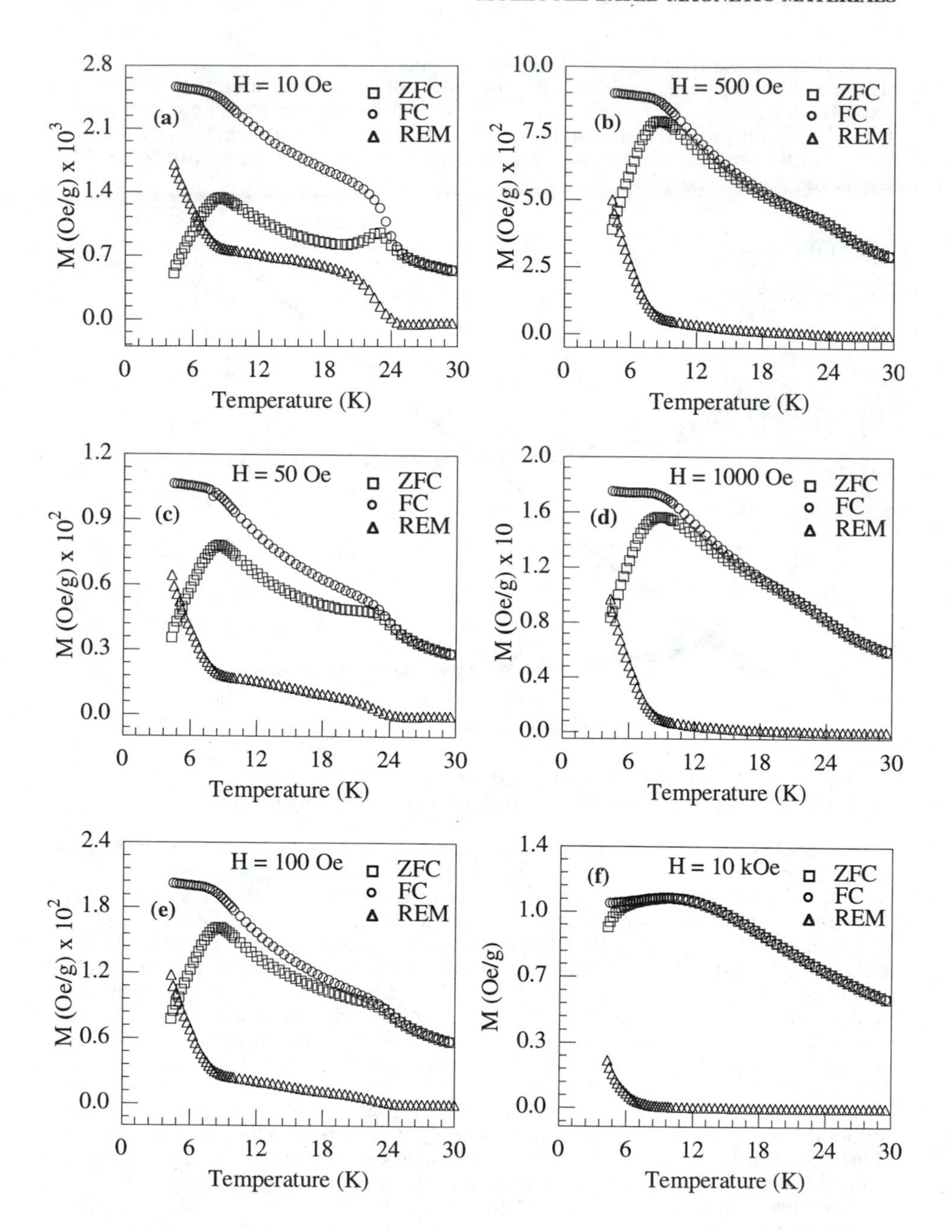

Figure 4. a-f) Field dependent magnetization vs. temperature data at applied field strengths indicated. Path dependent data were collected in the sequence ZFC-FC-REM as discussed in the text.

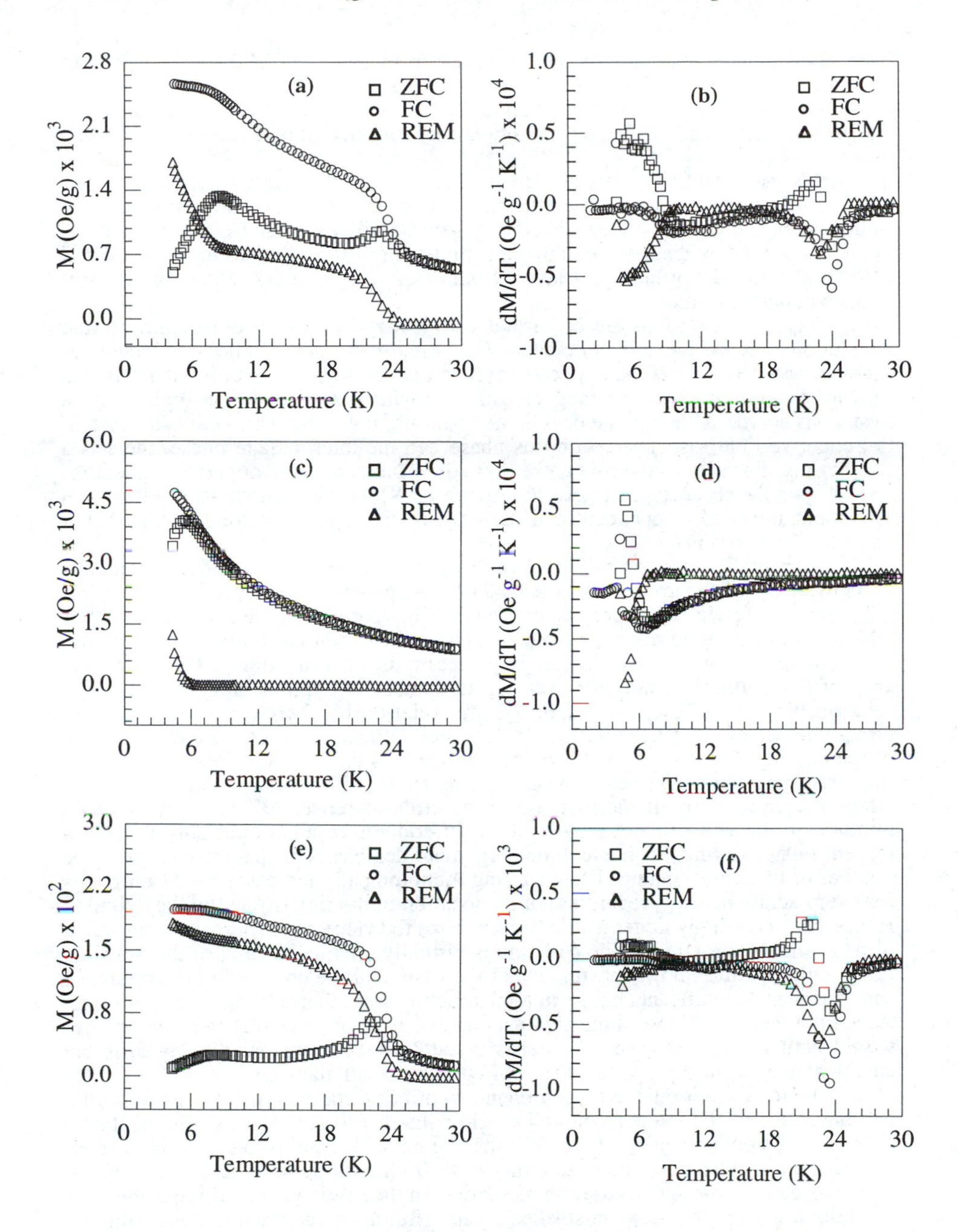

Figure 5. M vs. T and dM/dT vs. T data for the Harvard (a & b), Smithsonian (c & d), and Mineralogical Research (e & f) samples.

hand, the high temperature transition decreases from a value of 24.5 K at 0.01 kOe to a value of 23.25 K at 0.5 kOe. At field values greater than 0.5 kOe, the first derivative is no longer sensitive enough to ascertain the value of the phase transition. At fields greater than 1.0 kOe, the transition is no longer visibly detectable.

Magnetostructural Correlations. Upon completion of the data collection and analysis for all samples, certain trends in the data became evident. Most notably, the Smithsonian and UTEP samples are comprised primarily of layers from *small* radii cylinders and do not display the high temperature transition, whereas the Harvard and MR samples consist primarily of layers from *large* radii cylinders and exhibit a pronounced upper transition.

These observations are consistent with a magnetic model for cylindrite that calls upon three distinct magnetic behaviors. Field dependent studies conducted below 8 K and between -10 and 10 kOe confirm that the low temperature phase is that of a long range ordered ferromagnet with a remanent magnetization typified by a hysteresis curve. Although the data do not saturate, it is clear that there is a substantial coercive field associated with this phase. In the intermediate phase, there is a distinct non-linearity to the magnetization curve indicative of cooperative phenomena, despite the absence of hysteresis. Above 23 K, the magnetic susceptibility data are essentially field independent and show the linear dependence of M vs. H that is expected for a paramagnet.

A complete understanding of the magnetic behavior of cylindrite depends upon the successful correlation of the magnetic properties to the crystal chemistry. If the cylinders are viewed as perfect, concentric rings formed by the stacking of layers which comprise the minimum repeat sequence, then the total number of magnetic ions in a given ring (one layer repeat sequence) must be a function of the circumference of that particular ring. For example, in a cubic system, a simple mathematical relationship can be used to emphasize the relationship between the radius and circumference of any particular ring in "unit cell" dimensions. If the circumference of the nth ring, C_n, is given by $C_n = 2\pi r_n$, where r_n is the radius of the nth ring, then the circumference of the next concentric ring (n + 1) is given by $C_{n+1} = 2\pi\ r_{n+1}$, where $r_{n+1} = r_n + \Delta r$. If the increase in the circumference (ΔC) is expressed as a number of unit cells, then the n+1 ring must contain 2π additional unit cells with respect to the nth ring. Such a textbook argument demonstrates that the change in the number of unit cells contained in each ring monotonically increases by 2π unit cells for every additional ring, regardless of the location of the ring (n) within the cylinder. Hence for increasingly larger n values (as $n \rightarrow \infty$) the value of n approaches the value of n+1, and the structure of the n+1 ring is virtually identical to that of the nth ring. For smaller n values however, this is not the case, and the change in circumference can represent a significant change in total structure from that of the original nth ring. A qualitative way of describing this phenomena is to observe that very large rings would actually exhibit sheet-like behavior, and would therefore display different chemical and physical properties than rings of very small diameter.

To further this qualitative argument, an nth ring containing 100 unit cells with one magnetic Fe (II) ion per unit cell would contain 100 Fe (II) ions. The next successive ring would contain 106 Fe (II) ions, the next 112, and so on. This is in keeping with the intrinsic incommensurability of the cylindrite lattice. At large n values, this relationship corresponds to nothing more than the subtle effects of impurities in a crystalline lattice, i.e., between stacked layers. But again, at small n values, this incommensurability is significant and results in an amorphousness which must have an effect on the magnetic properties.

The question then arises as to what type(s) of magnetic behavior can be observed for a material whose microscopic structure systematically changes, exhibiting fundamental differences between the submicron level and the macroscopic centimeter scale. The paramagnetic regime (T > 30 K) of cylindrite is easily understood in terms of a six-coordinate Fe (II) d^6 ion which, in an ideal octahedral ligand field, would have a high spin (S = 2) $t_{2g}^4e_g^2$ ground state due to the intermediate field strength of the S ions. Such a $^5T_2$ ion is expected to show a spin only effective magnetic moment of 4.9 μ_B; typical Fe (II) compounds show effective moments in the range 5.0 - 5.5 μ_B. The observed high temperature limit to the effective magnetic moment in the range 4.9 - 5.1μ_B is in agreement with the expected results. Furthermore, the magnitude of the observed Curie constant (C = 2.83 emu mol^{-1} K^{-1}) yields an effective moment of 4.8 μ_B, which is in accord with the high temperature limit of the μ-effective data (Figure 3). Such a result is not unexpected in view of covalent effects in Fe-sulfides. Although the orbital momentum may be somewhat quenched due to covalent effects, it is feasible that Fe (II) displays single ion anisotropy (D) and that any exchange interactions (J) will deviate from isotropic Heisenberg character. For ionic Fe (II) systems, most observed anisotropies have been of the Ising type.

The crystallographic coordinates and the precise distribution of the Fe throughout the lattice are as yet undetermined. The current model suggests that the Fe is randomly distributed throughout the PH and PT layers. Such an arrangement of magnetic ions in a well-behaved crystal lattice would be expected to yield spin glass behavior with a distribution of intralayer exchange parameters. The stacking or curling of these layers on one another allows for additional interlayer exchange which leads to the 3-D ordering that we observe as the low temperature phase ($T < 8$ K). The question arises as to how the intra- and interlayer exchange values vary as a function of cylinder radius. Whereas the interlayer exchange pathways must change as a function of cylinder radius, it is not obvious why the distribution of the intralayer exchange pathways must vary as a function of cylinder radius. The multitude of variables regarding the total content and distribution of Fe (II) ions within the layers do not permit us to definitively assign a particular magnetic behavior to a bulk sample of cylindrite, i.e., the inner core of a cylinder is magnetically a different material than the periphery. This is particularly true in terms of long range order and the occurrence of domains. Cylinders only a few microns in diameter may not be large enough to sustain a single domain, whereas cylinders several millimeters in size could sustain single domains simply at the outer periphery. Therefore the bulk magnetic properties associated with the ferromagnetic long range order can be expected to be a function of sample selection.

Conclusions

The utility of SQUID-based magnetometry is emphasized by our need for extensive studies on a single material, such as cylindrite. The magnetic behavior of cylindrite results from having a magnetic ion whose crystallographic coordination environment varies and may therefore exhibit a range of single ion anisotropies ($|D| = 5 - 15$ cm^{-1}). This potential variation in D may result in exchange interactions that could range in anisotropy from Ising-like to near Heisenberg. Since the distribution of the Fe (II) among different lattice sites may be modulated throughout the lattice and may even exhibit clustering, a variation in the magnitude and nature (ferro- vs. antiferromagnetic) of the exchange could result in nanoscopic regions that exhibit localized ferrimagnetic ordering. Furthermore, any bulk ferrimagnetic character in cylindrite is likely to be enhanced due to the systematic increase of the Fe content in the consecutive, wrapping layers. Even if all exchange interactions throughout the crystal are assumed to be antiferromagnetic, this systematic increase would not permit net spin compensation. Intralayer exchange interactions in consecutive wrapping layers would be stronger than the interlayer exchange interactions and a crossover from 2-D to 3-D magnetism is likely associated with the low temperature phase transitions. With all of these variables, it is highly probable that the ferromagnetic ground state observed for cylindrite results from a quasi-low-dimensional assembly of spins for which the competing intra- and interlayer exchange interactions may be slightly anisotropic (Ising). The distribution of these interactions would result in a microscopic picture that corresponds to a correlated spin glass with ferrimagnetic attributes.

To further elucidate the nature of the magnetic behavior in cylindrite, future SQUID-based studies will focus on magnetic properties of single cylinders. The effects of radii and aspect ratios in the submicron to millimeter regime will be investigated in detail. Due to the Fe, Sb and Sn content, we anticipate that Mössbauer spectroscopy will provide information regarding the low temperature magnetic phases. If some of the low temperature properties are attributable to superparamagnetic or spin glass like behavior, AC susceptibility studies will reveal features of the spin dynamics which are not visible in the static studies carried out to date. Specific heat measurements should provide additional information about the role that dimensionality crossover plays in the low temperature transitions. Finally, it should be emphasized that all our cylindrite studies are currently limited to natural specimens. To enhance these studies, synthetic efforts are underway with an aim of controlling compositional variances and crystal growth.

Acknowledgments

We gratefully acknowledge the National Science Foundation, the Robert A. Welch Foundation and the Office of Naval Research for financial support of this work. We are indebted to Dr. Jeffrey E. Post and his associates at the Smithsonian NMNH, Department of Mineral Sciences, for providing samples and their valuable expertise. We also wish to express thanks to numerous students in our group, past and present, who have made contributions to our program. Special thanks go to Professor Philip Goodell for assistance in the mineralogical aspects of the research.

Literature Cited

1. Hatfield, W.E. in *Solid State Chemistry*; Cheetham, A.K. and Day, P., Eds.; Oxford University Press: Oxford, **1987**.
2. O'Connor, C.J. *Prog. Inorg. Chem.* **1982**, *29*, 203.
3. White, G. K. *Experimental Techniques in Low-Temperature Physics*, Clarendon Press: Oxford, **1987.**
4. Wilson, M.N. *Superconducting Magnets*, Clarendon Press: Oxford, **1983**.
5. Van Duzer, T. and Turner, C.W. *Principles of Superconductive Devices and Circuits,* Elsevier: New York, **1981**.
6. See for example *Superconductor Applications: SQUIDs and Machines*; Schwartz, B.B. and Foner, S. Eds.; Plenum Press, New York, **1976**.
7. Gallop, J.C. *SQUIDS, the Josephson Effects and Superconducting Electronics,* Adam Hilger: Bristol, **1991**.
8. Nelson, D.J. and ter Haar, L.W. *Inorg. Chem.* **1993**, *32*, 182.
9. Vaughan, D.J. and Craig, J.R. *Mineral Chemistry of Metal Sulfides*, Cambridge University Press: Cambridge, **1978**.
10. Wang, S. and Buseck, P.R. *American Mineralogist* **1992**, *77*, 758.
11. Hanson, S.L., et al. *Rocks & Minerals* **1992,** *67*, 113.
12. Trumbore, F.A. and ter Haar, L.W. *Chemistry of Materials* **1989**, *1*, 490.
13. Makovicky, E. *Neues Jahrbuch fur Mineralogie, Monatschefte* **1974**, *6*, 235.
14. Makovicky, E. and Hyde, B.G. *Structure and Bonding* 1981, 46, 101.
15. Makovicky, E. *Neues Jahrbuch fur Mineralogie, Monatschefte* **1971**, *2*, 404.
16. Makovicky, E. and Hyde, B.G. *Materials Science Forum* **1992**, *100*, 1.

Chapter 6

Continuous Wave and Fourier Transform Pulsed Electron Magnetic Resonance Spectroscopy in Organic–Molecular Magnetism

Theory and Applications

T. Takui[1], K. Sato[1], D. Shiomi[2], and K. Itoh[2]

[1]Department of Chemistry and [2]Department of Materials Science, Faculty of Science, Osaka City University, 3–3–138 Sugimoto, Sumiyoshi-ku, Osaka 558, Japan

CW electron spin resonance/electron nuclear multiple resonance (cw-Electron Magnetic Resonance: cw-EMR) has been used to examine microscopic details of high-spin molecules and molecular spin assemblies, focusing on electronic spin/molecular structural analyses and structure-magnetism relationships in solids. This paper presents an overview of recent developments in the rapidly growing interdisciplinary area of organic/molecular magnetism in terms of cw and FT-pulsed EMR spectroscopy. The main emphasis of this work is the spectral analysis of random orientation fine-structure ESR spectroscopy to identify molecular high-spin states with high sensitivity and to determine fine structure parameters with high precision. Some important recipes and guidelines for spectral simulation procedures are given and exemplified graphically. This paper discusses the potential capability of high-frequency (W-band) EMR spectroscopy in comparison with the X-band spectroscopy. This paper also deals with late breaking results from FT-pulsed ESR/ESTN (Electron Spin Transient Nutation) spectroscopy applied to high spin systems.

In the last decade organic/molecular magnetism *(1-10)* has become a rapidly growing multi-interdisciplinary field in the pure and applied natural sciences *(11-17,18-23)*. This is not only due to the rich variety of novel physical phenomena and properties which synthetic organomagnetic materials are anticipated to exhibit both macro- and meso-scopically, but also due to their underlying potential applications in future molecular-device technology and molecular quantum materials science based on both their multiple supramolecular functionality and "system" property *(18-23)*.

Among the diverse topics of organic/molecular magnetism, molecular high-spin chemistry continues to underlie organic/molecular magnetism, and the impressive progress in molecular design and synthesis of organic and inorganic molecular materials has led to various types of organic/molecular magnets *(11-17)*. Efforts to synthesize low (zero to two)-dimensional molecular magnetic systems which include extremely large spins (superparamagnets and super high-spin polymeric systems), are motivated not only by the device application capability

0097–6156/96/0644–0081$15.00/0

of novel "system" or "soft" magnetism, but also by interest in spin manipulation chemistry and in modulating spin structures of molecular orbital (crystal orbital) bands of neutral or ionic polymeric open-shell systems. More controlled attempts to generate extremely high-spin ground states of neutral or charged (polycationic or polyanionic) organic molecular systems continue, and extremely high-spin large clusters composed of transition metal ions have emerged (*11-17,24)*.

Practical Importance of Non-Oriented FS (Fine-Structure) ESR Spectroscopy in High-Spin Molecular Science

In order to characterize the spin structure of open-shell components or building units of molecular magnetic systems and assemblages in microscopic details, FS (Fine-Structure) ESR spectroscopy of randomly oriented or non-oriented media is a powerful spectroscopic technique and has been frequently used. Successful analyses of FS spectra intensify the investigation a great deal. Some fine-structure from high spin origins appears as structureless, apparently single absorption peaks subject to inhomogeneous line-broadening or exchange narrowing process, while others feature complex apparently bizzare lineshapes. Thus, a facile spectroscopic method and a practical easy-to-interpret approach are required to stimulate work in of spin manipulation science and organic/molecular magnetism.

The aim of this paper is to carry readers who are not specializing in EMR spectroscopy from a qualitative understanding of high spin FS ESR spectra to the stage where they are able to extract spin Hamiltonian parameters from FS spectra with the help of spectral simulation or directly, without simulation. The spin Hamiltonian parameters include spin quantum numbers (S), g values, zerofield splitting (ZFS) parameters (**D**), hyperfine splitting (HFS) parameters (**A**), and nuclear electric quadrupole splitting parameters (**P**), as given below:

$$\begin{aligned} H = \textstyle\sum_i \{ \beta \widetilde{\mathbf{B}} \cdot \mathbf{g}_i \cdot \mathbf{S}_i + \widetilde{\mathbf{S}}_i \cdot \mathbf{D}_i \cdot \mathbf{S}_i + J_{il} \widetilde{\mathbf{S}}_i \cdot \mathbf{S}_l \\ + \textstyle\sum_j [\widetilde{\mathbf{I}}_j \cdot \mathbf{A}_j \cdot \mathbf{S}_i - g_I \beta_N \widetilde{\mathbf{I}}_j \cdot \mathbf{B} + \widetilde{\mathbf{I}}_j \cdot \mathbf{P}_j \cdot \mathbf{I}_j + \textstyle\sum_k \widetilde{\mathbf{I}}_j \cdot \mathbf{J}_{jk} \cdot \mathbf{I}_k] \}, \quad (k \neq j,\ l \neq i)), \end{aligned} \qquad (1)$$

where group-theoretically allowed quartic or higher-order terms such as $BS_m{}^3$, $S_m{}^2S_n{}^2$, and $S_m{}^3S_n{}^3$ are neglected. It should be noted that these terms effect the interpretation of FS spectra if S and the spin-orbit interaction are large. The simulation procedure is not necessarily required in order to interpret the observed FS spectra and to extract, with high accuracy, the first four parameters mentioned above, as described later. For the ESR allowed transition ($\Delta M_S = \pm 1$ and $\Delta M_I = 0$), the other terms contribute to the second- and higher-order corrections in terms of perturbation theory.

Because of the above point and limited space, this paper does not deal with single-crystal EMR spectroscopy, only random orientation EMR spectroscopy. In general, random orientation spectroscopy is experimentally simple, but the spectra are not easily interpreted. This is particularly the case for FS spectra of randomly oriented non-crystalline media such as organic rigid glasses. We review mostly cw FS ESR spectroscopy from random orientation, exemplifying conventional cw X-band (~9 GHz) FS spectra. We also present cw W-band (~95 GHz) FS spectra for comparison's sake, illustrating what makes the X-band FS spectra complex and difficult to interpret.

Also, we present late breaking results from FT-pulsed ESR/ESTN (Electron Spin Transient Nutation) spectroscopy applied to high spin systems, showing what this new method arouses in high spin molecular science. A new methodological advance possibly intensifies the studies of high spin systems, organic or inorganic, and their quantum constrained effects in low-dimensional molecular magnetic systems, semi-macroscopically reduced dimensional magnetic systems, and charged high-spin polymeric systems created through laser excitation.

Approaches to Random Orientation X-Band FS Spectral Analyses

Exchange Coupled Systems and Quantum Spin Mixing. If the definition is broadly based, there are two kinds of molecular FS spectra. One arises from non-interacting high spin states without nearby electronic states where S is a good quantum number and no spin mixing takes place. The other arises from the exchange-coupling of spins and the resulting ESR transition fields and probabilities are interrelated between different spin states. Readers can refer to a comprehensive treatise on this issue by Bencini and Gatteschi *(25)*.

Figure 1 exemplifies the quantum spin mixing appearing in FS ESR spectra, shows typical FS ESR spectra at K-band (~25 GHz) calculated for an exchange-coupled triplet pair with the static magnetic field $\mathbf{B_0}$ along a given orientation, as described by the FS spin Hamiltonian, equation 1 for i = 1,2 and $\mathbf{D_1} = \mathbf{D_2}$. As the spin quantum mixing grows, new transitions arise with intensity borrowing and the transition fields shift appreciably. In the case of the complete mixing, the salient spectroscopic features arising from a quintet state and a triplet state disappear, and the spectra resemble a two triplet states case, where there are not two independent triplets, but where singlet-quintet complete mixing takes place due to the group-theoretical symmetry requirement for a pair of equivalent $S_i = 1$ spins and the triplet state is isolated because of symmetry requirements. (The permutation symmetry of Bose ($S_i = 1$) particles is symmetric, leading to no mixing between the triplet and the other spin multiplets).

Thus, a drastic spectral change is anticipated from complete spin quantum mixing near vanishing J_{il}, as illustrated in Figure 1. In X-band FS ESR spectroscopy, the transition probabilities which gain intensity when a good high-field approximation holds, as in high-frequency ESR spectroscopy, are reduced a great deal and, as a result, the intensity distribution characteristic of the high-field approximation with high spin states is changed. This decreases the practical advantage of using X-band ESR spectroscopy for the cases of quantum spin mixing. High-frequency ESR spectroscopy is desirable for the detection of spin quantum mixing in terms of the transition probability of intermediate spin states. The molecular design and fine tuning of quantum spin mixing for potential device application are current emphases of molecular magnetic science.

Salient Features of FS Spectra from Non-Interacting High Spins and Spectral Simulation. FS ESR spectral analyses for isolated molecular triplet states are well established. Assuming ZFS parameters not larger than the microwave transition energy ($h\nu$) employed at X-band and small g-anisotropy, there appear six group-theoretically allowed resonance peaks (singularities of absorption intensity) arising from canonical orientations where resonance fields correspond to $\mathbf{B_0}$ oriented along the principal axes (X, Y, and Z) of the $\mathbf{D}$ tensor, i.e., the number of the allowed transitions is given by 2S x 3 (X,Y,Z) for an arbitrary spin, S. For $|D| << h\nu$, these absorption peaks occur around the g~2 region of magnetic field. In addition, peaks due to group-theoretically forbidden transitions appear at low magnetic field and the maximum total number of possible peaks for an arbitrary S is given by S(2S-1) x 3.

Appearance of Off-Principal-Axis Extra Lines. In general, lineshapes of the forbidden transition peaks are anomalous, compared with those of the allowed ones. This is because the angular dependence (anisotropy) of the forbidden transitions causes angular anomalies, which correspond to stationary points with $\mathbf{B_0}$ oriented off the principal axis. A typical example is B_{min} appearing in FS spectra from triplet states. The resonance field of B_{min} is given by

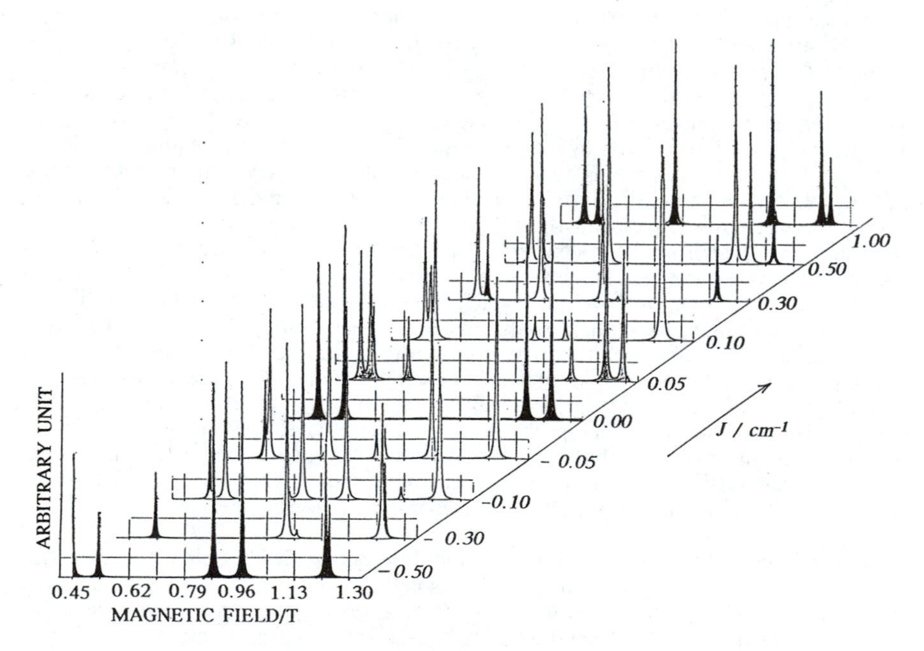

Figure 1. FS ESR spectra arising from quantum spin mixing in a weakly interacting triplet pair.

$$B \geq B_{min} = (2g\beta)^{-1}[(h\nu)^2 + 4(XY + YZ + ZX)]^{1/2}$$

$$= (2g\beta)^{-1}[(h\nu)^2 - (4/3)(D^2 + 3E^2)]^{1/2}. \quad (2)$$

$$X\sin^2\theta\cos^2\phi + Y\sin^2\theta\sin^2\phi + Z\cos^2\theta = XYZ/(g\beta B_{min})^2 \quad (3)$$

with $-X = -D/3 + E$, $-Y = -D/3 - E$, and $-Z = -2D/3$ and where $(\theta,\phi)_{min}$ is calculated by substituting B_{min} obtained from equation 2 into equation 3.

The B_{min} anomaly underlies the stationary behavior of the off-principal-axis orientation in FS spectra from high spins ($S_i > 1$) *(26,27)*. In order to understand straightforwardly the anomalous behavior of the forbidden transition $|M_S\rangle \leftrightarrow |M_S+2\rangle$, the general expression for the resonance field, B_0 and transition probability, $I_{M_SM_S+2}$ at an arbitrary coordinate-axis system are given explicitly (Iwasaki, M; Toriyama, K., unpublished. Takui, T.; Itoh, K. *Spectroscopy III*, Maruzen Publishers, Tokyo, 1993; pp 452).

Additional peaks due to angular anomalies corresponding to off-principal-axis orientations ($\mathbf{B_0} \nparallel$ X, Y, or Z) are called, in general, off-axis extra lines (or simply extra lines). The first HFS extra line observed was in the copper HFS spectrum of copper(II)phthalocyanine in H_2SO_4 glass *(28)*. Physical origins of HFS extra lines have been explained by several authors *(29,30)* .

FS extra lines have been studied extensively in terms of a higher-order perturbation treatment *(27)*. Angular anomalies do not show up for the allowed transitions of the triplet state within the framework of a second-order perturbation treatment *(27)*. Appearance of the extra lines features in FS ESR spectra arising from high spins larger than $S_i = 1$, inevitably from half-integer spins. In terms of perturbation theory, the angular anomaly originates in second- and higher-order corrections for resonance fields. For half-integer spins, first-order terms of the the resonance field corresponding to an $M_S = -1/2 \leftrightarrow M_S = +1/2$ transition vanish, leading to the condition where the second- and higher-order terms dominate *(31,32)*, as described in the following;

$$h\nu = E(M_S+1, M_I) - E(M_S, M_I)$$

$$= g\beta B_0 + (3/2)(\tilde{\mathbf{u}}\cdot\mathbf{D}\cdot\mathbf{u})(2M_S+1) + [K(M_S+1)M_I(M_S+1) - K(M_S)M_I(M_S)]$$

$$+ [\text{second- and higher-order terms}] \quad (4)$$

with

$$K^2(M_S) = \tilde{\mathbf{h}}\cdot\mathbf{K}^2(M_S)\cdot\mathbf{h}$$
$$\mathbf{K}(M_S) = (\tilde{\mathbf{A}}\cdot\mathbf{g}/g)M_S - g_I\beta_N B_0\mathbf{E}$$

where B_0 stands for the resonance field of the ESR allowed $|M_S, M_I\rangle \leftrightarrow |M_S+1, M_I\rangle$ transition. Thus, first-order treatment is not enough to generate simulated FS spectra nor to interpret observed FS spectra from spins larger than unity. Recipes for the analyses of FS extra lines and their appearance conditions have been given in detail *(27)*.

For the smaller ZFS parameters, first-order perturbation treatment has been used, for the sake of simplicity, in order to extract the ZFS parameters and g values. It turned out that most of the documented analyses based on first-order treatment of molecular high spins where π-π spin-spin interactions dominate the contribution to the ZFS parameters do not lead to the appearance of FS extra lines. During the simulation procedure, readers are strongly recommended to check the appearance condition by substituting the trial ZFS parameters and the microwave frequency employed into the equation given in the literature *(27)*. Whether or not extra lines appear also depends on the linewidth of a single transition used for the simulation. If the difference in resonance fields between extra lines and principal-axis lines is comparable to or smaller than the linewidth, the extra line will not be resolved, but asymmetry in the lineshape or intensity anomaly will be appreciable.

Higher-Order Perturbation Approach. In order to reproduce overall FS spectra including low-field absorption peaks from forbidden transitions, second-order, at least, or higher-order perturbation treatments are recommended with calculation of the angular dependence of all the resonance fields and corresponding transition probabilities. Taking the Boltzmann distribution into account for the peak intensities of the FS spectra observed at low temperature is trivial, but sometimes necessary even for X-band spectroscopy.

When the perturbation approach is applied in order to simulate the low-field X-band FS spectra, including forbidden transitions, attention must be paid to the appearance of somewhat peculiar-looking lineshapes (*e.g.*, out-of-phase lineshape) of the simulated spectra, which strongly indicates the possible breakdown of the perturbation approach.

All the mathematical expressions required for the spectral simulation are available in analytical forms with respect to an arbitrary coordinate-axis system *(27,31,32)* and complete FORTRAN program software packages based on the second-order perturbation approach including HFS terms have been made available by several authors for some times *(33,34)*. The perturbation approach has been developed to third-order only in terms of FS terms *(27)*. A program package based on the second-order perturbation approach to the spectral simulation for an arbitrary electron spin, S and arbitrary nuclear spins, I is commercially available, but its utility is hampered by the failure to simulate forbidden transition peaks *(35)*. The perturbation-approach programs are efficient in terms of computation time, but their weakness is still the failure to reproduce low-field peaks in X-band FS spectra. Programs to enable the simulation of complete FS spectra should, however, still be useful for the cases of $|D| << h\nu$ under the experimental conditions of X-band or high-frequency spectroscopy (K-, Q-, and W-band).

Program Software Packages. The program software packages which include ESR powder-pattern (random orientation) spectral simulation based on higher-order perturbation treatment are available on request *(36)*. The program package for random orientation ESR spectral simulation (second order for FS terms) with anisotropic linewidth variations is commercially available *(35)*.

Exact Numerical Diagonalization Approaches. The breakdown of the perturbation approach based on analytical solution can be avoided by using an exact numerical diagonalization of the spin Hamiltonian matrix, either the n x n eigenenergy matrix or the $n^2 \times n^2$ eigenfield matrix *(37,38)*, where $n = 2S+1$.

The former is subject to notorious non-convergence problems during the matching of $h\nu$ to the difference between the energies involved in the transitions, if the calculation process involves avoiding energy level crossing. This difficulty occurs in the simulation procedure of the transitions appearing in the low field region of X-band FS spectra.

Eigenfield Approach. The difficulty of the non-convergence can be overcome by the latter approach to the direct numerical diagonalization of eigenfield matrices. Note that the **exact analytical** solution of the eigenfield for triplet states can be derived in an arbitrary orientation of the static magnetic field, **Bo** *(39)*.

The only weakness of the eigenfield method is the sizable dimension of the $n^2 \times n^2$ eigenfield matrix. The eigenfield method requires solving generalized eigenvalue problems, which give rise to imaginary eigenfield values. During the numerical convergence procedure elaborate mathematical techniques are necessary. In addition, the method needs a large amount of the computation (CPU) time because the CPU time required for the numerical diagonalization usually increases in proportional to (dimension of the matrix)^3. Thus, the eigenfield method becomes impractical when the dimension of the original eigenenergy matrix is large, although we find the CPU time to be tolerable even for 2S+1 multiplets larger than quintet.

Once exact eigenfields are obtained, the corresponding transition probabilities can be numerically calculated by diagonalizing either the eigenfield transition probability matrix *(38,40)* or the eigenenergy transition probability matrix after substituting the exact transition (resonance) field already obtained and calculating the eigenfunctions *(41,42)*. This calculation procedure for the transition probability is a sort of a "hybrid approach" enabling us to save a great deal of CPU time without adding technically difficult calculations.

Figure 2 illustrates the X-band FS spectrum and the corresponding angular dependence of the transition fields and probabilities for a quintet state (S_i = 2) with relatively large ZFS parameters for which the higher-order perturbation approach breaks down; g = 2.003, D = 0.224 cm^{-1}, and E = 0.038 cm^{-1}.

The X, Y, and Z denote the three canonical orientations of the molecule, i.e., the orientations of $\mathbf{B_0}$ parallel to the three principal axes of the ZFS tensor, $\mathbf{D_i}$ with respect to each transition involved. The A designates the off-axis extra absorption peaks which occur from the stationary point (angular anomaly) with neither $\theta = 0, \pi/2$ nor $\phi = 0, \pi/2$, where θ and ϕ stand for the polar angles of $\mathbf{B_0}$ from the Z and X axis, respectively. Quite a few angular anomalous points and low transition probabilities spread out all over show that the high field approximation does not hold in this spin-quintet system. All the transition fields and probabilities have been computed by numerical direct diagonalization based on the eigenfield approach.

W-Band Quintet FS Spectrum. An intermediate case with $|D| \simeq h\nu$ gives rise to the most complex FS spectrum as seen in Figure 2. The perturbation approach will not yield accurate ZFS nor other spin Hamiltonian parameters from the observed FS spectra for the intermediate case. This is the reason why all documented ZFS data for spin-quintet nitrenes are incorrect, leading to the failure to interpret their electronic and molecular structures in a reasonable way (Fukuzawa, T. A.; Sato, K.; Kinoshita, T.; Ichimura, A. S.; Takui, T.; Itoh, K.; Lahti, P. M., submitted).

It has turned out that the W-band FS spectrum corresponding to the one in Figure 2 illustrates in a straightforward manner all the salient features anticipated for randomly oriented quintet states in the high field approximation. The allowed transitions are intense as expected, all the extra lines disappear, and the forbidden transitions appear in the low field region with only weak intensities. The spectral assignment is straightforward, and the spin Hamiltonian parameters can be extracted with high accuracy even by means of the perturbation approach.

It should be noted that the absolute sign of the D value can be determined experimentally, *i.e.*, the positive sign for the present case is concluded directly from the much stronger intensity of the outermost high field Z peak than the outlying low field Z peak.

Program Software Package. The program software package for the random orientation FS spectral simulation based on the eigenfield approach is available on request on diskette *(44)*. The package includes the 2D representation for the angular dependence of both transition fields and probabilities, and for transition

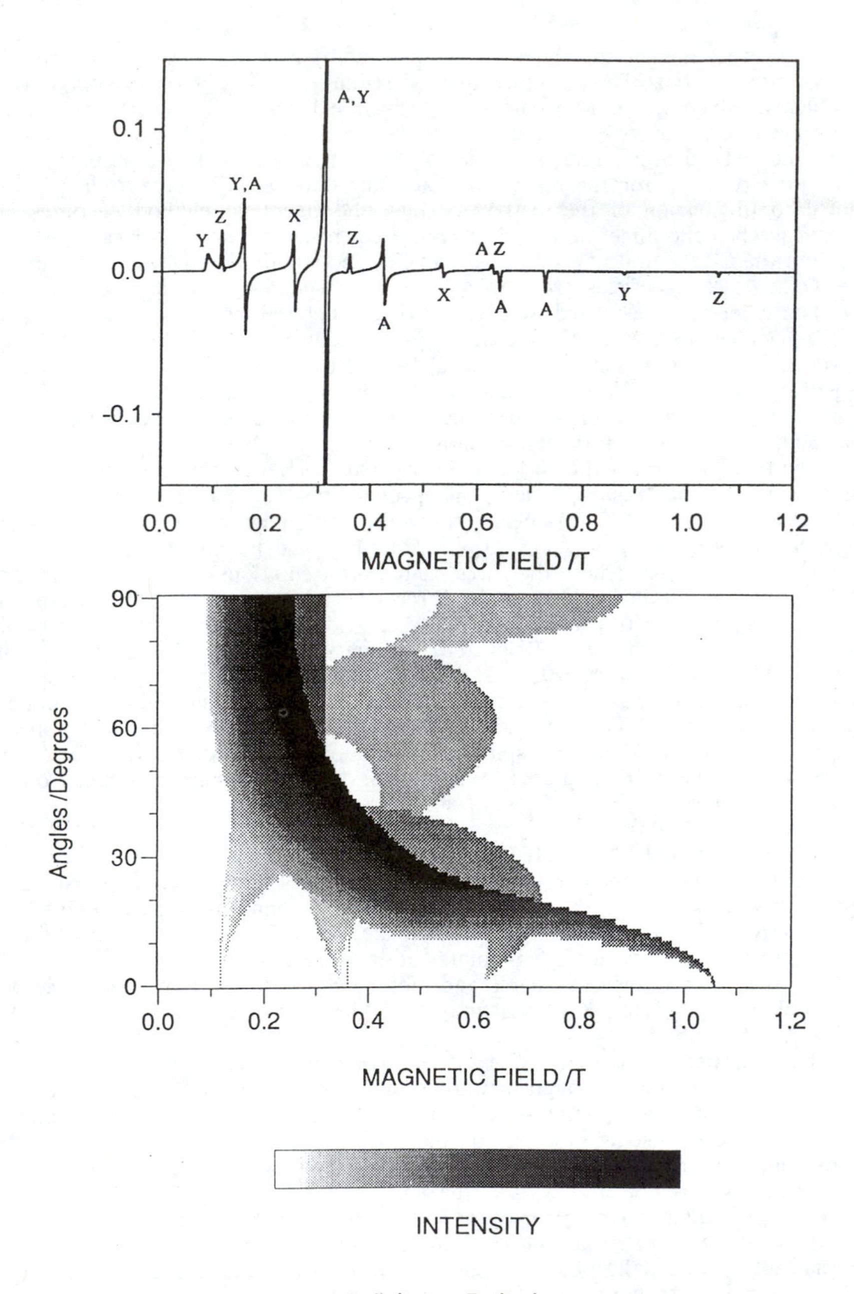

Figure 2. X-band FS spectrum (top) from a molecular quintet state and the angular dependence (bottom) of the transition fields and intensities.

intensities with Boltzmann's statistics taken into account. Anisotropic linewidth variations as a function of the orientation of **Bo** and the g and HFS tensors are incorporated in the program. The package (optional) incorporates the eigenenergy approach based on the numerical direct diagonalization of the energy matrix of the spin Hamiltonian, *i.e.*, equation 1 (i = 1).

Use of the Exact Analytical Expressions for the Resonance Fields and Eigenenergies of High Spins up to S = 3 and S = 4, Respectively. It is well known that equations up to quartic can be analytically solved whereas general analytical expressions for the exact solutions of equations higher than quartic are not available. Nonetheless, the eigenfields with **Bo** along the principal axes of the **D**i tensor can be solved exactly and the exact analytical expressions for the transition fields have been derived for the first time from the eigenfield FS equation up to $S_i = 3$, and the exact analytical expressions for the eigenenergies in the principal-axis orientation have been derived up to $S_i = 4$ *(40)*. The lineshapes characteristic of canonical orientations enable us to discriminate between the X, Y, and Z lines. Thus, the exact analytical expressions for the transition fields serve to determine the spin Hamiltonian parameters without numerical spectral simulation in many cases.

Figure 3 illustrates the transition assignment of the spin quartet FS spectrum (top) of Cr^{3+} in a cage complex of the ligand, 1,8-diamino-3,6,10,13,16,19-hexaazabicyclo[6,6,6]eicosane *(45)*. The angular dependence of the transition fields and probabilities are also shown (bottom). The complete spectrum has been simulated by the eigenfield approach. The spectral assignment has been carried out using the exact analytical solution which enables us to extract a set of the spin Hamiltonian parameters; S = 3/2, g = 1.99, D = 7134 MHz, and E = 422 MHz. The anomalous behavior of the transition fields and intermediate transition probabilities show up in Figure 3; The A's denote the off-axis extra absorption peaks and the F designates an apparent forbidden transition. Any perturbation approach would not enable the spin Hamiltonian parameters to be extracted with reasonable accuracy.

Figure 4 shows the simulated W-band FS spectrum of Cr^{3+} in the cage complex at 3 K assuming the positive D value, and exemplifies the inevitable appearance of the off-axis extra line from the $|Ms=+1/2\rangle \leftrightarrow |Ms=-1/2\rangle$ transition for a half-integer high spin even in the high field approximation. The strongest central peak on the high field side is attributable to the angular anomaly. The transition intensities for the forbidden transitions are diminished whereas the allowed transitions have gained intensity and dominate as expected, as seen in Figure 4 (bottom). In the W-band FS spectrum, the assignments for the canonical orientations are straightforward, facilitating the determination of the spin Hamiltonian parameters with good accuracy.

Program Software Package. The exact analytical expressions are tedious to write explicitly. They are available on diskette *(46)*. The program package generates the transition fields and probabilities, and eigenenergies in the principal-axis orientations of **Bo** for a given set of spin Hamiltonian parameters, *viz.*, a spin quantum number S ($\leq$ 4), g values, and ZFS parameters. The transition probabilities in the principal-axis orientations are calculated by means of the "hybrid" approach mentioned above *(42)*.

Statistical Molecular Structural Fluctuation and Linewidth Variation as a Function of the Orientation of Bo.

Statistical fluctuations of molecular structure show up prominently in organic rigid glasses and spectral simulations assuming random orientation do not repro-

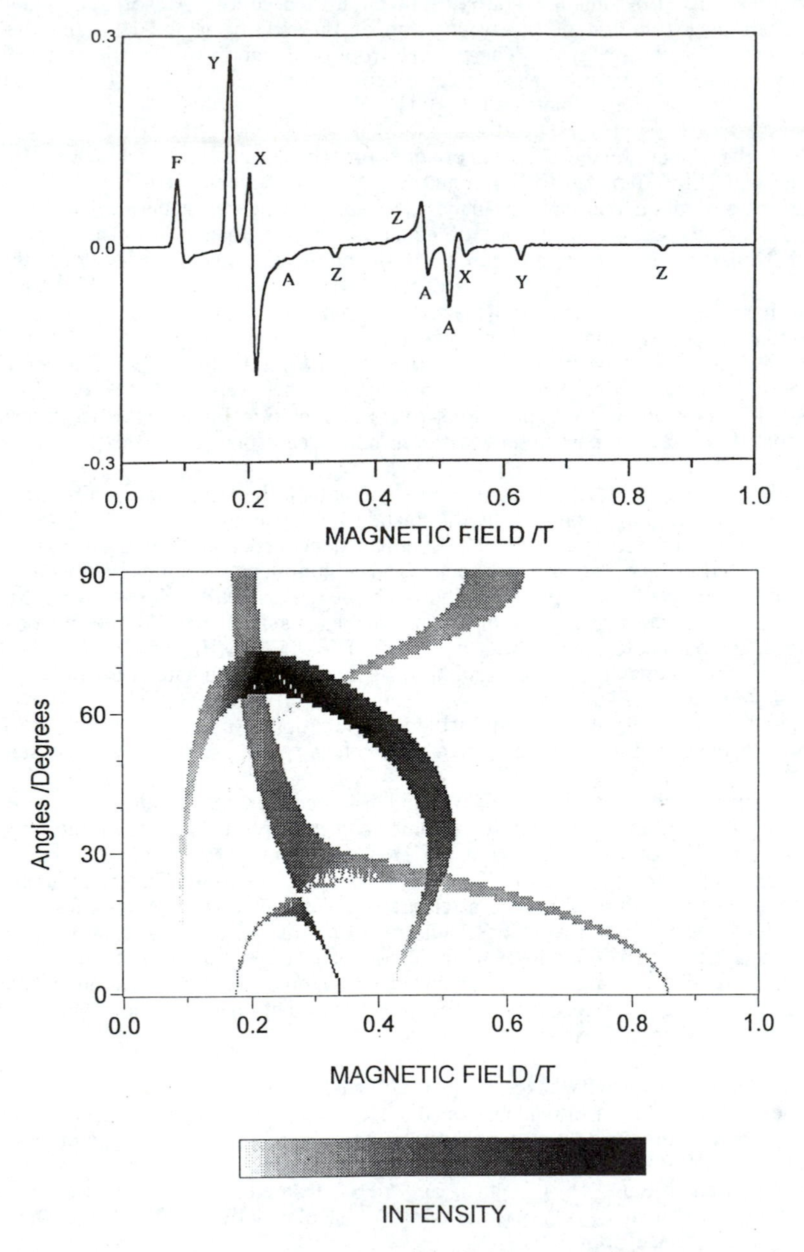

Figure 3. Transition assignments (top) of the quartet FS spectrum of Cr^{3+} in a cage complex and the angular dependence (bottom) of the transition fields and intensities.

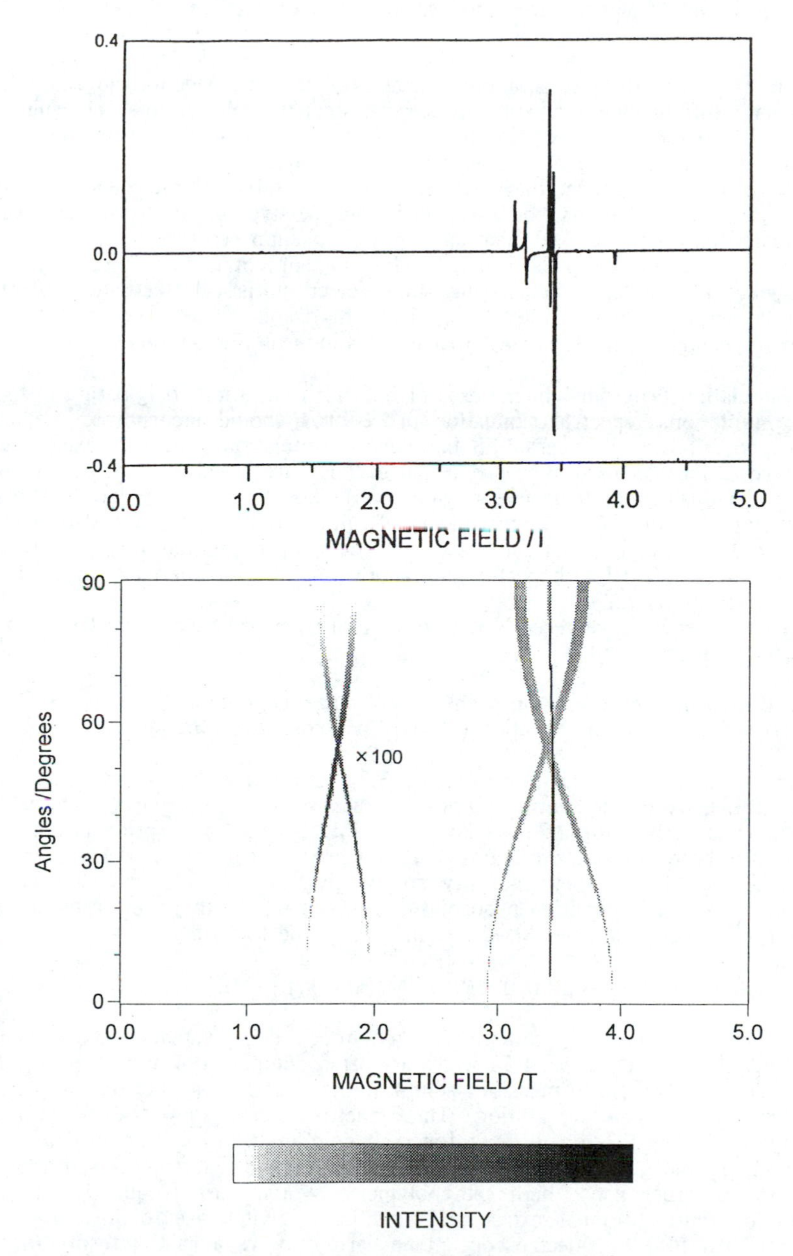

Figure 4. Simulated W-band FS spectrum (top) of Cr^{3+} in the cage complex and the angular dependence (bottom) of the transition fields and intensities.

duce observed FS spectra for this reason. The existence of geometrical isomers, conformers of high spin molecules, or various degrees of molecular clustering are the most common sources of such fluctuations. Particular absorption peaks arising from canonical orientations are subject to linebroadening in some cases. Thermally stimulated relaxation processes aroused by thermal annealing can change the ratio of the conformers or favor particular conformations if they persist during annealing of organic rigid glass states.

In order to interpret the occurrence of the statistical molecular fluctuation due to the continuous distribution of conformers, structurally reinforced molecular design has been invoked such as bridging to suppress the angular fluctuation of dihedral angles. Large or small, rigid planar or round molecules that do not experience intermolecular clustering show fewer statistical fluctuation effects in FS ESR spectra. On the other hand, linear high spin systems or hyperbranched π-aryl systems are made subject to various kinds of fluctuations.

Simulation Programs. In the case of linewidth variations originating in hyperfine interactions, spectral simulation procedures should incorporate the anisotropic contribution to overall FS lineshapes. Simulation programs executing the anisotropic linewidth variation are available. They are based on both the numerical direct diagonalization (eigenenergy approach) and the higher-order perturbation treatment. Phenomenological treatments of linewidth variations are also available. A program package based on a second-order approximation for fine structure terms and first-order in hyperfine interaction terms is commercially available with some restrictions *(35)*.

Pilbrow *et al.* have recently developed more general theoretical treatments of linewidth variations *(47)*.

New EMR Spectroscopic Approach to Molecular High-Spin Systems: FT-Pulsed Electron Spin Transient Nutation (ESTN) Spectroscopy; Motivation

With increasing spin quantum number, S, of molecular high-spin systems, the spectral density of the central regions increases rapidly ($\propto \sim S^2$) even if the spin-spin or spin-orbit interaction is kept constant. FS ESR absorption peaks spread out in the both outlying directions of the magnetic field. The peaks appearing at outermost field lose their intensity so rapidly. This arises from the $|M_s\rangle \leftrightarrow |M_s+1\rangle$ allowed transition probability, as given to zeroth-order transition probability for the $|M_s\rangle \leftrightarrow |M_s+1\rangle$ transition in the following;

$$I_{M_sM_{s+1}} \propto (2\beta B_1)^2 |G|^2 (1/4)[S(S+1)-M_s(M_s+1)]\, |\langle M_I(M_s)|M_I'(M_s+1)\rangle|^2 \quad (5)$$

When extremely large molecular or polymeric spin systems are designed and synthesized, their observed FS spectra are often composed of a mixture of different high spins or conformers. In addition, the spectra may be inhomogeneously broadened with poor resolution. These factors make spin discrimination and unequivocal identification difficult for high spin mixtures in non-oriented media such as organic glasses because of the spectral high density mentioned above.

On the other hand, high spin polymeric systems are frequently subject to inter- or intra-molecular exchange interactions, giving rise to structureless FS spectra. Cw ESR FS spectroscopy alone hardly offers a valid interpretation for such cases. These experimental difficulties encountered in studying the most complex molecular spin system can be eliminated by FT-pulsed ESR spectroscopy operating in the time domain instead of the cw frequency domain.

In the last part of this paper we concentrate on FT-pulsed ESR/ESTN (Electron Spin Transient Nutation) spectroscopy which has emerged recently *(48,49)* and has been developed for non-oriented molecular and polymeric high-spin systems *(50)*. FT-pulsed ESTN spectroscopy enables us to discriminate and unequivocally identify the spin quantum numbers of a system. The spin identification procedure is straightforward and does not require spectral simulation, nor a well-resolved cw FS spectra with high quality.

In the strict sense, ESR spectroscopy has been considered to be too specialized compared with NMR spectroscopy. This is in part due to many different terms to be handled in the spin Hamiltonian and in part due to the exploitation of the "notorious" static magnetic field-swept sheme instead of the energy(frequency)-swept sheme because of technical restrictions. The latter reason forces the spectral analyses and eigenvalue-eigenfunction problems to be mathematically complex. In this context, the spin discrimination analyses involved in ESTN spectroscopy are staightforward and it is easy to understand what happens with modern NMR-like ESTN experiments. It is natural that ESTN spectroscopy, based on FT-pulsed Electron Magnetic Resonance, gives rise to new aspects of EMR spectroscopy and the simplifications of the anisotropic nature intrinsic to the FS-HFS spectra and time-dependent phenomena of high spins or molecular high-spin assemblies, coupled or uncoupled.

Vectorial Picture of ESTN Spectroscopy and Nutation Frequency Obtained by Quantum Mechanical Treatment

What ESTN Spectroscopy Tells Us. ESTN spectroscopy is based on electron spin resonance to measure the spin Hamiltonian in terms of the rotating frame. The time evolution of the electron spin system in the presence of a microwave field $\mathbf{B}_1$ and a static magnetic field $\mathbf{B}_0$ is observed in the rotating frame. As a result, ESTN spectroscopy yields pseudo "zerofield" or pseudo "nearly zerofield" (as an offset parameter $x \sim 0$; see below for the physical meaning of x) FS spectra with high sensitivity due to the use of high-field ESR spectroscopy. The nutation spectroscopy provides us with extremely high resolution in terms of the spin multiplicity discrimination and magnitude of ZFS parameters (thus, the asymmetric nature of the ZFS tensor and surroundings) and directly with the anisotropic nature of FS spectra.

One of the advantages of FT-pulsed ESTN spectroscopy is the two-dimensional (2D) representation inherent in time-domain spectroscopy. The 2D representation of ESTN spectroscopy includes field-swept ($\mathbf{B}_0$) *versus* ESTN spectra and $\mathbf{B}_1$ *versus* ESTN phenomena. The resolution enhancement intrinsic to ESTN spectroscopy does not require high-quality cw FS spectra. On the contrary, the ESTN spectroscopy shows its real ability in the spectroscopic study of complex mixtures of high spins, displaying clear-cut experimental evidence of magnetic properties of spin systems under consideration in microscopic detail.

All aspects inherent in and characteristic of ESTN spectroscopy have not fully been disclosed until recently. This work is the first application of ESTN spectroscopy to non-oriented spin systems such as organic glasses and it has been developed from a methodological viewpoint. This work provides evidence for the existence of neutral high-spin organic polymers *(50)* and shows that ESTN spectroscopy enables us to experimentally determine and quantitatively evaluate extremely small ZFS parameters attributable to departure from a cubic or octahedral symmetry environment of the high spin state of a transition metal ion. The latter and its high resolution are owed to the intrinsic nature of pseudo zerofield spectroscopy of time-domain ESTN. Never before has such a small magnitude ($\sim 10^{-4}$ cm^{-1}) of the ZFS parameter been experimentally detected and evaluated. In addition, the nutation behavior of electron spin multiplets has been explained from the theoretical side, particularly suggesting the possible occurrence of multiple quantum nutation phenomena in high-spin systems which are useful in evaluating ZFS parameters (see Table I).

Vectorial Picture of ESTN Spectroscopy. Let us briefly describe the nutation of a single quantum transition in terms of a classical vectorial picture for the motion of the spin magnetization *(50)*. The magnetization $\mathbf{M_0}$ in the presence of a static magnetic field $\mathbf{B_0}$ precesses at the nutation angle ϕ from the initial direction around the effective field $\mathbf{B_e} = \mathbf{B_0} + \mathbf{B_1}$ when the microwave field ($\mathbf{B_1}$) pulse with width t_1 is applied. $\mathbf{M_0}$ in thermal equilibrium then undergoes free induction decay (FID) when the excitation pulse is turned off; $B_e = [(B_0-\omega/\gamma)^2 + B_1{}^2]^{1/2}$, $\tan\theta = B_1/(B_0-\omega/\gamma)$, and x is an offset parameter defined as $x = (B_0-\omega/\gamma)/B_1$. ϕ is defined as a nutation angle of $\mathbf{M_0}$ around $\mathbf{B_e}$ in time t and ϕ_0 is defined as the nutation angle of $\mathbf{M_0}$ in the same duration time t around $\mathbf{B_1}$, thus $\phi = \gamma B_e t$, $\phi_0 = \gamma B_1 t$ and $\phi = \phi_0(1+x^2)^{1/2}$ holds. ϕ_0 is the rotation angle on exact resonance ($x = 0$). In this classical vectorial picture, the nutation frequency ω_n is given as $\omega_n = \gamma B_e$ off exact resonance ($x \neq 0$) and $\omega_n = \omega_1 = \gamma B_1$ on exact resonance.

Quantum Mechanical Description of ESTN. In order to quantum mechanically describe ESTN phenomena, the equation of motion of the density matrix governed by the Liouville-von Neumann equation is invoked. For the vanishing FS term (H_D), the second term of equation 1 (i = 1), or $H_D << H_1$, the ensemble of high spins nutates at the frequency of $\omega_1 = -\gamma B_1$ under the on-resonance condition, where ω_1 is independent of S and H_1 stands for the interaction term between microwave field $\mathbf{B_1}$ and an electron high spin.

For non-vanishing H_D, the nutation is modified due to the presence of H_D in the rotating frame and not described in terms of a single frequency. In the weak extreme limit of $H_D >> H_1$, however, the nutation frequency ω_n is simply expressed as

$$\omega_n = \omega_1 [S(S+1)-M_s(M_s-1)]^{1/2} \tag{6}$$

where Ms denotes the electron spin sublevel involved in the ESR allowed transition *(48-50)*. The rotating-frame matrix element corresponding to the transition is given to first order as:

$$\langle S, M_s | H_{1,R} | S, M_s\rangle = -\omega_1 [(S+M_s)(S-M_s')]^{1/2} = -\omega_n \tag{7}$$

where $M_s' = M_s-1$ for the allowed transition, showing ω_n is equivalent to the transition frequency of coherent microwave excitation. Thus, for the weak extreme limit, $H_D >> H_1$, the nutation spectrum, ω_n depends on S and Ms.

For integer spins, S = 1,2,3,····, $\omega_n = \omega_1 [S(S+1)]^{1/2}$ for the $|S, M_s=1\rangle \leftrightarrow |S, M_s'=0\rangle$ or $|S, M_s=0\rangle \leftrightarrow |S, M_s'=-1\rangle$ transition. Therefore, even if the ESR transitions involving the $|S, M_s=0\rangle$ level overlap due to the small ZFS parameters, the spin quantum number S can be discriminated in the nutation spectrum by time-domain ESTN spectroscopy. Practically, the effect of the offset frequency on the nutation must be carefully considered in some cases in carrying out $\mathbf{B_0}$-swept 2D nutation spectroscopy.

For half-integer spins, S = 3/2, 5/2,···, the fine structure term $\omega_D(2M_s-1)$ to first order for allowed $M_s \leftrightarrow M_s-1$ transitions vanishes for the $|S, M_s=1/2\rangle \leftrightarrow |S, M_s'=-1/2\rangle$ transition, and higher order corrections due to the fine structure term contribute only as off-axis extra absorption peaks in the random orientation FS spectrum if ω_D is large *(27)*. The corresponding nutation frequency ω_n is simply given as $\omega_n = \omega_1(S+1/2)$. Thus, the nutation spectrum is distinguishable between S=1/2 and other S values even if the fine structure splitting is not seen in the cw ESR spectrum because of inhomogeneous linebroadening, large ω_D values, *etc.* For intermediate cases, *i.e.*, $H_D \sim H_1$, the nutation spectrum appears more complicated, but is interpretable in terms of the rotating frame spin Hamiltonian.

In addition, multiple quantum transitions are observable in the nutation spectroscopy in the weak extreme limit. Theoretical treatments show that the nutation frequency arising from the multiple quantum transition is considerably reduced due to the scaling effect of the effective field that spin ensembles experience in the rotating frame *(50)*. In Table I are summarized nutation frequencies ω_n on resonance for various cases of S and H_D *vs.* H_1.

Table I. On-Resonance Nutation Frequencies for Various Cases

	ω_n, ω_n^{dq}, ω_n^{tq}
$H_D = 0$	$\omega_n = \omega_1$
$H_D << H_1$	$\omega_n \sim \omega_1$
$H_D \sim H_1$	not a single ω_n
$H_D >> H_1$	$\omega_n = \omega_1[S(S+1)-M_SM_S']^{1/2}$ with $(M_S'=M_S-1)$
	$\omega_n = \omega_1(S+1/2)$ for the $M_S=1/2 \leftrightarrow M_S'=-1/2$ transition $(S = 3/2, 5/2, 7/2, \cdots)$
	$\omega_n = \omega_1[S(S+1)]^{1/2}$ for the $M_S=0 \leftrightarrow M_S'=-1$ or $M_S=1 \leftrightarrow M_S'=0$ transition $(S = 1, 2, 3, \cdots)$
	$\omega_n^{dq} = \omega_1(\omega_1/\omega_D)$ for $S = 1$
	$\omega_n^{dq} = \omega_1(7\omega_1/4\omega_D)$ for $S = 3/2$
	$\omega_n^{tq} = \omega_1(3\omega_1/8\omega_D)^2$ for $S = 3/2$

ω_n^{dq} and ω_n^{tq} denote the nutation frequency for double and triple quantum transitions ($S \geq 1$), respectively.

Experimental Method The nutation experiment can be made by either observing the FID or electron spin (or rotary) echo (ESE) signal, $s(t_1, t_2)$ as time-domain spectroscopy (time axis t_2) by incrementing the time interval t_1 of the microwave pulse excitation with the field strength $\propto \omega_1$ ($B_1=-\gamma\omega_1$) parametrically, as depicted schematically *(48-50)*. The two time variables t_1 and t_2 are independent. In the echo-detected scheme, a typical two-pulse sequence can be invoked for the detection period with exploiting particular phase cyclings in order to detect the desired echo signal. The signal $s(t_1, t_2)$ measured as a function of t_1 and t_2 is converted into a 1D or 2D frequency domain spectrum, *i.e.*, $S(f_1, t_2)$ (or $S(t_1, f_2)$) or $S(f_1, f_2)$ by Fourier transformation. The nutation spectrum $S(f_1, t_2)$ measured also as a function of the static magnetic field $\mathbf{B_0}$ gives a 2D $\mathbf{B_0}$-swept nutation spectrum, which enables us to discriminate directly between ESR transition assignments in FS spectra and to separate different spins spectroscopically.

2D ESTN Spectroscopy: An Example. Figure 5 exemplifies 2D ESTN spectroscopy and shows the direct spin discrimination and ESR transition assignments for a spin-quartet state observed in an organic glass, where other lower spin states (S = 1/2, 1) from intermediate oxidation states or by-products are expected as contaminants during the chemical reaction. The central peak (ω_b) and the two "outermost" peaks (ω_a and ω_c) along the magnetic field B_0 are unequivocally assigned to the $|S=3/2, M_S=+1/2\rangle \leftrightarrow |S=3/2, M_S=-1/2\rangle$ and the $|S=3/2, M_S=\pm3/2\rangle \leftrightarrow$

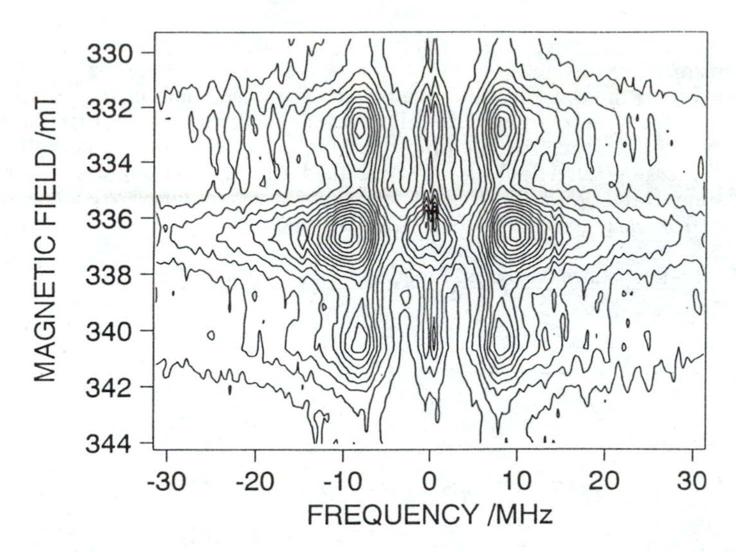

Figure 5. 2D representation for static magnetic field swept *versus* ESTN of a quartet spin state

|S=3/2, Ms=±1/2> transitions, respectively. This is because ω_b : ω_a (or ω_c) = 10.0 MHz : 8.1 MHz $\simeq 2\omega_1 : \sqrt{3}\omega_1$, the ratio expected theoretically from equation 6 for S = 3/2 in the weak extreme limit.

Also, Figure 5 apparently shows that no trace peak attributable to S=1/2 nor S=1 species was observed at 7K, demonstrating that the cw FS spectrum at 7K originates entirely in the highest spin-quartet state which is in the electronic ground state with no neaby excited states of lower spins. It should be emphasized that the central peak, whose transition (nutation) frequency is ω_b, does not contain any appreciable amount of spin-doublet character and thus the FS spectrum is purely from the quartet state. The FS-HFS spectral simulation has been carried out by incorporating hyperfine interaction terms only for the pure spin-quartet state. The excellent agreement enables us to extract the spin Hamiltonian parameters with high accuracy.

A comprehensive treatment will be published on both experimental and theoretical techniques for random orientation ESTN spectroscopy for high-spin transition ions, molecular high-spin clusters, ionic high-spin polymers, and low dimensional molecular magnetic assemblies.

Acknowledgments. This work has been supported by Grants-in-Aid for Scientific Research on Priority Area "Molecular Magnetism" (Area No.228/04 242 103 and 04 242 105) from the Ministry of Education, Culture and Science, Japan and also by the Ministry of International Trade and Industries, Japan (NEDO Project "Organic Magnets"). The authors (K. S. and D. S.) acknowledge the Ministry of Education, Culture and Science, Japan for Grants-in-Aid for Encouragement of Young Scientists (Grant No. 07740468 and 07740553, respectively).

Literature Cited.

1. Morimoto, S.; Tanaka, F.; Itoh, K.; Mataga, N. *Preprints of Symposium on Molecular Structure, Chem. Soc. Japan* **1968**, 67.
2. Mataga, N. *Theor. Chim. Acta* **1968**, *10*, 372.
3. Itoh, K. *Bussei* **1971**, *12*, 635.
4. Itoh, K. *Pure Appl. Chem.* **1978,** *50*, 1251.
5. Ovchinnikov, A. A. *Theor. Chim. Acta* **1978,** *47,* 297.
6. Takui, T. *Dr. Thesis (Osaka University)* **1973**.
7. McConnell, H. M. *J. Chem. Phys.* **1963**, *39*, 1910.
8. McConnell, H. M. *Proc. R. A. Welch Found. Chem. Res.* **1967,** *11*, 144.
9. Itoh, K. *Chem. Phys. Letters* **1967,** *1,* 235.
10. Wasserman, E.; Murray, R. W.; Yager, W. A.; Trozzolo, A. M.; Smolinsky, G. *J. Am. Chem. Soc.* **1967,** *89,* 5067.
11. For a recent overview, see the following references, 11-17 Miller, J. S.; Dougherty, D. A., Eds.; *Mol. Cryst. Liq. Cryst.* **1989,** *176,* 1-562.
12. *Advanced Organic Solid State Materials*; Chiang, L. Y.; Chaikin, P. M.; Cowan, D. O., Eds.; MRS, 1990, 1-92.
13. *Molecular Magnetic Materials*; Gatteschi, D.; Kahn, O.; Miller, J. S.; Palacio, F., Eds.; Kluwer Academic Publishers: Dortrechdt, Netherland, 1991.
14. Miller, J. S.; Epstein, A. J., Eds.; *Mol. Cryst. Liq. Cryst.* **1993,** *232/233*, 1-724.
15. Miller, J. S.; Epstein, A. J., Eds.; *Mol. Cryst. Liq. Cryst.* **1995**, *271/272,* 1-222/1-216, and *ibid., 273/274,* 1-218/1-226.
16. Miller, J. S., Epstein, A. J.; Reiff, W. M. *Acc. Chem. Res.* **1988,** *22*, 144 and references therein.
17. Miller, J. S.; Epstein, A. J.; Reiff, W. M. *Chem. Rev.* **1988,** *88*, 201 and references therein.
18. Takui, T.; Itoh, K. *Polyfile* **1990,** *27*, 49.
19. Takui, T., Itoh, K. *J. Mat. Sci. Japan* **1991**, *28*, 315.
20. Takui, T.; *Polyfile* **1992,** *29*, 48.
21. Takui, T.; *Chemistry* **1992,** *47*, 167.
22. Miller, J. S.; Epstein, A. J. *Chemtech* **1991,** *21*, 168.
23. Landee, C. P.; Melville, D.; Miller, J. S. In *Molecular Magnetic Materials*; Gatteschi, D.; Kahn, O.; Miller, J. S.; Palacio, F., Eds.; Kluwer Academic Publishers: Dortrechdt, Netherland, 1990.
24. *Suprafunctionality: Magnetism*; Kahn, O. Eds.; Kluwer Academic Publishers: Dortrechdt, The Netherlands, 1996, pp 1-442.
25. Bencini, A.; Gatteschi, D. *EPR of Exchanged Coupled Systems*; Springer-Verlag: Berlin, Heidelberg, New York, 1990: 1-287.
26. de Groot., M. S.; van der Waals, J. H. *Mol. Phys.* **1960,** *3*, 190.
27. Teki, Y.; Takui, T.; Itoh, K. *J. Chem. Phys.* **1988,** *88*, 6134.
28. Neiman, R.; Kivelson, D. *J. Chem. Phys.* **1961,** *35*, 156.
29. Ovchinnikov, V.; Konstantinov, V. N. *J. Magn. Reson.* **1978,** *32*, 179.
30. Weltner Jr., W. *Magnetic Atoms and Molecules*; Scientific and Academic Editions, New York, NY, 1983; pp 1-422.
31. Iwasaki, M. *J. Magn. Reson.* **1974**, *16,* 417.
32. Rockenbauer, A.; Simon, P. *J. Magn. Reson.* **1973**, *11,* 217.
33. Toriyama, K.; Iwasaki, M. *personal communication,* **1974**.
34. Shimokoshi, K. *personal communication,* **1975**.
35. Bruker Report; Bruker Analytishe Messtechnik GMBH, D-7512 Rheinstetten, FRG.

36. Sato, K.; Takui, T.; Itoh, K. *Program Software Package (ESR Spectral Simulation Based on High-Order Perturbation Approach)*; Osaka City University.
37. Banwell, C. N.; Primas, H. *Mol. Phys.* **1963**, *6,* 225.
38. Belford, G. G.; Belford, R. L.; Burkhalter, J. F. *J. Magn. Reson.* **1973**, *11,* 251.
39. Wasserman, E.; Snyder, L. C.; Yager, W. A. *J. Chem. Phys.* **1964**, *41,* 1763.
40. Sato, K. *Dr. Thesis (Osaka City University)* **1994**.
41. Teki, Y.; Fujita, I.; Takui, T.; Kinoshita, T.; Itoh, K. *J. Am. Chem. Soc.* **1994**, *116,* 11499.
42. Sato, K.; Takui, T.; Itoh, K. *Program Software Package (ESR Spectral Simulation Based on Eigenfield Approach)*; Osaka City University.
43. Strach, S. J.; Bramley, R. *J. Chem. Phys.* **1988**, *88,* 7380.
44. Sato, K.; Takui, T.; Itoh, K. *Program Software Package (Exact Analytical Expressions for Eigenfields up to $S = 3$ and Eigenenergies up to $S = 4$)*; Osaka City University.
47. Pilbrow, J. R. *Abstracts of Twelfth Conference of ISMAR*; ISMAR: Sydney, 1995; L 18.1.
48. Isoya, J.; Kanda, H.; Norris, J. R.; Tang, J.; Bowman, M. K. *Phys. Rev.* **1990**, *B41,* 3905.
49. Astashkin, A. V.; Schweiger, A. *Chem. Phys. Lett.* **1990**, *174,* 595.
50. Sato, K.; Shiomi, D.; Takui, T.; Itoh, K.; Kaneko, T.; Tsuchida, E.; Nishide, H. *J. Spectrosc. Soc. Japan* **1994**, *43,* 280.

Chapter 7

Heat-Capacity Calorimetry of Molecule-Based Magnetic Materials

Michio Sorai

Microcalorimetry Research Center, Faculty of Science, Osaka University, Toyonaka, Osaka 560, Japan

Based on precise heat-capacity measurements at low temperatures one can elucidate various aspects arising from spin-spin interactions in magnetic materials. Cooperative long-range spin ordering effect manifests itself as a characteristic heat-capacity peak due to phase transition phenomena. Dimensionality of the magnetic lattice and the type of spin-spin interaction sensitively reflect on the heat capacity curve. Zero-field splitting of a single magnetic ion brings about a Schottky heat-capacity anomaly. Heat capacity at extremely low temperatures is described by spin-wave excitation. This paper demonstrates representative heat-capacity data for various kinds of molecular-based magnetic materials and shows how these informations are derived from the adiabatic calorimetry.

Physical quantities obtained from thermodynamic measurements reflect macroscopic aspects of materials in the sense that they are derived as the ensemble averages in a given system. However, since those quantities are closely related to the microscopic energy schemes of all kinds of molecular degrees of freedom in statistical manner, one can gain detailed knowledge on the microscopic level on the basis of precise calorimetry. Among various thermodynamic measurements heat-capacity calorimetry is an extremely useful tool to investigate thermal properties of materials at low temperatures. We shall discuss what kinds of microscopic and macroscopic information are derived from actual calorimetries for molecular-based magnetic materials.

Thermodynamic Quantities Derived from Heat Capacity

Heat capacity is usually measured under constant pressure and designated as C_p. It is defined as "the enthalpy H required for raising the temperature of one mole of a given

0097–6156/96/0644–0099$15.00/0

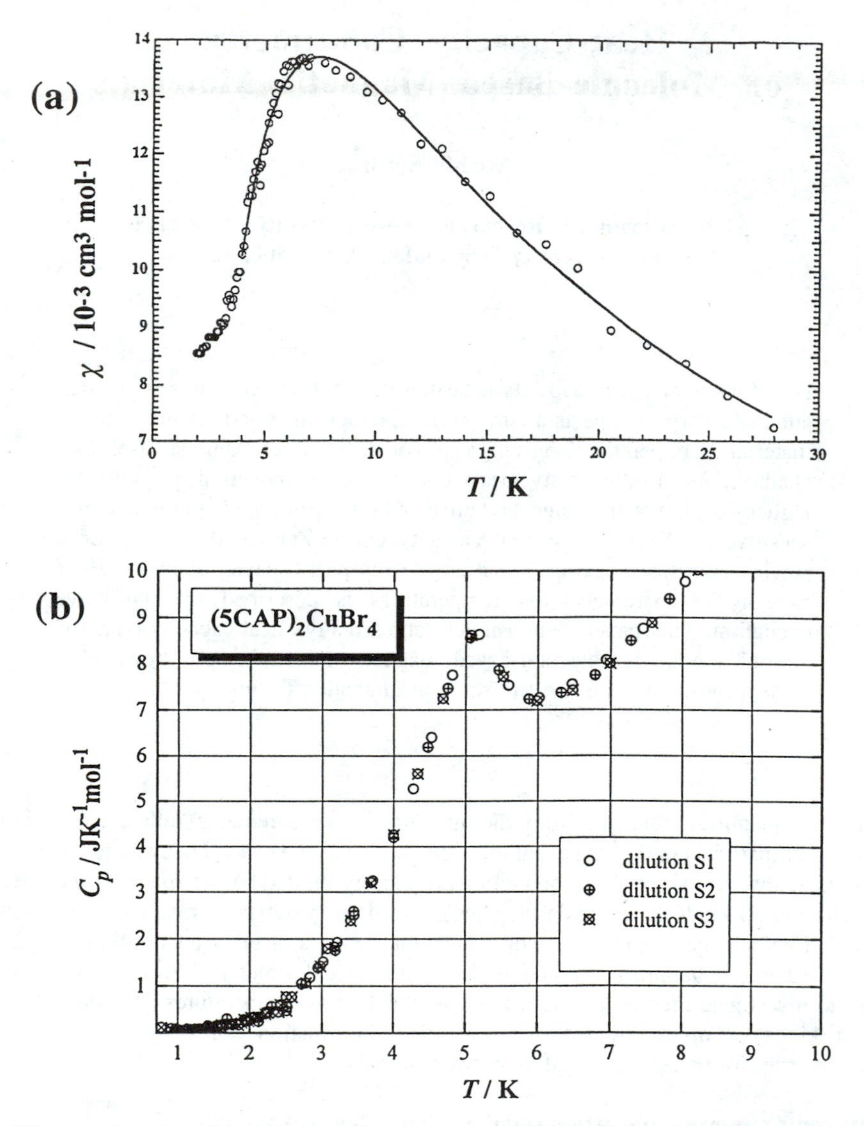

Figure 1. 2D Heisenberg antiferromagnet, $(5CAP)_2CuBr_4$ (*1*): (a) magnetic susceptibility and (b) molar heat capacity. The solid line in (a) is a theoretical curve determined for two-dimensional spin S=1/2 Heisenberg antiferromagnet.

substance by 1 K". From this definition, $C_p \equiv (\partial H/\partial T)_p$, enthalpy increment is determined by integration of C_p with respect to temperature, T :

$$H = \int C_p \, dT \qquad (1)$$

Since C_p is alternatively defined as $C_p \equiv T\,(\partial S/\partial T)_p$, entropy S is also obtainable by integration of C_p with respect to $\ln T$:

$$S = \int C_p \, d \ln T \qquad (2)$$

Now that both enthalpy and entropy are derived from C_p measurements, one can estimate the Gibbs energy G as follows:

$$G = H - TS \qquad (3)$$

Heat capacity is a very valuable and efficient physical quantity in that three fundamental thermodynamic quantities (H, S and G) can be simultaneously determined solely from C_p measurements.

What Can One Learn from Heat Capacity ?

When there exists magnetic interaction between spins of unpaired electrons, the spin energy-levels otherwise degenerate becomes either less degenerate, or completely non-degenerate. Heat capacity of the system clearly exhibits either cooperative phase transition or non-cooperative anomaly depending on the magnetic interaction acting on the spins.

Phase Transitions. Heat capacity is sensitive to a change in the degree of the short-range or long-range ordering. Thus, heat-capacity calorimetry is the most reliable experimental tool to detect the existence of phase transition originating in the onset of a long-range ordering. In case of magnetic materials, the onset of the long-range spin-ordering appears as a sharp C_p peak at Curie temperature T_C for ferromagnets and at Néel temperature T_N for antiferromagnets. Contributions from all the degrees of freedom are principally involved in experimental heat capacity. Therefore, for discussion of the nature of phase transition, the contribution gained by the relevant degrees of freedom should accurately be separated from observed heat capacity. In the case of magnetic phase transitions, contribution from spin-spin interaction appears as an excess heat capacity ΔC_p beyond a normal (or lattice) heat capacity. The excess enthalpy ΔH and entropy ΔS arising from the phase transition are determined by integration of ΔC_p with respect to T and $\ln T$, respectively. Based on these quantities, one can gain an insight into the mechanism of phase transition.

An example to show sensitiveness of heat capacity is seen in Figure 1, where magnetic susceptibility and heat capacity are compared for two-dimensional spin S=1/2 Heisenberg antiferromagnet, *bis*(5-chloro-2-aminopyridine) copper(II) tetrabromide, (5-CAP)$_2$CuBr$_4$ (*1*). As can be seen in Figure 1 (a), the magnetic susceptibility data seem to be well reproduced by a theoretical curve calculated for 2D S=1/2 Heisenberg antiferromagnet (2). As far as the magnetic susceptibility is concerned, this complex seems not to exhibit a phase transition, indicating an ideal 2D Heisenberg magnet. Contrary to this, the heat capacity shown in Figure 1 (b) clearly reveals the existence of a phase transition caused by weak 3D interaction. Although heat capacity and magnetic susceptibility are physical quantities similarly derived as the ensemble averages in a given system, the former is often more sensitive to the onset of a long-range ordering than the latter.

Many magnetic phase transitions are of a second-order and hence often regarded as "critical phenomena". For those phase transitions, magnetic heat capacities C in the vicinity of the critical temperature T_C are reproduced by the following equation:

$$C \propto \varepsilon^{-\alpha} \quad (\varepsilon > 0) \qquad \text{or} \qquad C \propto (-\varepsilon)^{-\alpha'} \quad (\varepsilon < 0) \tag{4}$$

where $\varepsilon \equiv (T - T_C) / T_C$, and α and α' are critical exponents, from which one can get a clue as to the magnetic lattice structure, magnetic interaction involved in the system and so on. In case of molecular-based magnetic materials, dominant spin-spin interaction is very anisotropic and is often realized in a low dimensional manner. In such cases, heat capacity cannot simply be described by these equations.

Boltzmann's Principle. Although entropy is a physical quantity characteristic of macroscopic aspect of a system, the Boltzmann principle relates such macroscopic entropy with the number of energetically equivalent microscopic states:

$$S = k\, N_A \ln W = R \ln W \tag{5}$$

where k and N_A are the Boltzmann and Avogadro constants, R the gas constant, and W stands for the number of energetically equivalent microscopic states. If one applies this principle to the transition entropy, the following relationship is easily derived because ΔS is the entropy difference between the high- and low-temperature phases:

$$\Delta S = R \ln (W_H / W_L) \tag{6}$$

where W_H and W_L mean the number of microscopic states in the high- and low-temperature phases, respectively. In many cases the low-temperature phase corresponds to an ordered state and hence $W_L = 1$. Therefore, when a phase transition is concerned only with the spin-ordering, one can determine the spin quantum number S, on the basis of experimental ΔS value, by the relationship $W_H = 2S + 1$.

Lattice Dimensionality. Spin ordering crucially depends on the magnetic lattice structure, resulting various short-range and long-range order effects upon heat capacity curves. When paramagnetic species form clusters magnetically isolated from one another, the magnetic lattice should be regarded as being of zero-dimension (0D), whose spin energy scheme consists of a bundle of levels. In such a case there exists no phase transition: heat capacity exhibits a broad anomaly characteristic of the cluster geometry.

On the other hand, what happens when spins are arranged on lattice sites spread infinitely ? For a linear-chain structure (one-dimensional lattice: 1D), no phase transition is theoretically expected because fluctuation of the spin orientation is extremely large. Only a broad C_p anomaly characteristic of 1D structure is observed. Stepping up to two-dimension (2D), dramatic change appears: a phase transition showing remarkable short-range order effect takes place when the interaction is of the Ising type, while no phase transition for the Heisenberg type. Contrary to this, three-dimensional structure (3D) gives rise to a phase transition with minor short-range order effect, independently of the type of spin-spin interaction. In Figures 2a and 2b are illustrated magnetic heat capacities C_m of different dimensionality for the spin S=1/2 Ising and Heisenberg models, respectively (*3*). Temperature is reduced by the transition temperature θ of the mean-field model.

In actual magnetic materials, interaction paths, through which super-exchange spin-spin interaction takes place, are often much more complicated than being classified into a homogeneous single dimension. As the result, it happens that apparent dimensionality seems to change with temperature. This feature is called "dimensional crossover". Figure 3 shows the dimensional crossover between 1D and 2D lattices for the spin S=1/2 ferromagnetic Ising model (*4*). The dashed curve indicates the heat capacity anomaly arising from a linear chain with the intra-chain interaction J and inter-chain interaction J' = 0. The dot-dashed curve is the heat capacity of a square planar lattice (J = J') showing a phase transition, corresponding to the exact solution by Onsager (*5*). The solid curve shows the dimensional crossover between 1D and 2D lattices occurring in an anisotropic square lattice, in which the inter-chain interaction J' is assumed to be as small as 1/100 of the intra-chain interaction J. At high temperatures the heat capacity curve asymptotically approaches the 1D curve, while at low temperatures the existent two-dimensional inter-chain interaction, though weak, diminishes the fluctuation of the system and leads to a phase transition.

Type of Spin-Spin Interaction. Fluctuation concerning the spin alignment strongly depends not only on the lattice dimensionality but also on the type of spin-spin interaction. Effective spin Hamiltonian accepted for the exchange interaction between spins 1 and 2 is given by the following generalized equation:

$$\mathcal{H} = -2J\,[\alpha(S_1{}^xS_2{}^x + S_1{}^yS_2{}^y) + \beta S_1{}^zS_2{}^z\,] \qquad (7)$$

$\alpha = 1,\ \beta = 1$: Heisenberg model

$\alpha \neq \beta\ (\neq 0)$: Anisotropic Heisenberg model

$\alpha = 1,\ \beta = 0$: XY model

$\alpha = 0,\ \beta = 1$: Ising model

where S_1 and S_2 are the spin operators and J means the exchange interaction parameter.

In the case of a magnetic system characterized by Ising- or XY-type interaction, heat capacity curves calculated for ferromagnetic and antiferromagnetic interactions are identical, unless the lattice dimensionalities are different. However, this is not the case for the system showing Heisenberg-type interaction. This feature is demonstrated in Figure 4 for one-dimensional S=1/2 case (*3*): the heat capacities of ferromagnetic (curve (d)) and antiferromagnetic (curve (c)) Heisenberg chains are quite different from each

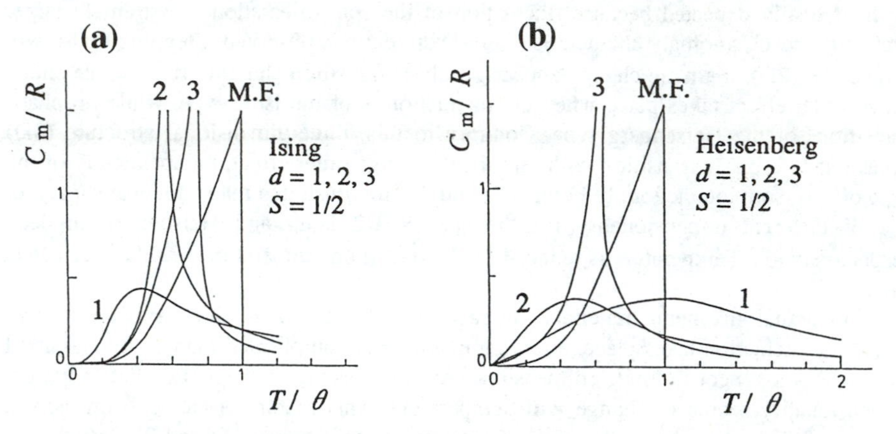

Figure 2. Magnetic heat capacity curves C_m of different dimensional lattices of spin S=1/2. "d " indicates the lattice dimension. The heat capacity derived from the mean-field approximation is labeled "M.F.". Temperature is reduced by the transition temperature θ of the mean-field model. (a) Ising model: 2d and 3d are calculated for square-planar and simple cubic lattices. (b) Heisenberg model: 1d for antiferromagnetic chain, 2d for ferromagnetic square-planar, and 3d for ferromagnetic body-centered-cubic lattices (*3*). Reproduced with permission from reference 3. Copyright 1974. Taylor & Francis, Ltd.

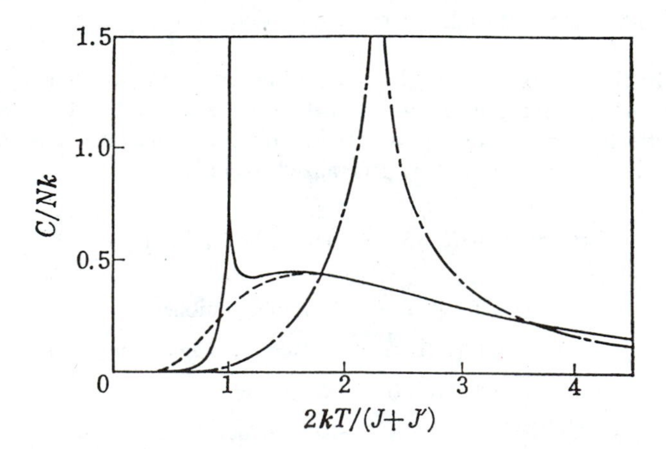

Figure 3. Dimensional crossover between 1D and 2D lattices for the spin S=1/2 ferromagnetic Ising model. Dashed and dot-dashed curves correspond to 1D and 2D models, respectively, while solid curve indicates the dimensional crossover between them. J and J' mean the intra- and inter-chain interaction parameters, respectively (*4*). Reproduced with permission from reference 4. Copyright 1960. Taylor & Francis, Ltd.

other. By comparing experimental heat capacity with these theoretical curves one can easily estimate the type of spin-spin interaction operating in the material.

Schottky Anomaly. Since spin is a physical quantity characterized by quantization, a system containing spins gives rise to various non-cooperative heat-capacity anomalies in addition to cooperative phase transitions. A system consisting of finite number of energy levels brings about a Schottky anomaly. Such systems are encountered in paramagnetic clusters, the Zeeman splitting, tunnel splitting, zero-field splitting and so on. As an example, let's consider a simple two-level system consisting of g_0-degenerate ground state separated by δ from g_1-degenerate excited state. Figure 5 shows Schottky anomalies of such two-level systems as a function of the degeneracy ratio g_1/g_0. The peak value of the heat-capacity anomaly increases when the ratio is increased, while the temperature giving the peak is decreased. By comparing experiment with theory one can elucidate the degeneracy ratio.

Zero-field splitting of a single ion arising from crystalline-field anisotropy is a good example for the Schottky anomaly, whose spin Hamiltonian is written as follows:

$$\mathcal{H} = D\{S_z^2 - S(S+1)/3\} + E(S_x^2 - S_y^2) \tag{8}$$

where D and E are uni- and bi-axial zero-field splitting parameters, respectively. A representative compound described by this equation is nickel(II) complexes. Figure 6 shows the molar heat capacity of *trans*-bis(ethylenediamine)bis(isothiocyanato)nickel (II), $[Ni(en)_2(NCS)_2]$, below 20 K (*6*). This complex exhibits a broad anomaly centered around 3.5 K. This anomaly can be attributed to a magnetic origin arising from the zero-field splitting of the Ni^{2+} ion. As shown in Figure 7, the magnetic heat capacity of this complex was well accounted for by equation (8) with $D/k = 8.98$ K and $|E|/k = 1.62$ K. It should be remarked here that a different set of D and E for negative D also provides the same energy-level scheme, that is $D/k = -6.92$ K and $|E|/k = 3.67$ K. Although, solely from heat capacity data, one cannot judge which set is appropriate, reliable values for the crystalline-field splitting parameters D and E can be estimated only if the sign of D is determined by other experiment.

Spin Wave Excitation. Spin-wave (magnon) theory has been proved to be a good approximation to describe the low-temperature properties of magnetic substances. The limiting low-temperature behavior of heat capacity due to spin-wave excitation (C_{SW}) is conveniently given by the following formula (*3*):

$$C_{SW} \propto T^{d/n} \tag{9}$$

where d is the dimensionality and n is defined as the exponent in the dispersion relation. For antiferromagnetic magnons $n = 1$, while for ferromagnetic magnons $n = 2$. Thus the spin-wave heat capacity of a 3D ferromagnet goes with $T^{3/2}$ and of a 2D antiferromagnet with T^2, and so on. However since actual low-dimensional magnets contain, more or less, weak three-dimensional interaction, the dimensionality sensed by spin at extremely low temperatures should be $d = 3$, as exemplified below.

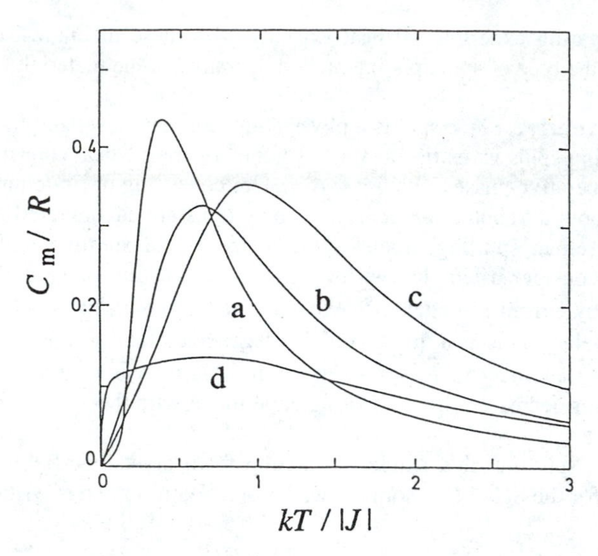

Figure 4. Theoretical heat capacities of magnetic chains with spin S=1/2: (a) Ferro- or antiferromagnetic Ising model, (b) ferro- or antiferromagnetic XY model, (c) antiferromagnetic Heisenberg model, (d) ferromagnetic Heisenberg model (*3*). Reproduced with permission from reference 3. Copyright 1974. Taylor & Francis, Ltd.

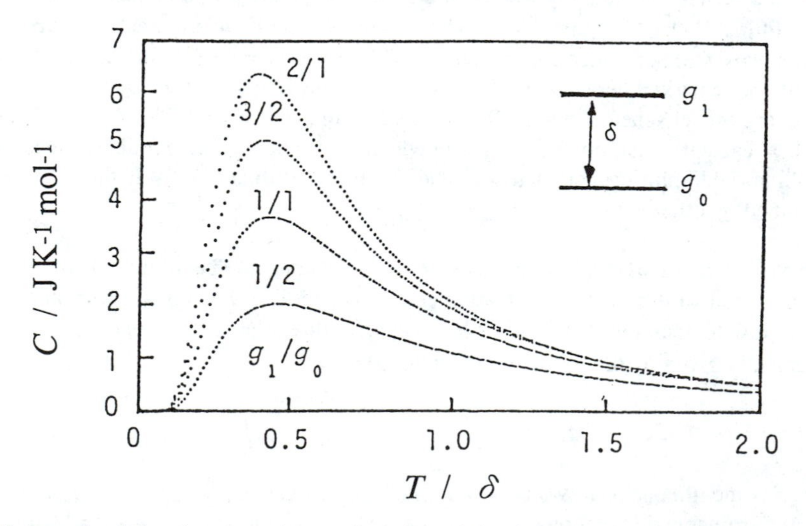

Figure 5. Schottky anomaly as a function of the degeneracy ratio of the ground (g_0) and excited (g_1) states. Temperature is reduced by the energy separation δ between the ground and excited levels.

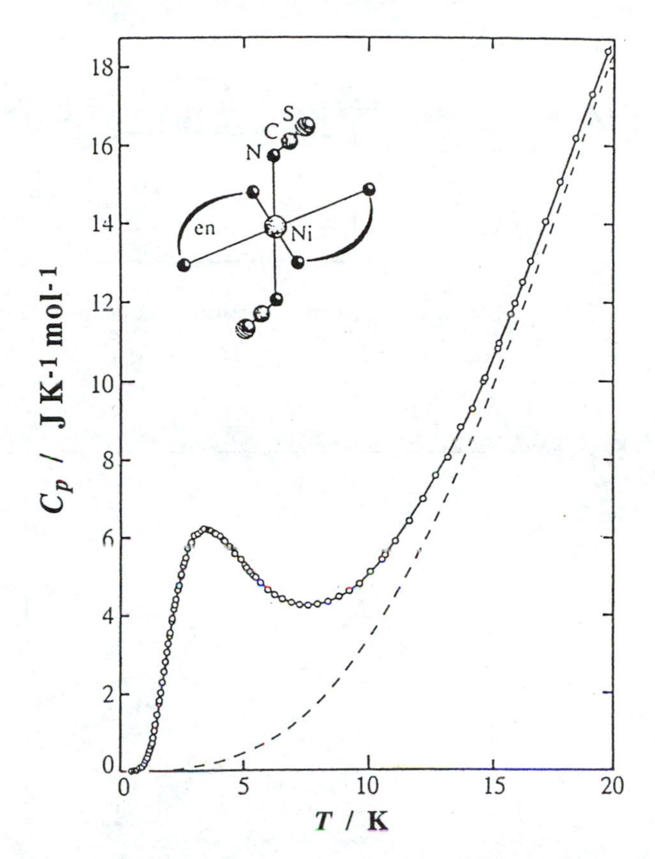

Figure 6. Molar heat capacity of $[Ni(en)_2(NCS)_2]$. Open circles are experimental values and the broken curve indicates the lattice (or normal) heat capacity (*6*). Reproduced with permission from reference 6. Copyright 1987. Academic Press, Ltd.

Heat Capacities of Molecular-Based Magnetic Materials

In this section we shall discuss how the tutorial given above may be applied to actual molecular-based magnets. The materials studied calorimetrically are organic free radical crystals, charge-transfer complexes and mixed-metal complexes, which have been obtained according to the representative strategies to realize molecular-based magnets.

Organic Free Radical Crystals. 4-Methacryloyloxy-2,2,6,6-tetramethylpiperidine-1-oxyl (MOTMP), 4-acryloyloxy-2,2,6,6-tetramethylpiperidine-1-oxyl (AOTMP) and 4-methacryloylamino-2,2,6,6-tetramethylpiperidine-1-oxyl (MATMP) are pure organic magnets having similar molecular structures (see Figure 8). The magnetic susceptibility from 2.5 to 300 K of MOTMP, for instance, indicates the evidence of ferromagnetic interaction between the neighboring radicals (*7*). As reproduced in Figure 9, its heat

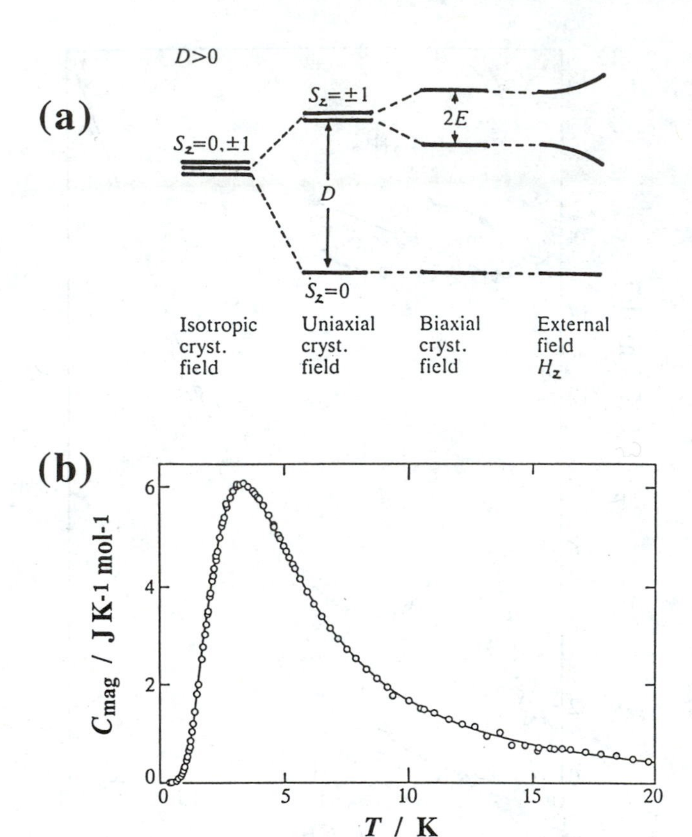

Figure 7. (a) Energy-level splitting of the spin-state of a Ni^{2+} ion with respect to uni- and bi-axial crystalline-field anisotropies. The right-hand column shows magnetic-filed dependence of the spin levels. (b) Magnetic heat capacity C_{mag} of $[Ni(en)_2(NCS)_2]$. The theoretical curve has been calculated for D/k = 8.98 K and $|E|/k$ = 0.18 K (*6*). Reproduced with permission from reference 6. Copyright 1987. Academic Press, Ltd.

capacity measured down to 0.07 K shows a λ-type transition due to the onset of long-range ordering at 0.14 K (*7*). Unexpectedly, a remarkable broad anomaly was observed above the transition temperature. This anomaly was well accounted for in terms of the short-range ordering in the ferromagnetic Heisenberg chains with the interaction parameter of J/k = 0.45 K (*8*). As far as its crystal structure is concerned, there seems to be no indication of one-dimensional favorable exchange paths. This fact evidently indicates how sensitive to dimensionality the heat capacity is. This finding of one-dimensional behavior was confirmed by magnetic susceptibility measurements done below 1 K (*9*). As discussed above, the initial slope of the heat capacity at the lowest

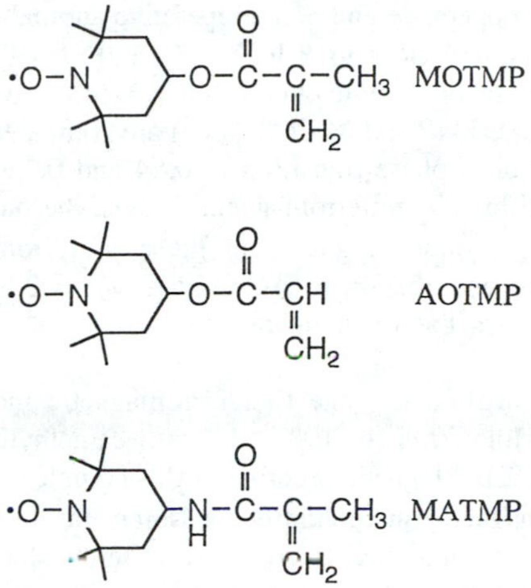

MOTMP

AOTMP

MATMP

Figure 8. Molecular structure and abbreviation of the organic free radicals studied.

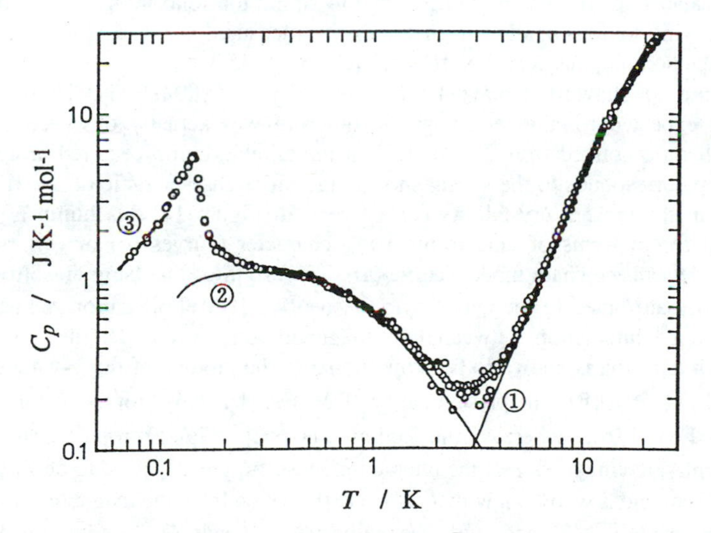

Figure 9. Molar heat capacity of MOTMP. ① Lattice heat capacity; ② 1D Heisenberg ferromagnet ($J / k = 0.45$ K); ③ Spin-wave contribution.

temperatures corresponds to the spin-wave excitation. In case of MOTMP, temperature dependence given by equation (9) was d/n = 1.53. This value is well approximated by 3/2, indicating three dimensional ferromagnet below the transition temperature. The entropy gain due to the cooperative and non-cooperative anomalies was ΔS = 5.81 $JK^{-1} mol^{-1}$. Since this value is very close to $R \ln 2$ = 5.76 $JK^{-1}mol^{-1}$ expected for the spin S=1/2 system, it turns out that one mole of spin is involved in MOTMP crystal.

Heat capacities of AOTMP and MATMP (*10*) are compared with that of MOTMP in Figure 10. They exhibit a phase transition at 0.64 and 0.15 K, respectively. Since AOTMP is characterized by 1D antiferromagnetic chains, the broad C_p anomaly due to the short-range ordering is much higher than those of ferromagnetic MOTMP and MATMP. The spin-wave contribution to AOTMP is d/n = 2.98, coinciding with the value 3 for 3D antiferromagnet at low temperatures.

Charge-Transfer Complexes. The first ferromagnetic molecular charge-transfer complex prepared by Miller *et al.* in 1987 (*11*) is decamethylferrocenium tetracyanoethenide, [DMeFc][TCNE]. Magnetic ordering in this complex was intensively studied by magnetization and magnetic susceptibility measurements (*11,12*), ^{57}Fe Mössbauer spectroscopy (*11,13*), and neutron diffraction (*12*). Below 4.8 K the charge-transfer complex displays the onset of spontaneous magnetization in zero applied field consistent with a 3D ferromagnetic ground state. However, ^{57}Fe Mössbauer spectroscopy exhibited zero-field Zeeman split spectra even up to 20 K. This was a great mystery because at that time there had been reported no conclusion about the question, whether the Mössbauer line-width broadening is due to single ion relaxation or to a cooperative effect involving solitons in a 1D magnet (*11*). We measured heat capacity of this complex (*14*) and the homologous complex [DMeFc][TCNQ] (*15*).

Figure 11 shows the magnetic heat capacity of [DMeFc][TCNE]. The phase transition expected for the ferromagnetic ordering was actually observed at 4.74 K. Another feature noticed from Figure 11 is a remarkable hump centered around 8.5 K. This hump corresponds to the strong short-range order characteristic of the 1D stacking structure in the present crystal. As can be seen in Figure 11, this hump is favorably accounted for in terms of anisotropic Ising character (curves (b) or (c)) rather than isotropic Heisenberg chain model (curve (a)). The origin of this Ising anisotropy can be satisfactorily attributed to the anisotropic g-tensor of [DMeFc]$^+$ cation radical. For the superexchange interaction between two magnetic ions A and B, the anisotropy of magnetic interaction is known to be proportional to the product of the g-values, $J_\perp/J_\parallel \approx (g_\perp^A g_\perp^B)/(g_\parallel^A g_\parallel^B)$. The present cation [DMeFc]$^+$ has anisotropic g-values ($g_\parallel$=4.0 and $g_\perp$=1.3) (*16*), whereas the anion [TCNE]$^-$ is isotropic ($g_\parallel$=$g_\perp$=2.0). Consequently, the magnitude of the interaction anisotropy is expected to be $J_\perp/J_\parallel$=0.33. This value coincides well both with $J_\perp/J_\parallel \approx$0.30 derived from the magnetic susceptibility for single crystals (*12*) and with the calorimetric result (*14*) shown in Figure 11 suggesting a very strong Ising character of the magnetic interaction, $J_\perp/J_\parallel < 0.5$ (curve (b)).

The total entropy gain due to the phase transition at 4.74 K and the non-cooperative anomaly around 8.5 K, ΔS = 12±1 J $K^{-1}mol^{-1}$, is well approximated by $2R$

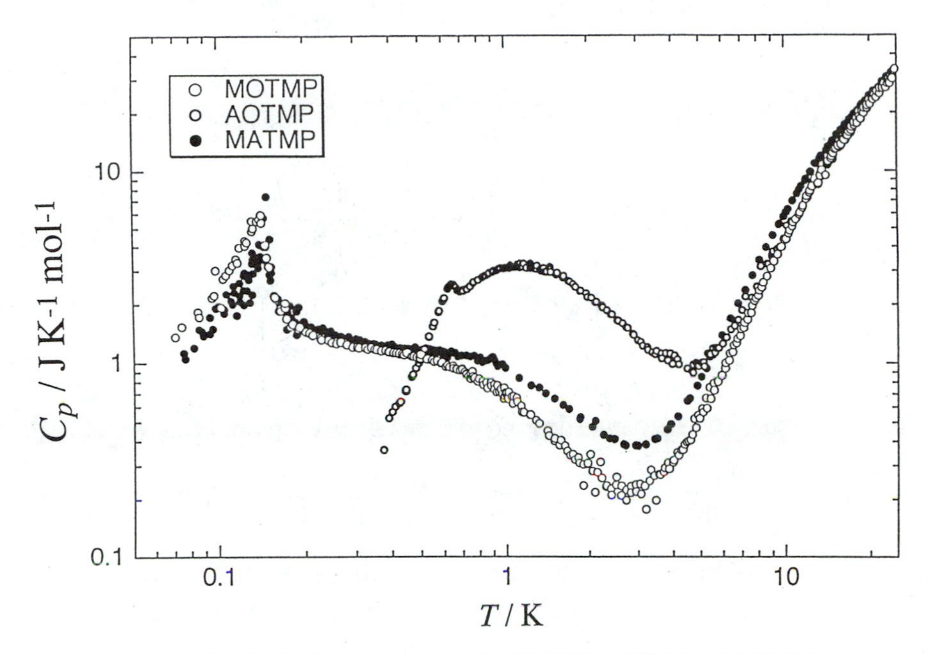

Figure 10. Molar heat capacities of MOTMP, AOTMP and MATMP.

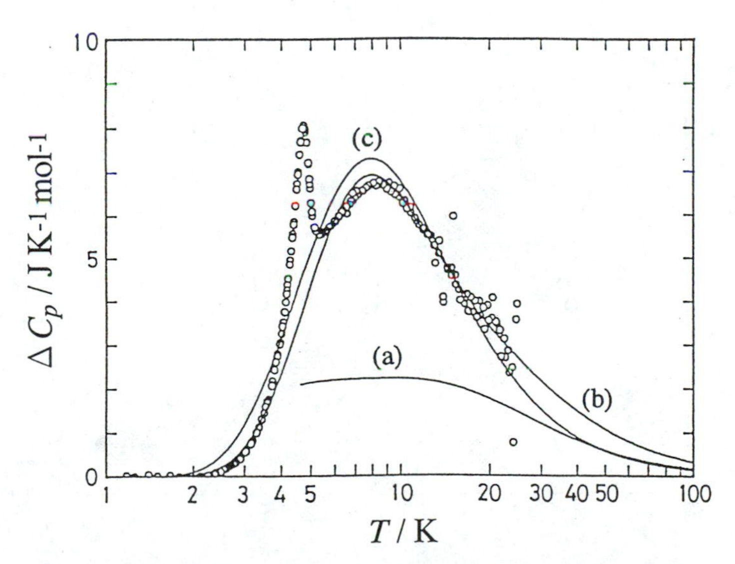

Figure 11. Magnetic heat capacity of [DMeFc][TCNE]. (a) Ferromagnetic Heisenberg chain ($J_{\perp}/k = J_{\parallel}/k = 13$ K), (b) Anisotropic Heisenberg chain ($J_{\parallel}/k = 25$ K; $J_{\perp}/J_{\parallel} = 0.5$), (c) Ising chain ($J_{\parallel} = 19$ K; $J_{\perp} = 0$) (*14*). Reproduced with permission from reference 14. Copyright 1990. Elsevier Science - NL.

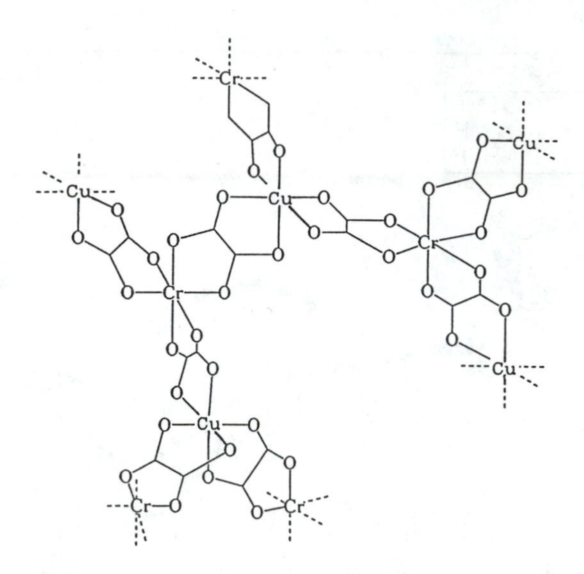

Figure 12. Structure of $\{[CuCr(ox)_3]^-\}_n$ ion.

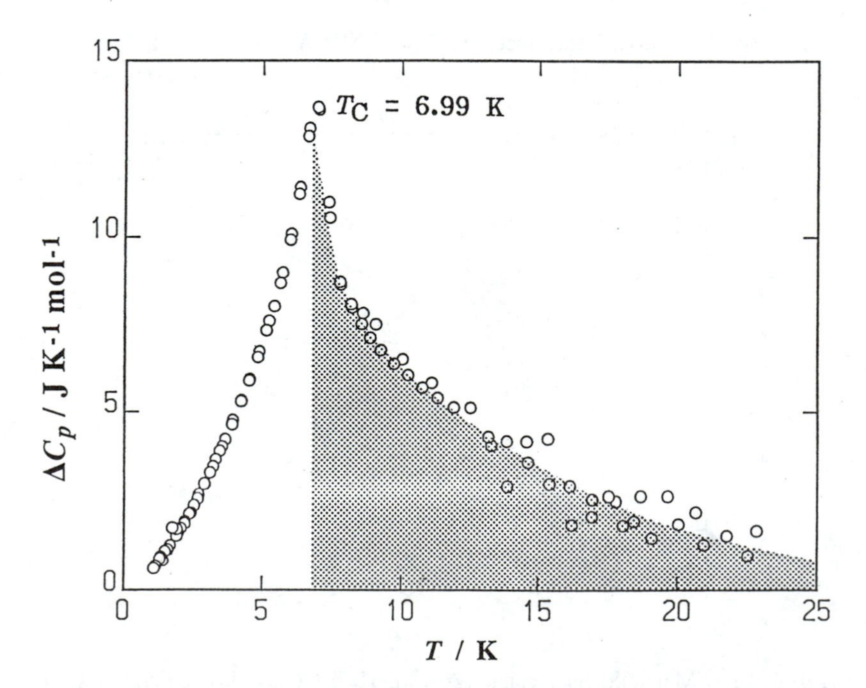

Figure 13. Excess heat capacity ΔC_p of $\{NBu4[CuCr(ox)_3]\}_n$. Shaded area corresponds to the short-range order effect persisting in the temperature region above the Curie point.

ln 2 = 11.53 $JK^{-1}mol^{-1}$. This fact confirms that the charge transfer from the donor to the acceptor is complete and that the present complex consists of a spin 1/2 cation and a spin 1/2 anion.

Mixed-Metal Complexes. Many molecular-based magnets are formed by a bundle of 1D paramagnetic chains as the constituents. Okawa *et al.* (*17*) reported a series of mixed-metal assemblies $\{NBu_4[MCr(ox)_3]\}_n$ (NBu_4^+ = tetra-*n*-butylammonium ion; ox^{2-} = oxalate ion; M = Mn^{2+}, Fe^{2+}, Co^{2+}, Ni^{2+}, Cu^{2+}, Zn^{2+}) (see Figure 12). All these complexes except for the zinc homologue exhibit spontaneous magnetization at low temperatures. On the basis of molecular model considerations it is suggested that these complexes form either a 2D- or 3D-network structure extended by the Cr(III)-ox-M(II) bridges (*17*). Since they are prepared as fine powder, single crystal X-ray diffraction study has not been successful. As heat capacity is sensitive to lattice dimensionality, we tried calorimetry of $\{NBu_4[CuCr(ox)_3]\}_n$ (Asano, K.; Ohmae, N.; Tamaki, H.; Matsumoto, N.; Okawa, H.; Sorai, M., Osaka University, unpublished data). Figure 13 represents the excess heat capacity ΔC_p of this complex beyond the normal heat capacity. A ferromagnetic phase transition was observed at T_C= 6.99 K. A remarkable feature of this phase transition is a large C_p tail above T_C, the shaded area in Figure 13. This type of short-range ordering effect is characteristic of two-dimensional structure. Therefore, from a thermodynamic viewpoint, we predict that the crystal structure might be two-dimensional. It is worth while to remark here that recently Decurtins *et al.* (*18*) reported the single crystal X-ray study of an analogous complex $\{P(Ph)_4[MnCr(ox)_3]\}_n$ (Ph = phenyl) and concluded a 2D network structure.

The heat capacities of $\{NBu_4[CuCr(ox)_3]\}_n$ in the spin-wave excitation region are well reproduced by $T^{1.51}$. Since the exponent 1.51 can be approximated by d/n = 3/2 in equation (9), the ordered state at low temperatures is proved to be of ferromagnet.

Concluding Remarks

Heat capacity is a physical quantity containing contributions from all kinds of molecular degrees of freedom at the rate of the Maxwell-Boltzmann distribution. This makes a sharp contrast with various spectroscopies in which particular nuclides and/or particular modes are selectively sensed. In other words, thermodynamic measurements are not restricted by any selection rules or selectivity. Consequently they are applicable to a wide range of materials and phenomena. The only shortcoming of heat capacity would be the ambiguity involved when the total heat capacity is separated into individual contribution. In case of molecular-based magnetic materials, dominant contribution at low temperatures is the spin degrees of freedom and molecular vibrations (or phonon). When the magnetic phenomena occur at low temperatures, separation of these two contributions is successfully made because the phonon contribution is extremely small.

Modern sciences are greatly specialized in an individual field of study. Therefore, better understanding of materials and/or natural phenomena would be achieved by employing experimental techniques which can play complementary roles between experiments leading to microscopic aspects and those to macroscopic aspects. Calorimetry is a comprehensive keeper of energetic and entropic aspects.

Acknowledgments. This work was supported by a Grant-in-Aid for Scientific Research on Priority Area "Molecular Magnetism" (Area No. 228/04242103) from the Ministry of Education, Science and Culture, Japan. Collaboration with Professors M. Kamachi, C. P. Landee, H. Okawa and N. Matsumoto is greatly appreciated. Heat capacity measurements were made by the author's research group, to whose members thanks are due. Contribution number 119 from the Microcalorimetry Research Center.

Literature Cited

1. Albrecht, A. S.; Landee, C. P.; Matsumoto, T.; Miyazaki, Y.; Sorai, M. *the March 1996 Meeting of the American Physical Society*, 1996.
2. Baker Jr., G. A.; Gilbert, H. E.; Eve, J.; Rushbrooke, G. S. *Phys. Lett.* **1967**, *25A*, 207.
3. de Jongh, L. J.; Miedema, A. R. *Adv. Phys.* **1974**, *23*, 1.
4. Dom, C. *Adv. Phys.* **1960**, *9*, 149.
5. Onsager, L. *Phys. Rev.* **1944**, *65*, 117.
6. Murakawa, S.; Wakamatsu, T.; Nakano, M.; Sorai, M.; Suga, H. *J. Chem. Thermodynamics* **1987**, *19*, 1275.
7. Sugimoto, H.; Aota, H.; Harada, A.; Morishima, Y.; Kamachi, M.; Mori, W.; Kishita, M.; Ohmae, N.; Nakano, M.; Sorai, M. *Chem. Lett.* **1991**, 2095.
8. Kamachi, M.; Sugimoto, H.; Kajiwara, A.; Harada, A.; Morishima, Y.; Mori, W.; Ohmae, N.; Nakano, M.; Sorai, M.; Kobayashi, T.; Amaya, K. *Mol. Cryst. Liq. Cryst.* **1993**, *232*, 53.
9. Kobayashi, T.; Takiguchi, M.; Amaya, K.; Sugimoto, H.; Kajiwara, A.; Harada, A.; Kamachi, M. *J. Phys. Soc. Japan* **1993**, *62*, 3239.
10. Ohmae, N.; Kajiwara, A.; Miyazaki, Y.; Kamachi, M.; Sorai, M. *Thermochim. Acta*, **1995**, *267*, 435.
11. Miller, J. S.; Calabrese, J. C.; Rommelmann, H.; Chittipeddi, S. R.; Zhang, J. H.; Reiff, W. M.; Epstein, A. J. *J. Am. Chem. Soc.* **1987**, *109*, 769.
12. Chittipeddi, S. R.; Selover, M. A.; Epstein, A. J.; O'Hare, D. M.; Manriquez, J.; Miller J. S. *Synth. Met.* **1988**, *27*, B417.
13. Zhang, J. H.; Reiff, W. M. *Hyperfine Interactions* **1988**, *42*, 1099.
14. Nakano, M.; Sorai, M. *Chem. Phys. Lett.* **1990**, *169*, 27.
15. Nakano, M.; Sorai, M. *Mol. Cryst. Liq. Cryst.* **1993**, *233*, 161.
16. Duggan, D. M.; Hendrickson, D. N. *Inorg. Chem.* **1975**, *14*, 955.
17. Tamaki, H.; Zhong, Z. J.; Matsumoto, N.; Kida, S.; Koikawa, K.; Achiwa, N.; Hashimoto, Y.; Okawa, H. *J. Am. Chem. Soc.* **1992**, *114*, 6974.
18. Decurtins, S.; Schmalle, H. W.; Oswald, H. R.; Linden, A.; Ensling, J.; Gütlich, P.; Hauser, A. *Inorg. Chim. Acta* **1994**, *216*, 65.

Chapter 8

Use of Mössbauer Effect Spectroscopy in the Study of the Low-Dimensionality Magnetism and Long-Range Order of Magnetic Systems of Contemporary Interest

William M. Reiff

Department of Chemistry, Northeastern University, 360 Huntington Avenue, Boston, MA 02115

The onset of nuclear Zeeman splitting resulting from the growth of internal hyperfine fields owing to exchange fields and ultimate long range magnetic order is directly observable in Mössbauer effect spectra of appropriate metal nuclides, e.g. Fe-57. On the other hand, slow paramagnetic relaxation resulting in part from the (single ion) zero field splitting of the ground electron spin manifold of paramagnetic oxidation states of a metal can also lead to readily observed hyperfine splitting of Mössbauer spectra, particularly for high-spin and to a lesser extent low-spin FeIII. These two effects are contrasted with emphasis on applications in the study of long range and local magnetic order in certain chain and layered magnets as well as dimers and complex paramagnetic salts. Specifically, the consequences of domain wall movement (soliton effects) and zero point spin reduction are considered relative to the Mössbauer spectra of 1-D magnets. The application of large external fields as an approach to ascertaining the nature (ferri-, ferro-, or antiferromagnetic) of the 3-D ordered ground state from Mössbauer spectra is also discussed. The correlation of the determination of critical ordering temperatures (T_{Curie} or $T_{Néel}$) via Mössbauer spectroscopy with other more classical methods (heat capacity, bulk ac or dc susceptibility) is also considered.

The study of magnetic interactions in solids has traditionally focused primarily on susceptibility, magnetization, neutron diffraction, and heat capacity measurements. The discovery of the Mössbauer effect in 1958 has led to a

0097–6156/96/0644–0115$16.50/0

welcome spectroscopic technique for the direct observation single ion and cooperative magnetic behavior in solids. In this tutorial, we review some of the relevant features of iron-57 Mössbauer spectroscopy in the study of slow paramagnetic relaxation and three-dimensional magnetic ordering of inorganic and organometallic solids having iron atoms as their paramagnetic metal ion centers. In what follows, we outline the basic features of magnetic hyperfine splitting (i.e. origins) owing to slow paramagnetic relaxation vs. cooperative ordering. Aspects of applied field Mössbauer spectroscopy are then considered. Finally, a number of different examples of Mössbauer spectroscopy in the study of a variety of magnetic behavior from this author's and colleague's research are presented. A basic familiarity with Nuclear Gamma Resonance (Mössbauer effect) spectroscopy is assumed. In any event, the reader is directed to a number of important resources (*1,2,3*) dealing with detailed basic aspects of NGR. A very brief primer on NGR based largely on ideas discussed in references (*1,2,3*) is now presented.

Brief Primer on NGR

In a nutshell, a Mössbauer effect experiment is typically based on the recoil free emission and subsequent recoil free, resonant absorption of low energy gamma rays ($E\gamma$ generally < 100 kev) in the solid state for identical isotopes of a given element. The requirement for essentially any optical resonance phenomena (infra-red, uv-vis, etc.) to occur is summarized succinctly in Figure 1 wherein the resonant overlap (dark-shaded area) of Gaussian source and absorber energy distributions is shown. The fundamental source transition energy (E_T)(shown for convenience as a δ-function), is actually broadened to varying degrees depending on the natural excited state lifetime, $\tau_{(0)}$, of the source according to the Heisenberg uncertainty principle such that $\Gamma_{S(0)}$ (natural full width at half maximum intensity) = $h/\tau_{(0)}$. The centroids of the pertinent distributions are separated by $2E_R$ where E_R (the recoil energy) = $E_T/2MC^2$ for the case of a free gas phase atom photon emitter of mass M. Clearly the recoil energy will be larger for gamma-ray transitions (transitions between different nuclear spin manifolds (I_i)) than say uv-visible or infra-red quanta (transitions between different electronic or vibrational states respectively) owing to the inherently much larger values of E_T. At the same time gamma ray transitions for convenient Mössbauer effect isotopes typically have small natural line widths (10^{-8} to 10^{-9} ev). These facts combine to limit resonance overlap in the case of gamma ray transitions to ~0.

Alternatively, one can say that the emitted gamma photon's energy is highly degraded from the actual value of the nuclear spin state energy (change) , i.e. E_T, by recoil energy loss. This makes it impossible for the transition to occur between the same nuclear spin states for the particular isotope starting in its ground state, i.e. via resonant absorption when this ground state form of the isotope is in reasonable proximity to the emitter.

These "problems" are largely overcome when a gamma ray emitter nucleus is incorporated in a solid state lattice. In this situation the effective mass (M) of the emitter is some 10^{15} times that of its individual isotopic mass (even for small polycrystals of the source matrix). Hence, $E_R = E\gamma^2/2MC^2$ and $<E_D> = E\gamma\ (2<E_K>/MC^2)^{1/2}$ are clearly now negligible where $<E_D>$ is the average Doppler broadening energy (see Figure 1), the other factor contributing to the energy degradation of the emitted gamma-photon. $<E_D>$ is seen proportional to the square root of the average (classical Boltzmann) kinetic energy, $<E_K>$, where $E_k=k_BT$ per translational degree of freedom.

Even though the recoil energy is apparently substantially reduced in view of the above "classical considerations" for nuclei bound in a solid, the possibility of truly recoil free emission and absorption processes arises from consideration of the quantized nature of the vibrational states of a solid. The vibrational energy eigen-function for a solid viewed as a collection of simple one-dimensional harmonic oscillators is $E_n = (n + 1/2)\ h\upsilon$ where $n = 0$ is an allowed and is, in fact, the lowest energy vibrational (phonon) state. (Recall that at $T = 0°K$ such a system still possesses $1/2\ h\upsilon$ zero point vibrational energy, a direct result of the Heisenberg uncertainty principle.) By comparison, for example, the energy of the electronic levels of the hydrogen atom or hydrogen-like ions is $E_n \propto n^{-2}$ which clearly precludes the $n = 0$ state since otherwise the energy would diverge (become infinite). These observations are particularly important in the context of emission and absorption of gamma rays in solids in that for sufficiently low energy (so called "soft") gamma rays,a fraction, f, of such emissions or absorptions can occur with no vibrational excitation of the lattice at all, i.e. as zero phonon events. Momentum, however, must still be conserved. This is achieved through manifestation of recoil as infinitely small translational motions of the (entire) infinitely massive (relative to individual nuclei) collective source atom lattice. Summarizing, then, in these circumstances one has the emission of highly mono-chromatic undegraded gamma photons of narrow line width. Their spectroscopic resolving power expressed as the fractional line width, $\Gamma_{S(0)}/E_T$, is the finest available, by far, for any known spectroscopic technique. This allows the measurement of <u>energy changes</u> varying between a few parts in 10^{13} (Fe^{57})to a few parts in 10^{16} (Zn^{67}). Thus, the previous conventional wisdom that: *the modulation of the energy separations between <u>different</u> nuclear spin-states (i.e., gamma ray transition energies) owing to variations in the nature of the chemical and magnetic environment or their symmetry is negligible and hence, unmeasurable* is clearly no longer valid.

The extremely small difference between E_T (source) vs E_T (absorber) owing to any of differences in chemical ligation of the Mössbauer isotope nucleus, local ligation symmetry, oxidation or spin state or local magnetic exchange interaction and degree of magnetic order are now in fact measurable using Mössbauer effect (NGR) gamma rays.

The basic (typical) experimental approach to measuring the foregoing differences consists of the restoration of resonance (now destroyed by any of the above environmental disparities between the source and absorber) through simple Doppler shifting (typically electro-mechanically) of the source gamma ray energy relative to the absorber: E_γ (resonance) = $E_T(1 + V_X/C)$ where V_X is an extremely small Doppler transducer velocity relative to the speed of light, C. The necessary Doppler shifting and velocity separations of the observed resonance transition(s) can be related to fundamental nuclear-extra-nuclear electron hyperfine interaction parameters such as: (A) the chemical isomer shift (δ), (B) the electric quadrupole interaction (ΔE_Q), and (C) the development of an internal hyperfine field (H_n). In addition to fundamental nuclear parameters (A) is a function of the difference between the "s" electron density at the Mössbauer effect nucleus for the absorber and source, i.e. $\delta \propto \{|\Psi_{ABS}(0)|^2 - |\Psi_{SOURCE}(0)|^2\}$. Thus the magnitude and sign of δ has implications related spin and oxidation state, coordination number and degree of ionicity for the local Mössbauer isotope-ligand bonding. (B) reflects the interaction of the nuclear quadrupole moment of the Mössbauer isotope nucleus with an electric field gradient arising from a low symmetry distribution of valence shell electrons and/or nearby ligand charges or dipole moments and thus gives detailed information on the symmetry of the coordination environment. Finally (C) results in nuclear Zeeman splitting which is directly observable in the case of $\Delta E_Q = 0$, simply as the overall velocity separation of the resonances of a zero applied field Mössbauer spectrum. This situation will arise from the development of a large Weiss molecular exchange field on achievement of a 3D-magnetically ordered state (at some critical temperature) for a given paramagnetic material and, as such, is the primary thrust of the rest of this chapter.

Origins of Nuclear Zeeman Splittings

Application of External Magnetic Fields (H_0) to Nuclei in Otherwise Diamagnetic Environments. This is, of course, the most obvious method for realizing nuclear Zeeman splitting and is of central importance to NMR (rf induced intraspin manifold transitions) e.g. through the use of superconducting magnets. Since iron-57 has significantly smaller nuclear moments (*4*) for its ground (I = 1/2) and first excited (I = 3/2) nuclear spin states($\mu 1/2$ = 0.0906 β_N, $\mu 3/2$ = -0.1539β_N) than the proton ($\mu 1/2$ = 2.7927β_N) seemingly large values of H_0 are not particularly useful in producing spectroscopically resolvable (vis à vis Mössbauer spectra) nuclear Zeeman splittings except in certain special cases to be considered subsequently.

Internal Hyperfine Fields Arising from an Atom's Unpaired Electron Spins. Since the ratio μ(electron)/μ(proton) is ~ 660, it is the atom's own electrons, by far, that have the greatest potential for producing internal hyperfine fields and concomitant large nuclear Zeeman splittings that are readily resolved in typical NGR spectra, of say iron-57. The contributions to the internal hyperfine field are considered in some detail subsequently. Let it be said now that there are two distinct phenomena intimately involving electron spin fluctuations, and magnetic exchange interactions that ultimately lead to very large (static) internal hyperfine fields at NGR nuclei. These are:

(a) Slow Paramagnetic Relaxation. This is basically a single ion effect requiring (1) low T (but not always), (2) dilution of paramagnetic ion centers and (3) zero field splitting of single ion electron spin manifolds.

(b) Cooperative (3D) Magnetic Order which requires (1) "low kT" relative to the strength of magnetic exchange interactions, and (2) direct contact or atom-bridged (super) exchange pathways. Factors (1) and (2) lead to the "sudden" development of an internal molecular field (H_n) such that dH_n/dT is usually large near $T_{critical}$ and where H_n, results in large resolvable nuclear Zeeman splittings. These can be as large as 62T for high spin Fe^{3+} (vide infra) and (in contrast to slow relaxation) are often accompanied by an anomaly in susceptibility (χ_m) vs T, magnetic heat capacity (C_m) vs T, or the observation of "magnetic reflections" in neutron diffraction studies.

Effective Magnetic Fields. In addition to the sign (*5,6*) of ΔE (the quadrupole splitting) it is also possible to determine the effective magnetic field (H_{eff}) from perturbed Mössbauer spectra. This is of considerable use in iron coordination chemistry in that H_{eff} is related to features of electronic structure such as oxidation and spin state as well as degree of covalency and nature of magnetic anisotropy. Measurement of effective magnetic fields also allows determination of the presence or absence of intramolecular antiferromagnetism in coordination systems that are otherwise magnetically dilute.

In general, the effective magnetic field, H_{eff}, at a Mössbauer nucleus is related to the applied field (H_0) and internal hyperfine field (H_n) by

$$H_{eff} = H_0 + H_n \qquad (1)$$

where all H_i are of course vector quantities.

We are interested in spectra of quadrupole split diamagnetic and rapidly relaxing paramagnetic systems as well as the cases of slow relaxation and pure Zeeman interaction which are considered first. The following situations arise where ΔE is the quadrupole splitting.

(a) $\Delta E = 0$, $H_n \neq 0$. In this instance, the characteristic six-line pure Zeeman pattern (Figure 2(a),(b)) of a magnetically ordered iron system is observed. A similar spectrum is also observed for a slowly relaxing paramagnet in the limit of infinitely long relaxation time between the components of its Kramers' doublets. The spectra correspond to the normally allowed transitions ($\Delta M_I = 0, \pm 1$) among the magnetically split I = 1/2 ground and I = 3/2 excited states with magnetic field relaxation time long relative to the nuclear Larmor precession frequency. The growth of such spectra (in $H_0 = 0$) as a function of decreasing T is shown in several examples (vide infra) and for cooperative systems is a fairly direct, highly visual way of determining critical ordering temperatures.

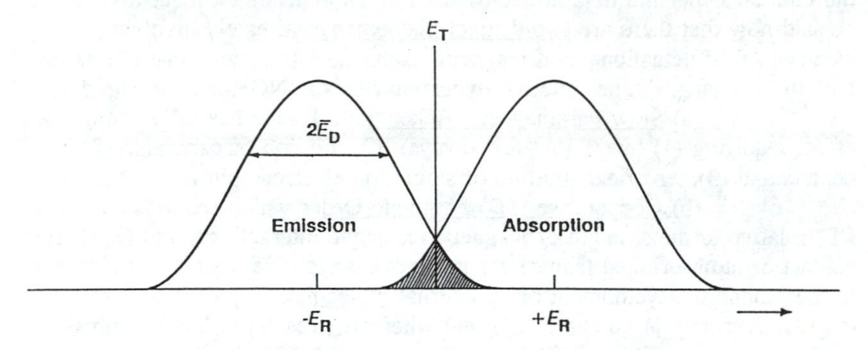

Figure 1. Resonance overlap of source and absorber energy distributions.

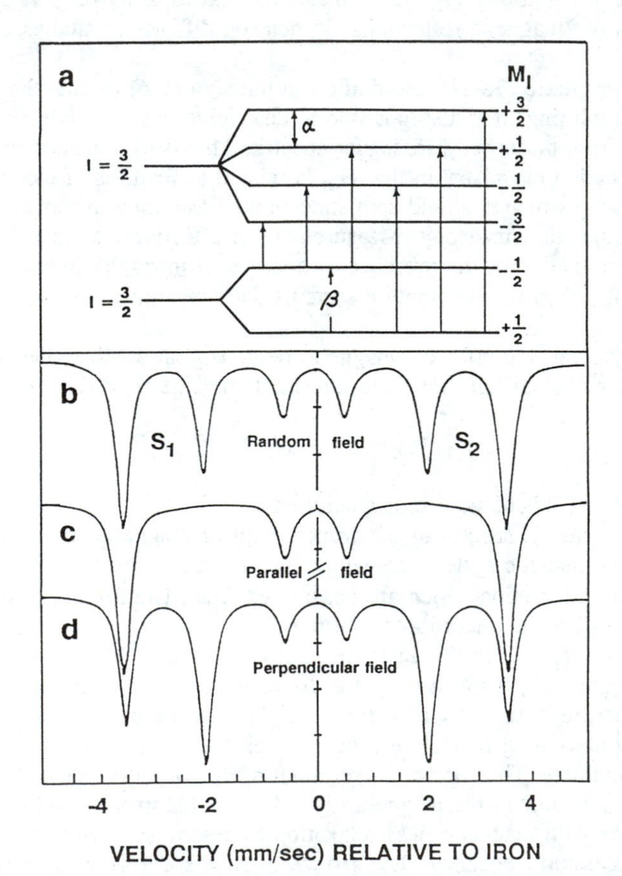

Figure 2. Pure Zeeman Splitting.

(b) $\Delta E = 0$, $H_n = 0$, $H_0 \neq 0$ and parallel to the γ-ray. This situation corresponds to an effective magnetic field equal to the applied, as in the case of a diamagnet or a rapidly relaxing paramagnet (whose metal ions are in sites of cubic symmetry) in the limit of large H_0. The result is the spectrum of Figure 2(c). One sees that transitions 2 and 5 vanish owing to nuclear magnetization and their $\sin^2 \Theta$ dependence(*3*). The angle Θ between the direction of γ-ray propagation and H_0 is zero for a longitudinally (axially) applied field. The energy separation of transitions 1-6 is $3\alpha + \beta$ and is, of course, proportional to the applied field. The quantities $\alpha = g_{3/2}\beta_N H$ and $\beta = g_{1/2}\ \beta_N H$ where $g_{3/2}$ and $g_{1/2}$ are the excited and ground state gyromagnetic ratios and β_N is the nuclear magneton. For iron-57, the ratio $\beta/\alpha = g_{1/2}/g_{3/2}$ has been determined (*4*) using Mössbauer spectroscopy and has the value -1.715. The ground state gyromagnetic ratio has been measured (*7*) independently using NMR and is 0.1828. Thus with the preceding values of α,β etc. and a known applied field, the total splitting $3\alpha + \beta$ is determined. In the case of the standard Mössbauer calibrant α-iron, the splitting is 10.626 mm/sec, corresponding to H_n = 330 kGauss (33 Tesla). These numbers correspond to the nearly saturated ferromagnetic state at ambient temperature. Single-line ($\Delta E = 0$) diamagnetic absorbers such as $K_4Fe(CN)_6$ and TiFe alloy (*8*) also exhibit spectra such as that illustrated in Figure 2(c) in large external (axial) fields.

(c) $\Delta E = 0$, $H_n = 0$, $H_0 \neq 0$ and perpendicular to the γ-ray. The comments for this case are similar to those for case (B), except that transitions 2 and 5 are intensified relative to 1, 3, 4, and 6 which are all expected to be weaker from their $1+\cos^2\Theta$ dependence (*3*) (Figure 2(d)).

(d) Combined Nuclear Zeeman and Quadrupole Splitting, $H_n \neq 0$, $\Delta E \neq 0$, $H_0 = 0$

The spectrum in this case under the assumption $H_n >> \Delta E$ is like that of Figure 2(a) except that the center of the inner four transitions is shifted toward either lower or higher energy relative to the center of transitions 1 and 6 because of the non-zero quadrupole interaction. For axial symmetry, this shift may be related to the angle Θ between H_n and the principal axis of V_{zz} (the principal component of the electric field gradient tensor) by the relation (*3*):

$$S_1 - S_2 = -\Delta E(3\cos^2\Theta - 1) \tag{2}$$

where S_1 is the separation of transitions 1 and 2 and S_2 that of transitions 5 and 6, Figure 2(B), and ΔE is the quadrupole splitting determined from the paramagnetic phase. Thus for pure Zeeman splitting($\Delta E = 0$), as in Figures 2(a)-(d), S_1-S_2 = 0 and a symmetric pattern is evident.

Contributions to the Internal Hyperfine Field (H_n). The extremely large internal hyperfine fields exerted on the nuclei of atoms in exchanged-coupled cooperatively-ordered or slowly-relaxing systems ultimately arise from the paramagnetism of the atom's electrons and can be divided into three components: (*3*)

(a) $H_S \propto \{\psi_\uparrow(0)^2 - \psi_\downarrow(0)^2\}$ (the Fermi contact term);
(b) $H_L \propto \langle r^{-3}L_Z \rangle$ or $H_L \alpha \langle r^{-3} \rangle (g-2) \langle S_Z \rangle$ (the orbital moment contribution) and
(c) $H_D \propto \langle r^{-3}(3 \cos^2\Theta - 1)S_Z \rangle$ or alternatively $H_D \alpha$: $V_{ZZ}/e \langle S_Z \rangle$ (the dipolar interaction).

Thus, $H_{eff} = H_0 + H_n = H_0 + H_S + H_L + H_D$. It is seen that (a) is the result of an imbalance of "s" electron α and β spin density at the nucleus. This imbalance results from the polarization caused by the differential interaction of the core "s"(α) and "s"(β) with unpaired valence shell "d" and "f" electrons. On the other hand, (b) is directly related to the magnitude of orbital angular momentum (L) for a particular unpaired valence shell electron of radius r. It is clear that (c) is operative only for sites of less than cubic symmetry and is the result of a through space interaction of the valence shell spin angular momentum with that of the nucleus. The terms H_L and H_D can either oppose or add to H_S. For the case of iron III, H_S is ~ 12T/unpaired electron (1T = 10 kiloGauss). Thus for high spin iron III, ^{6}A ground term, in cubic symmetry where L, H_L, and H_D are all ~ 0, limiting low temperature values of H_n ranging between 40 to 60T can be observed. For high spin ironII, somewhat smaller values of H_n (to ~ 35T) are observed owing to a combination of the smaller value of $\langle S \rangle$, namely two, and the greater likelihood of an orbital contribution, H_L that generally opposes H_S. Reduction of H_n for any oxidation state is also related to variable covalency and delocalization effects depending on the ligands (3). This can lead directly to delocalization of metal ion "s" spin density and an increase of $\langle r^3 \rangle$ for the valence shell electrons. These are effects that result in a decrease of any of H_S, H_L or H_D in view of their respective equations. An additional fundamental and rather interesting effect leading to substantially reduced values of H_n is the so-called "zero point spin reduction" very much evident for certain low dimensional magnetic systems(*9*) (vide infra).

Single Ion Zero Field (ZFS) Splitting and Slow Paramagnetic Relaxation versus Cooperative Three Dimensional Order. Slow paramagnetic relaxation-hyperfine splitting is a dynamic single ion effect resulting in part from the zero field splitting of the ground (electronic) spin manifold. This is shown in Figure 3 for the spin sextet of high-spin iron III. Since this is a Kramers ion, the ground state of the zfs ion must be the doubly degenerate $M_S = \pm 1/2$ (D>O) or $M_S = \pm 5/2$ (D<O) in zero applied magnetic field. Paramagnetic relaxation in the former is rapid while that in the latter as well as $M_S = \pm 3/2$ is slow. We will consider only the case of high spin iron III with D large and negative and at low temperatures. In this situation, the dominant relaxation mechanism is spin-spin relaxation (interatom exchange of S_z values) as opposed to spin-lattice that is more important to $L \neq 0$ ions such as high spin Fe^{2+} or low-spin Fe^{3+}. If the metal ions are closely situated so as to allow for rapid interatom spin flips via direct dipolar interactions (but not close enough for direct magnetic exchange or superexchange) there will be no spectral broadening even though a "slowly" relaxing ($M_S = \pm 5/2$) doublet is being populated. However, dilution (*10*) of the metal ions to distances

typically ≥7.5 Å (for high-spin FeIII) leads to longer spin-spin relaxation times whose reciprocal corresponds to a dynamic, temperature dependent frequency that eventually becomes comparable to the Larmor precession frequency of the nuclear moment (*11*). This leads to the gradual development of a non-zero time averaged value of H_n and gradual Zeeman splitting of the Mössbauer spectrum as the $M_S = \pm 5/2$ doublet is progressively populated. One observes a single six line pattern in the limit of very low T since the $M_S = \pm 3/2$ at 4D is little populated when D is large.

The limiting (zero field) low temperature spectra corresponding to three dimensional magnetic ordering processes (antiferromagnetism. ferro- and rerrimagnetism) are generally identical in appearance to those for slow relaxation. The difference is that H_n at the individual metal ion sites now corresponds to and is generally thought of as collinear with a spontaneous magnetization developing in the bulk sample. The latter originates from exchange interactions (either direct e.g. α-Fe (12), or super-exchange e.g. α-$Fe_2 0_3$) between the metal ions that become comparable to or greater than the thermal spin randomization effects as the temperature is decreased and a molecular exchange field spontaneously develops. In general and in contrast to slow relaxation, the hyperfine splitting process occurs "suddenly" over a small temperature interval reflecting the cooperative (usually second order) phase transformation nature of magnetic ordering. In addition, the internal hyperfine field is very temperature dependent in the vicinity of $T_{critical}$ and only levels off as magnetic saturation is reached at low temperatures, often as T → O°K. Finally, the temperature dependence of the spectral line widths differs for the two extremes of slow paramagnetic relaxation vs. on-set of long range magnetic order (see for instance the following discussion of soliton line width broadening effects in chain magnets just above $T_{critical}$).

The Néel or Curie temperature as measured via extrapolation of $H_n \rightarrow 0$ is usually in reasonable agreement with susceptibility results. Actually, at values for critical temperatures as determined by more precise methods such as classical measurements of the temperature dependence of magnetic heat capacity, Mössbauer spectra sometimes exhibit a substantial non-zero value of H_n. This problem is not dealt with here. The dimensionality of magnetic order (1, 2, or 3) is related to the so-called "critical exponent" which can be determined from fits to functions such as:

$$H_n(T)/H_n(O°K) = B(T_c - T)^{\beta}/T_c \qquad (3)$$

in temperature ranges as close as feasible to $(T_c\text{-}T) = 0$ where β is the critical exponent. It should be mentioned that hyperfine splitting and ordering may not be sharp but can actually be spread over larger temperature intervals as exemplified in finely divided superparamagnetic materials or highly defect structures. Rather characteristic Mössbauer spectra are observed in these cases. Finally, no hyperfine splitting may occur even though other techniques give clear evidence of magnetic order. This is rare but can occur when the various contributions to H_n, namely H_S, H_L, H_D fortuitously cancel e.g. as for the case of anhydrous ferrous chloride (*3*) for which magnetic susceptibility studies clearly indicate $T_{Néel} = 23$ K.

Intensities - Single Crystal Studies. So far, all of the discussion has referred to the Mössbauer spectra of isotropic, polycrystalline powders or powders whose paramagnetic (nuclear and electron) moments are polarized to the direction of an applied field, H_0. We now focus on oriented single crystals in zero and non-zero magnetic fields. For convenience throughout the rest of this section, should a field be applied, it will either be longitudinal (axial), i.e. parallel to the direction of gamma ray propagation, $E\gamma$, or transverse, i.e. perpendicular to $E\gamma$. The orientation will then refer to the angle (θ) between the easy axis or plane of magnetization and $E\gamma$ for a three dimensionally ordered material. Hopefully, the foregoing axes will be collinear with a convenient laboratory crystallographic axis of a favorable unit cell, e.g. orthorhombic or tetragonal. As with other spectroscopies, the utility of the Mössbauer effect spectroscopy is significantly enhanced when applied to single crystal samples with an excellent example being the case of Fe_2O_3 (*13*).

For the transitions of the polycrystalline powder form of a 3D-ordered ferro- or antiferro-magnet and thus a random orientation of magnetic domains in zero applied field, one expects the familiar 3:2:1:1:2:3 pattern of integrated intensity as in Figure 2b for an idealized texture free, thin absorber. Otherwise for $E\gamma$ parallel or perpendicular to an easy axis (or plane) of magnetization (H_0 again equal to zero) one sees 3:0:1:1:0:3 (Figure 2c) or 3:4:1:1:4:3 (Figure 2d) respectively. The $\Delta M_I = \pm 2$ transitions are normally forbidden or of very low intensity and are not considered further herein. The application of these results in obtaining useful magneto-structural correlations for a number of specific examples is now considered.

Zero and High Field Mössbauer Spectra of Locally and Long Range Ordered Systems and Determination of the Three Dimensional Ground State

Weak Antiferro- and Ferromagnetic Interactions - Ionic Fluorides of High Spin Fe(III). We now consider the weak super-exchange exhibited in ionic fluorides of high-spin iron III recently studied by this author. The exchange involves close contacts of delocalized electron spin density between $[FeF_6]^{3-}$ or $[FeF_5{\cdot}H_2O]^{2-}$ polyhedra via hydrogen bonds or lithium atom bridging.

(1) $K_2FeF_5{\cdot}H_2O$

The structure of this material is shown in Figure 4 while pertinent Mössbauer spectra are given in Figure 5. Clearly, $K_2FeF_5{\cdot}H_2O$ is ordered and exhibits a low field spin-flop transition in a transverse field of ~ 1 T for $T < T_N$ (~ 0.9 K). (Note: diminution of the $\Delta M_I = 0$ transitions in Figure 5). This agrees with a 3D-AF ground state suggested by a susceptibility study (*14*). Above T_N, the field dependence is quite different (*31*) and suggests the possibility of substantial 1D-antiferromagnetic correlation.

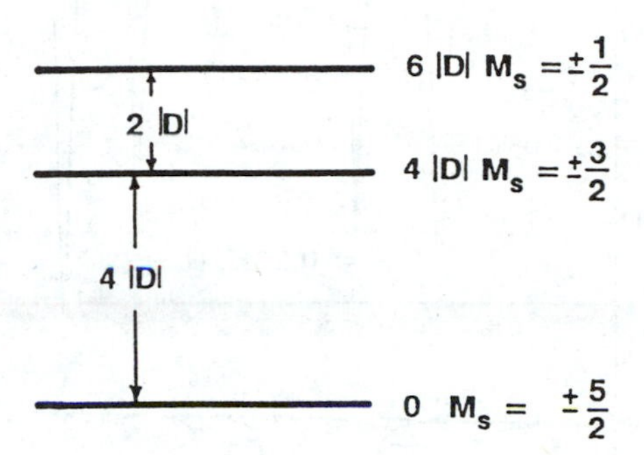

Figure 3. Negative axial zero field splitting for high spin Fe(III).

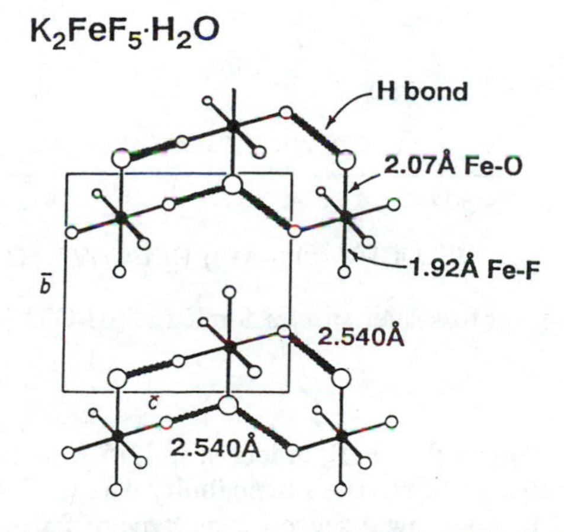

Figure 4. Local structure and hydrogen bonding of the 1-D chain $K_2FeF_5{\cdot}H_2O$.

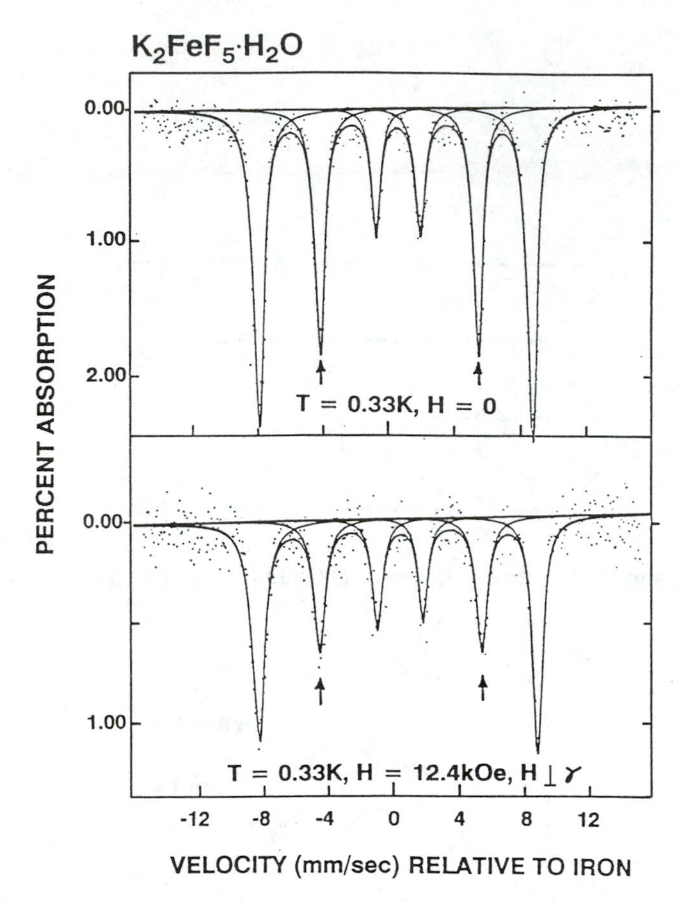

Figure 5. Mössbauer spectra for $K_2FeF_5{\cdot}H_2O$ at 0.33 K.

(2) β-Li_3FeF_6

We have found (*15*) that β-Li_3FeF_6 orders near 1.15 K as shown via zero field Mössbauer spectroscopy. The susceptibility data (d.c. squid results at H_0 = 30 Gauss to 2 K; not shown) suggest some type of ferromagnetic behavior (the value of $\Theta = +1$ K and there is a very rapid rise in moment). We are, however, unfortunately unable to reach temperatures < 1.25 K with our susceptibility equipment (d.c. squid magnetometry and a.c. susceptometry) in order to verify the ferromagnetic ground state suggested by the applied field Mössbauer spectroscopy results. Clearly more detailed susceptibility study is necessary. In any event, the <u>transverse field spectra</u> at 0.51 K (Figure 6) and 0.33 K (not shown) -- i.e. well below $T_{critical}$ -- show clear polarization of the sample moments (for a random magnetically ordered polycrystalline powder sample) <u>perpendicular</u> to the direction of γ-ray propagation fully consistent with a 3D-ferromagnetic ground state. (Note: The expected enhancement of

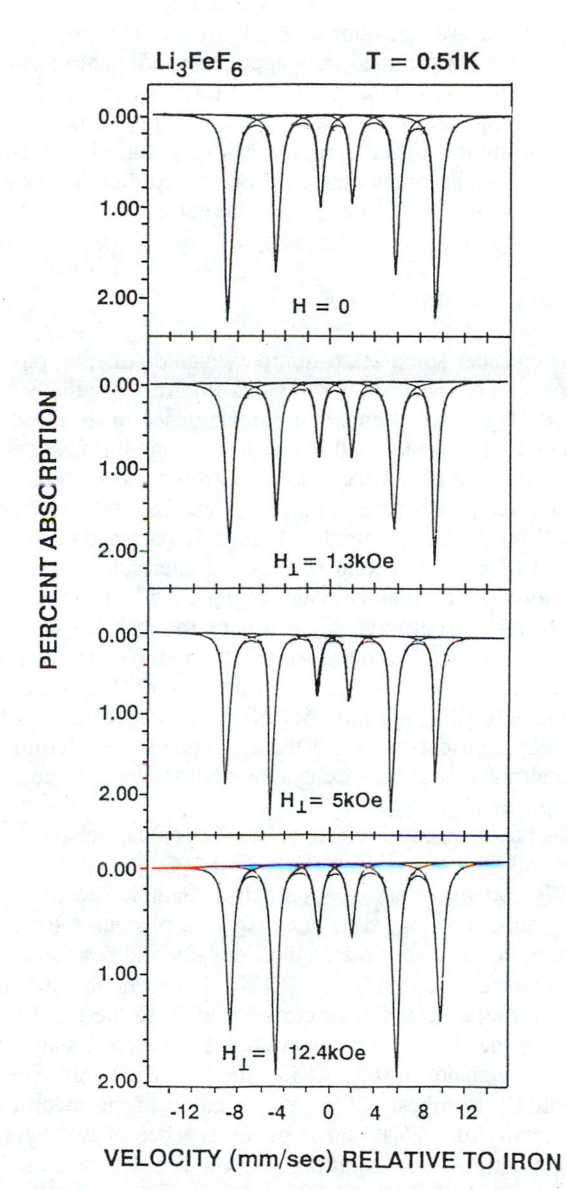

Figure 6. Mössbauer spectra for β-Li_3FeF_6 at 0.51 K in various transverse fields.

the $\Delta M_I = 0$ transitions in Figure 6). This behavior is not only unexpected but the exchange interactions are surprisingly strong relative to, say, $K_2FeF_5 \cdot H_2O$ which has hydrogen bonding (Figure 4). The magnetic exchange in β-$Li_3[FeF_6]$ is likely transmitted (and apparently efficiently) through LiF_4 tetrahedra (2/9 of the total Li) that bridge the ferric octahedra (*16*). It might be expected that α-$Li_3[FeF_6]$ (the elpasolite polymorph) orders at a somewhat lower temperature in view of its ferric ion bridging entirely via LiF_6 octahedra implying weaker individual Li-F bonds. Additional, detailed studies of both the α and β forms of Li_3FeF_6 are anticipated.

Ferromagnetically Coupled Dimers

We now consider some 2:1 cation:polycyanide dianion phases whose charge transfer polymer structure is shown in Figure 7 for the example of $[Fe(Cp^*)_2]_2^+[C_4(CN)_6]^{2-}$. The temperature dependence of its Mössbauer spectra is clearly unprecedented and to our knowledge the first example in which hyperfine splitting grows, reaches a maximum intensity and then, with further decreasing temperature, is observed to vanish totally (*17,18*). $[Fe(Cp^*)_2]_2^+[C_4(CN)_6]^{2-}$ as is pictured in Figure 7, corresponds to parallel chains of pairs of (S = 1/2) $[Fe(Cp^*)_2]^+$ cations alternating with single (S = 0) $C_4(CN)_6^{2-}$ di-anions. Its characteristic singlet Fe-57 Mössbauer spectrum at 293 K (Figure 8, top) incontrovertibly confirms the fact that all iron is present as low-spin Fe(III), i.e. decamethylferrocenium cations. (Ferrocene and decamethylferrocene formally contain iron(II) and always exhibit well resolved quadrupole doublet spectra with $\Delta E > 2$mm/sec) (*3*). Thus from electrical neutrality considerations and the 2:1 stoichiometry confirmed by x-ray structure determinations, the hexacyanobutadiene moiety must be present as a dianionic species, $C_4(CN)_6^{2-}$.

The Mössbauer spectra indicate unusual dynamic behavior and the presence of a novel triplet species in $[Fe(Cp^*)_2]_2^+[C_4(CN)_6]^{2-}$ as ferromagnetically coupled dimeric cation pairs. Sample spectra are shown in Figure 8. On gradual cooling, slow paramagnetic relaxation-broadening and hyperfine splitting develop with maximum intensity and resolution for the Zeeman split component occurring at ~ 10 K. However, below this temperature the strong central singlet corresponding to the rapidly relaxing paramagnetic phase begins to grow again until a broadened singlet reminiscent of 300 K is all that remains at 0.53 K, i.e. the hyperfine split component of the spectrum entirely vanishes. Closer examination of the spectra (not shown) in the vicinity of ~ 25 K shows the emergence of two hyperfine patterns. The multiple Zeeman splitting patterns are observed as the result of highly anisotropic magnetic hyperfine interactions, i.e. $g_\parallel$ and $g_\perp$ spectra. These interactions undoubtedly arise from the large g factor anisotropy of the

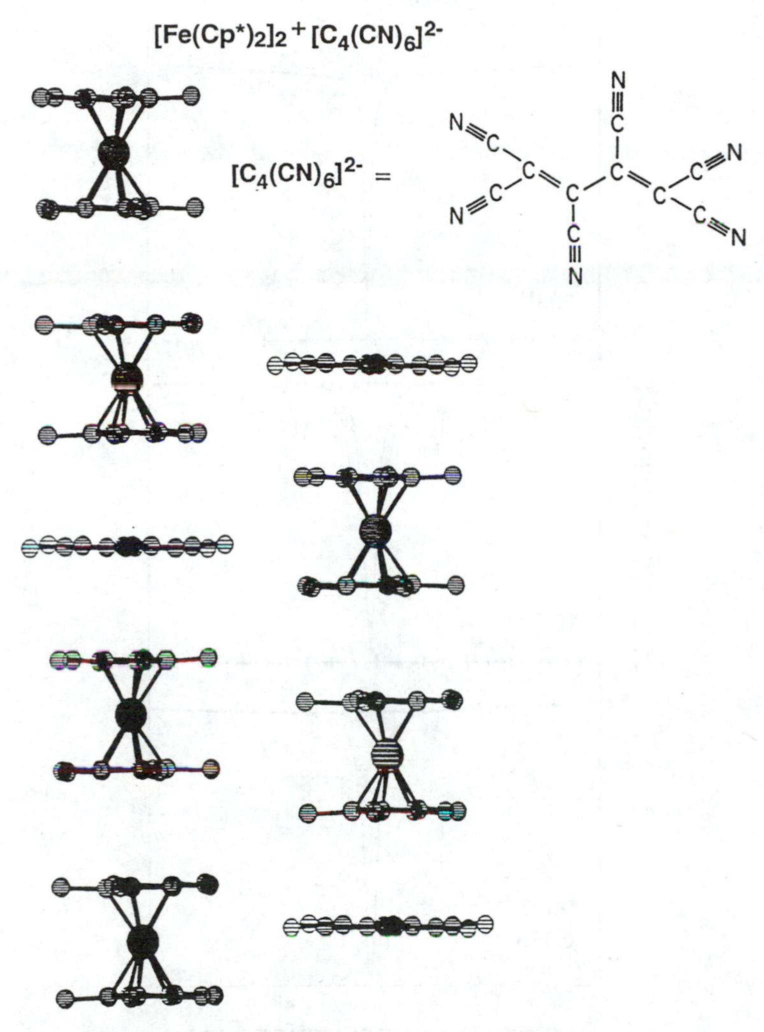

Figure 7. Schematic of the 2:1 charge transfer polymer $[Fe(Cp^*)_2]_2{}^+[C_4(CN)_6]^{2-}$.

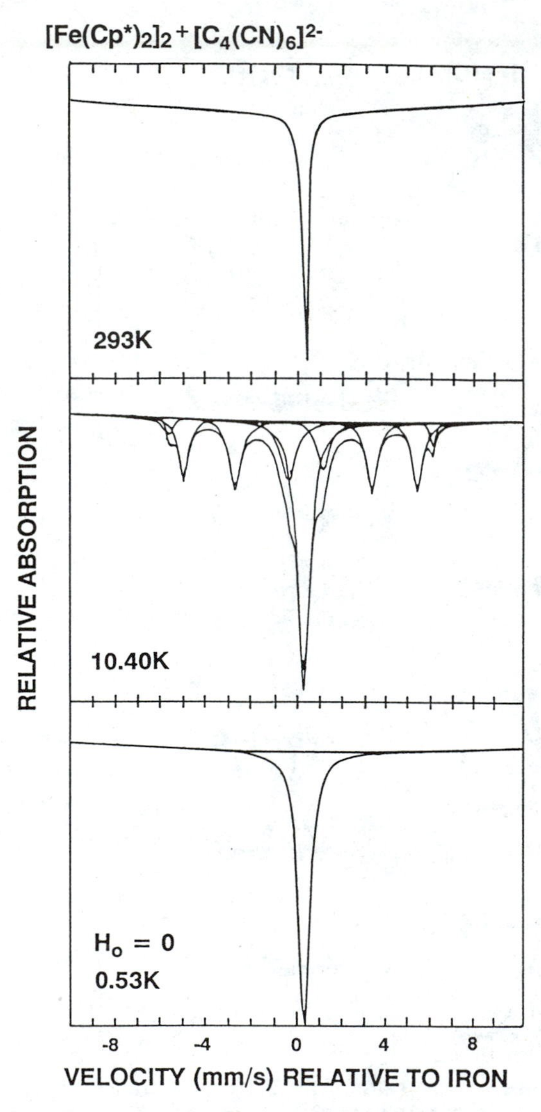

Figure 8. Sample spectra showing the growth and vanishing of hyperfine splitting for $[Fe(Cp^*)_2]_2{}^+[C_4(CN)_6]^{2-}$.

low-spin Fe(III) ions and the strongly anisotropic ferromagnetic exchange between them. We emphasize that the observed behavior has been reproduced and is entirely reversible. In addition, it is also found for other dianionic species such as $C_6(CN)_6^{2-}$ and iso-$C_4(CN)_6^{2-}$

The explanation of our observations for the 2:1 cation:dianion phases is suggested by the structure of these systems, i.e. one containing cation pairs (Figure 7). At sufficiently low temperatures, such pairs can act as magnetic exchange coupled dimers (very anisotropically along the cationic dimer axis). The "magnetic" literature is replete with such s=½, s=½ dimers, particularly s_{total} = 0 (singlet) ground state antiferromagnetically coupled dimers of Cu(II). For such AF dimers, a maximum in the temperature dependence of the magnetic susceptibility (χ_m) is typically observed (for samples free of paramagnetic impurities) and is theoretically predicted (*19*). The exchange interaction J is negative with the singlet-triplet separation being 2 J. There are, however, well characterized examples of s_{total} = 1 ground state (J > 0) ferromagnetically coupled Cu^{2+} dimers for which a simplified (spin only) energy level diagram is given in Figure 9. This is undoubtedly the case for the present $[Fe(Cp^*)_2]_2^+$ pairs. Such a dimer's s_{total} = 1 (nominal) ground state can be (positive) zero field split to an excited $m_{s(total)}$ = ±1 and a true $m_{s(total)}$ = 0 ground state, as shown on the right side of Figure 9. Large zero field splittings of this type can result from the highly anisotropic magnetic exchange operative in these systems. Decreasing the temperature for such a system would then initially lead to population of the slowly relaxing $m_{s(total)}$ = ±1 (since $\Delta m_{s(total)}$ = ±2 electronic transitions are highly forbidden) and hyperfine splitting via slow (intra and inter) dimer paramagnetic relaxation. Ultimately exclusive population of the nonmagnetic $m_{s(total)}$ = 0 ground state singlet occurs at very low T. Concomitant with the latter would be the vanishing of all hyperfine splitting effects as observed. *Finally, and consistent with our observations, J positive dimers are not expected (19) to exhibit a maximum in* χ_m. Good fits to the variation of the effective magnetic moment μ vs T for the combined effects of ferromagnetic intradimer exchange and zero field splitting are obtained for 2J and D values of + 180 cm^{-1} and + 6 cm^{-1} respectively. That D is, in fact, positive is clearly confirmed by the Mössbauer data, i.e. a non-magnetic, non-hyperfine split singlet ground state. Note, a value of D = 6 cm^{-1} is equivalent to ~ 9 K. This is consistent with the observed temperature, ~ 10 K, of maximum intensity for the hyperfine background that eventually vanishes.

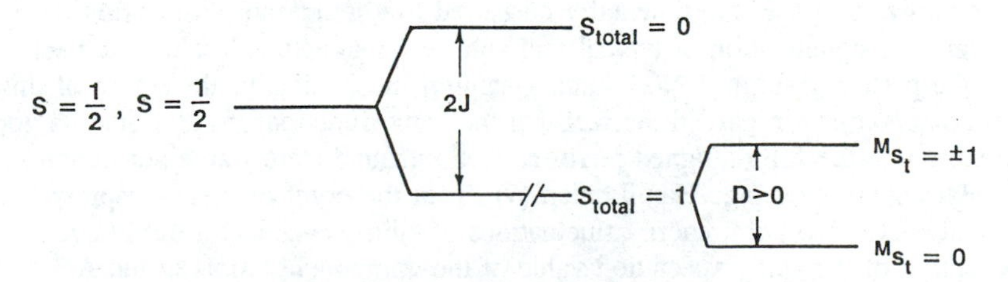

Figure 9. Energy level diagram for a feromagnetically coupled (J > 0) dimer including positive axial zero-field splitting (D > 0) of the triplet state, to give a non-magnetic M_S = 0 ground state.

Non-linear Low Dimensionality Effects - Solitons

It is clear that structurally induced magnetic one dimensionality is an important aspect of many of the systems discussed herein. This effect reinforces the already significant anisotropic single ion magnetic behavior typical of high spin Fe(II) and certain low-spin Fe(III) ferrocenium ions and can potentially lead to highly anisotropic one and three dimensional exchange behaviors intermediate to the Heisenberg and Ising extremes. It is also known that the motions of domain walls (kinks) in antiferro- or ferromagnetic chains obey the sine-Gordon equation and thus can be viewed as soliton excitations. The spin flips accompanying domain wall boundary movement in a highly 1D-magnet can lead to reversals of the internal hyperfine field at Mössbauer active nuclei with observable line width broadening effects. Their observation depends on the rate of soliton passage vs. the reciprocal of the excited state lifetime of the Mössbauer effect nuclide. Such soliton broadening effects have in fact been observed in detailed line width analysis just above and below T(critical) for chain magnets based on high spin iron II by Thiel an co-workers (*20, 21*) and 1-D high spin iron III fluoride systems by Johnson and co-workers (*9*). In all of these systems the spin resides exclusively on the ferrous or ferric ions. A good example (Figure 10) of these "soliton line width broadening effects" is seen in the specific cases hydrazinium ferrous and cobaltous sulfate whose linear chain structure is shown in Figure 11 where T_C for the iron system is 6.0 K.

It seems likely that the line width broadening effects observed near $T_{critical}$ for certain ferrocenium based electron donor-acceptor compounds (*22*) (the donor cation being low-spin ferric) are soliton related rather than solely single ion or super paramagnetic relaxation effects. In any event, this is certainly a rich area for further study of donor-acceptor chain magnetic materials. Such investigations are expected to be more complicated for the low spin Fe(III) systems owing to larger, highly sensitive orbital contributions to the internal field and the fact that both the anion and cation of the donor-acceptor chain magnet systems possess spin. In this situation, simple dilution experiments (*21*) with diamagnetic $s=0$ species such as deca-methylcobalticenium may not clearly distinguish single ion low paramagnetic relaxation broadening versus cooperative soliton effects.

Zero Point Spin Reduction Effects in Low-D Antiferromagnets

It is known that the experimentally measured low temperaure saturation sublattice magnetization of typical real anti-ferromagnets is less than expected for the perfectly alligned Néel state. Quantum mechanically, the origin of this phenomenon lies in part in the fact that the wave function of latter state is not (contrary to the fully alligned ferro-magnetic ground state) an eigen-function of Heisenberg exchange Hamiltonian.(*9*) From the point of view of spin-wave (*23*) theory, zero-point energy fluctuations of spin-waves lead to effective reduction of the spin expectation value of the component atoms in the AF coupled lattice relative to the value, S, for the "free" ion. For such an AF lattice, one has an $\langle S \rangle_0$ (average spin expectation value at absolute zero) such that there results a zero-point spin reduction $\Delta S = S - \langle S \rangle_0$ that can

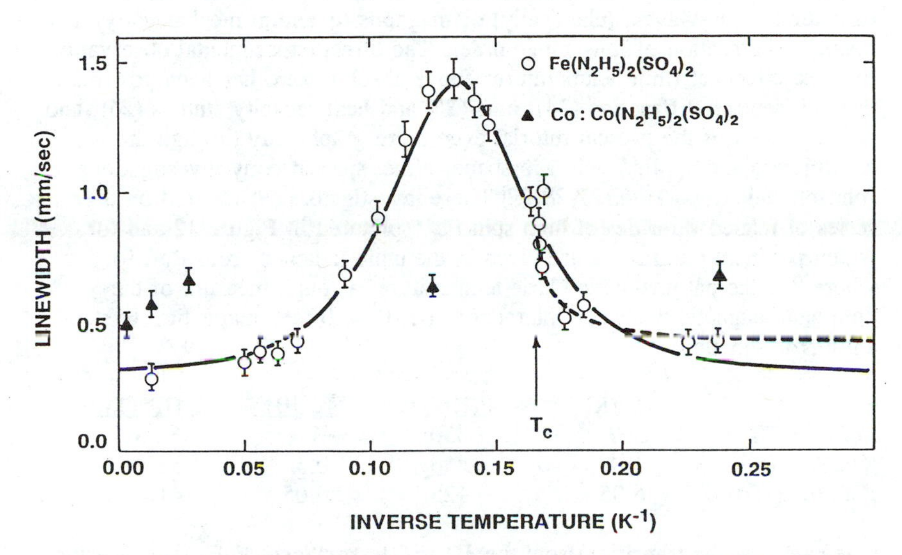

Figure 10. Linewidth versus inverse temperature for the $Fe(N_2H_5)_2(SO_4)_2$ (o) and $Co(N_2H_5)_2(SO_4)_2$ (Δ) compounds. (Adapted from ref. 21).

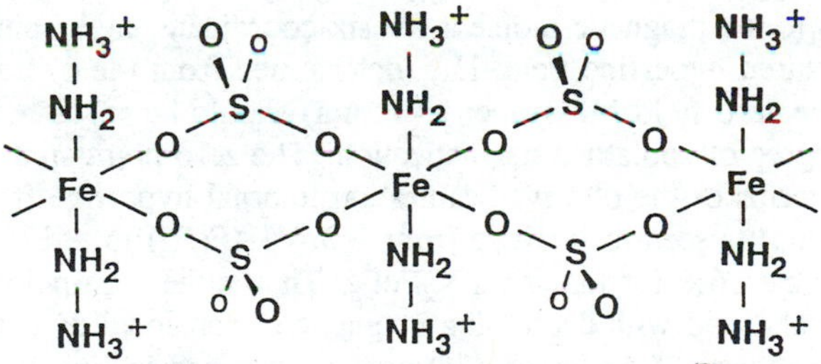

Figure 11. Chain structure of hydrazinium metal (II) sulfates.

vary from a few percent for a 3D magnet to ~35% for certain 1D-chain systems (vide infra) and is a sensitive (inverse function) of the magnetic anisotropy. Reduced magnetization then follows since: $M = Ng_B\mu<S>$ is now replaced by $M_0 = Ng\mu<S>_0$. Zero-point spin reduction is a manifestation of the Heisenberg uncertainty principle. Thus the zero-point energy fluctuation in an AF ordered array can be thought of as leading to excitations, spin-waves, (classically) or magnons (quantum mechanically) that result in destruction of long range order. The direct experimental observation of these effects at finite temperatures above absolute zero has been possible through neutron diffraction (24), nmr (*25*) and heat capacity studies (*26*) and in the context of the present tutorial even more graphically through the careful zero and applied field Mössbauer effect spectroscopy investigations of Johnson and co-workers.(*27,28,29*) These investigators (*9*) focused on the series of related fluorides of high spin Fe^{3+} pictured in Figure 12 and for which pertinent parameters are given in the table (adapted from Ref. 9): where θ is the paramagnetic Curie temperature, a rough measure of the <u>dominant</u> magnetic exchange interaction, i.e. $\theta > 0$ ferromagnetic, $\theta < 0$ antiferromagnetic.

	T_N(K)	θ(K)	$T_N/\lvert\theta\rvert$	H_n^{sat}(T)
FeF_3 (3D)	362	~-350	~1	61.8
$KFeF_4$ (2D)	134	~-250	0.5	53.4
K_2FeF_5 (1D)	6.95	~-125	0.05	41.0

One can view the transition from the 3D anti-ferromagnet FeF_3 (face sharing FeF_6 octahedra) to the 2D layered magnet $KFeF_4$ (edge sharing FeF_6 octahedra) to the linear chain (*30*) K_2FeF_5 (corner sharing octahedra) via successive magnetic dilutions of FeF_3 with one mole of KF and where the K^+ cations act as a diamagnetic diluent separating layers or chains. Since orbital contributions to the magnetic moment for six-coordinate, high spin Fe^{3+} are ≈ 0, the measured hyperfine field, H_n, (determined from the overall Zeeman splitting of the zero field Mössbauer spectrum) should be $\propto <S>$ and directly reflect the degree of sublattice magnetization. The zero point spin-reduction measured in terms of the observed saturation internal hyperfine field (right side of the table) is seen to increase from ~ 3.4% (FeF_3) to ~ 17% to ($KFeF_4$) to some 36% for the chain K_2FeF_5. That is the zero-point spin reduction is enhanced with decreasing magnetic dimensionality as predicted by the spin-wave theory of Anderson (*23*) and summarized in Figure 13 where the ratios J'/J and T_N/θ (see table) reflect the "degree" of low dimensionality and the anisotropy ratio $\alpha = H_A/H_E$. The quantities H_A and H_E are the anistropy and exchange fields respectively such that

$$H_{spinflop}=\sqrt{(2H_AH_E)}$$

For zero, zero point spin reduction, i.e., with the full spin of high spin Fe^{3+}, $S = 5/2$, H_n is assumed to be 64 Tesla, i.e., some 12.8T/spin. (*9,27*)

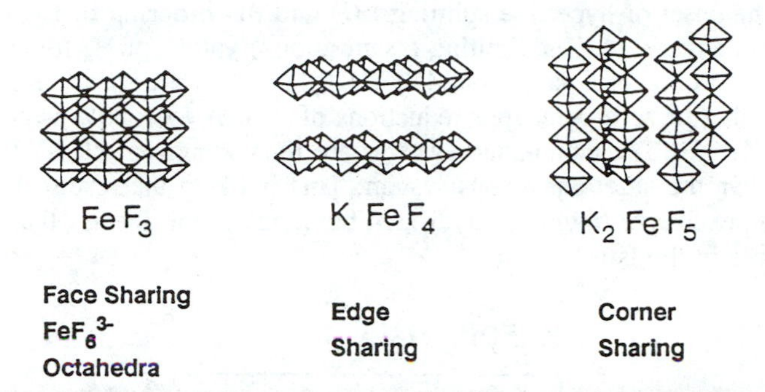

Figure 12. $(FeF_6)^{3-}$ octahedra for FeF_3 (3D) $KFeF_4$ (2D) and K_2FeF_5 (1D). (Adapted from ref. 9).

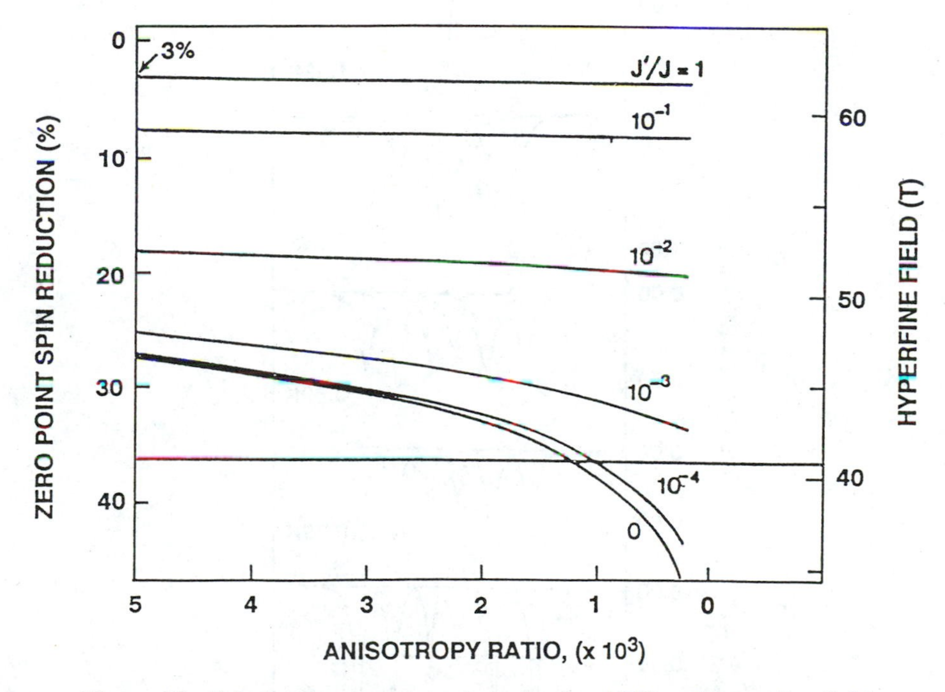

Figure 13. Calculated zero point spin reduction $\Delta S(0)$ and hyperfine field as a function of the anisotropy ratio B_A/B_E and J'/J. (Adapted from ref. 9).

We have studied the related $M_2FeF_5{\cdot}H_20$ chain series (*31,32,33,34*)(Figure 4), $M^+ = K^+$, $NH_4{}^+$, Rb^+ and Cs^+ with T_N ranging from ~ 1.0K(K^+) to 4.0K (Rb^+). The onset of hyperfine splitting (*31*) and 3D ordering of $K_2FeF_5{\cdot}H_20$ is shown in Figure 14. The limiting ("saturation") values of H_n for these chain anti-ferromagnets vary from 45T ($NH_4{}^+$) to 52T (K^+,Cs^+, Rb^+) corresponding to zero-point spin reductions of ~ 30% to ~ 20% versus ~ 36% for K_2FeF_5.The mono-aquo systems are thus somewhat "less" 1-D chain magnets than the latter anhydrous system. This is likely the result of stronger inter-chain exchange (owing to hydrogen bonding) in the former that is simply not possible in the latter.

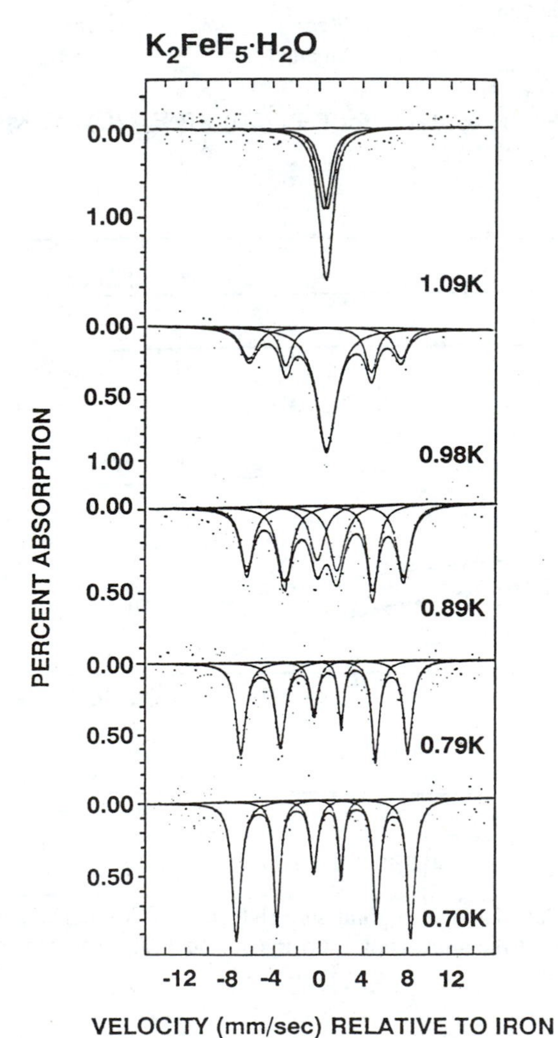

Figure 14. Temperature dependence of the Mössbauer spectrum of $K_2FeF_5{\cdot}H_2O$). (Adapted form ref. 31).

We (Reiff, W. M. and co-workers, unpublished results) observe a comparable zero-point spin reduction in the five coordinate high spin Fe^{3+} system Fe(2,2'-bipyridine)Cl_3 which has a structure similar to that shown in Figure 15 for the related Fe(2-(pyridin)-2-yl)benzothiazole)Cl_3. This is arguably a "very" molecular magnet being composed of 5-coordinate high spin Fe^{3+} monomers articulated to double chains that are held together solely by close Cl-Cl contacts and π-π stacking interactions, i.e., no formal covalently bonded bridging super exchange pathways. Nevertheless Fe(2,2'-bipyridine)Cl_3 exhibits unusually well resolved 1D and 3D magnetic effects in its D.C. susceptibility (H_0=100 Oe, Figure 16), i.e. $T\chi_{max}$ 1D and 3D values of ~6K and ~3K respectively or T_N ~ 3K. The zero field Mössbauer spectrum at 1.52K (Figure 17), i.e., well below T_N corresponds to H_n=34T,

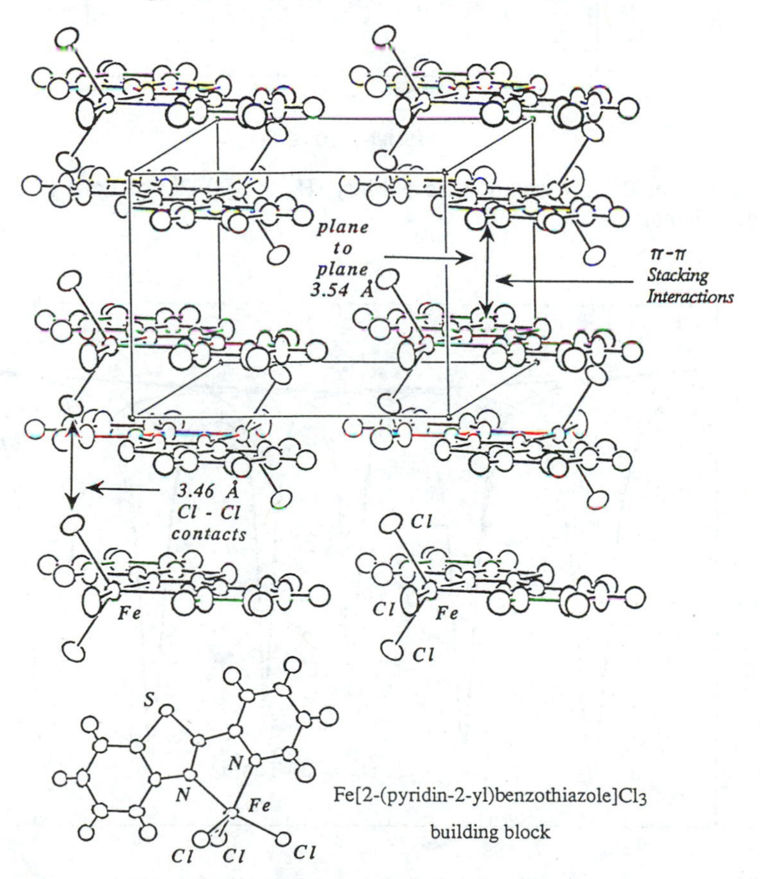

Figure 15. Double chains of five coordinate monomers of Fe(2-(pyridine)-2-yl)-benzothiazole)Cl_3.

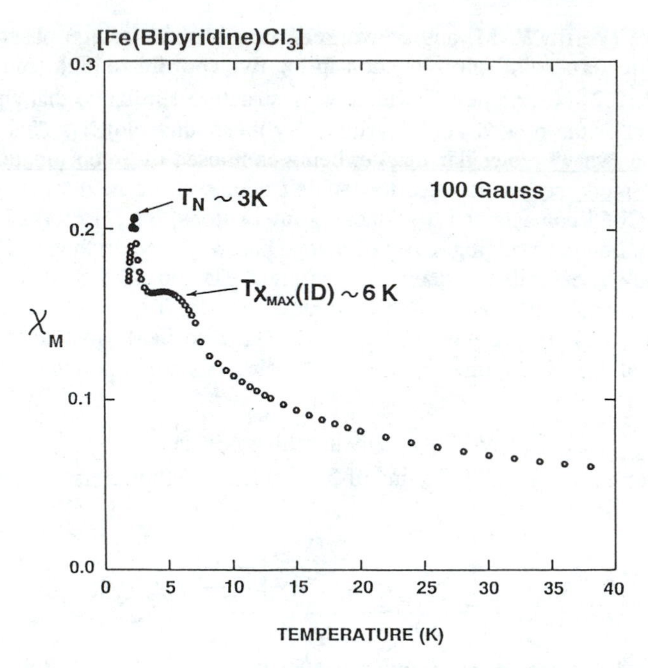

Figure 16. D.C. - "squid" susceptibility (H_o = 100 Oe) of FeIII (2,2′-bipyridine)Cl_3.

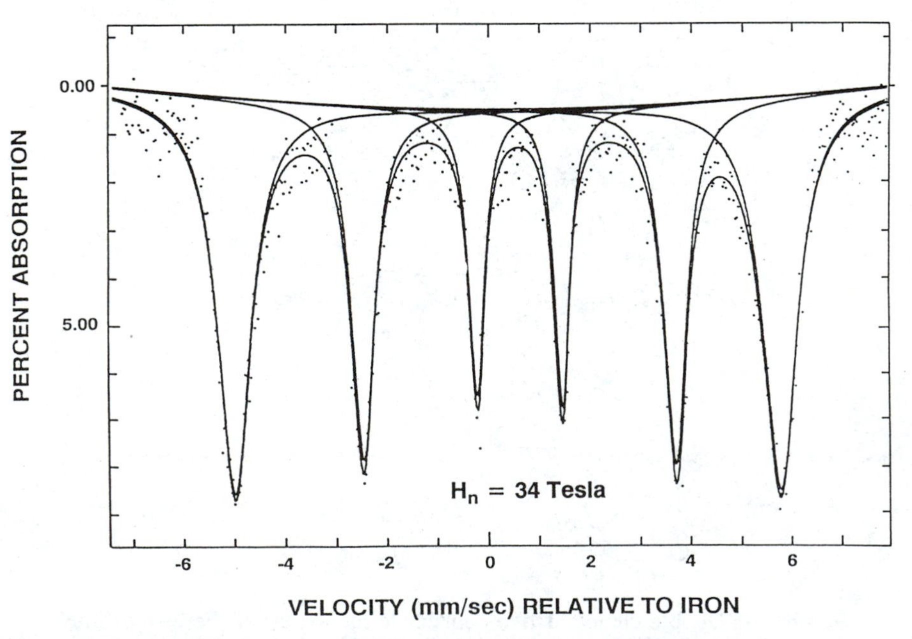

Figure 17. Mössbauer spectrum of FeIII (2,2′-bipyridine)Cl_3 at 1.52 K.

some ~ 30% zero-point spin reduction by comparison to $FeCl_3$ (H_n (0°K) = 48.7 T)(3). The latter salt is chosen so as to eliminate covalency delocalization effects in the comparison, where such effects can also lead to a decrease in H_n(3). Essentially the same value of H_n is observed to 0.6K, indicating that the material is clearly magnetically saturated and that we are observing a remarkably large zero-point energy spin reduction effect in a molecular magnetic material. The observed saturation hyperfine fields of K_2FeF_5 and Fe(2.2'-bipyridine)Cl_3 correspond to effective 3.2 and 3.5 spins respectively vs the expected 5 spins of high spin ferric.

Conclusion

It is clear that nuclear gamma resonance (Mössbauer effect) spectroscopy is a powerful tool in the study of a variety of magnetic behavior of contemporary interest. The reader is directed to other interesting topics such as the field dependence of $T_{Néel}$ (*35*) and spin reorientation transitions (36) that are particularly amenable to Mössbauer effect study.

Acknowledgments

The author thanks the organizers of this symposium for the opportunity to present this tutorial. He is also indebted to Professors C. E. Johnson (Liverpool University) and R. C. Thiel (Leiden University) for permission to freely use the results of their investigations of zero point spin reduction and soliton effects.

Literature Cited

1. *Chemical Application of Mössbauer Spectroscopy*, Goldanskii, V. I.; Herber R. H. Eds.; Academic Press, New York, 1968.
2. Wertheim, G. K. *Mössbauer Effect: Principles and Applications*, Academic Press, New York, 1964.
3. Greenwood, N .N.;.Gibb T. C., "Mössbauer Spectroscopy," Chapman and Hall, Ltd., London, 1971.
4. Preston, R. S.; Hanna, S. S.; Heberle, *J. Phys. Rev.* , **1962**, *128*, pp.2207.
5. Reiff, W. M. *Coord. Chem. Rev.* **1973**,*10*, pp. 37.
6 Collins, R .L. *J. Chem. Phys.*, **1965**, *42*, pp. 1072.
7. Ludwig, G. W.; Woodbury, H. H. *Phys. Rev.*, **1960**, *117*, pp. 1286.
8. Swartzendruber, L. J.; Bennett, L. H. *J. Res. Nat. Bur. Stand.*, **1970**, *Sect. A, 74*, pp. 691.
9. Johnson, C. E. "Proceedings of the International Conference on the Applications of the Mössbauer Effect," Jaipur, India, **1981**, p. 72.
10. Wignall, J. W. G. *J. Chem. Phys.*, **1966**, *44*, pp. 2462.
11. Wickman, H. H.; Wertheim, G. K. 'Spin Relaxation in Solids and After Effects of Nuclear Transformations" in *Chemical Applications of Mössbauer Spectroscopy*, Goldanskii, V.I.; Herber, R.H. (Eds.), Academic Press, New York, 1968.

12. Perlow, G. J.; Hanna, S. S.; Hamermesh, M.; Littlejohn, C.; Vincent, D. H.; Preston, R. S.; Heberle, S. *Phys. Rev. Letters*, **1960**, *4*, p. 74.
13. Frankel, R. B.; *Mössbauer Effect Methodology* **1974**,*9*, 151 and references therein.
14. Carlin, R. L.; Burriel, R.; Rojo, J. A.; Palacio, F. *Inorg. Chem.***1984**, *23*, p. 2936.
15. Kwiecien, M. J.; Takacs, L.; Reiff, W. M. *Inorg. Chem. Paper No. 170*, 193rd ACS Meeting, Denver, CO, April 1987.
16. Massa, W. *Z. Kristallogr.* **1980**, *153*, p. 201, and private conversations.
17. Reiff, W. M. *Hyp. Interact.* **1988**, *40*, 195.
18. Enz, E. Helv. *Phys. Acta.* **1964**, *37*, p. 245.
19. Carlin, R. L. *Magnetochemistry* , Springer-Verlag, Berlin, 1986.
20. Thiel, R. C.; deGraaf, H.; deJongh, L. J. *Phys. Rev. Lett.* **1981**, *47*, p. 1415.
21. deGroot, H. J. M.; deJongh, L. J.; Elmassalami, M.; Smit, H. H. A.; Thiel, R. C. *Hyp. Interact.* **1986**, *27*, p. 93.
22. Reiff, W. M. in *Mössbauer Spectroscopy Applied to Magnetism and Materials Science*; Long, G. J.; Grandjean, F., Ed.; Plenum Press: New York, 1993, Vol. 1, 205.
23. Anderson, P.W. *Phys. Rev.* **1952**, *86*, 694-701.
24. Steiner, M.; Villain, J.; Windsor, C. H. *Adv. Phys.* **1976**, *25*, 87-209.
25. Hirakawa, K.; Yamada, I.; Kurogi, Y. *J. Physique* **1970**, *32*, C1 890-1.
26. Copla, J. H.; Sieberts, E. G.; Van der Linde, R. H. *Physica* **1971**, *51* 573-87.
27. Gupta, G. P.; Dickson, D. P. E.; Johnson, C. E.; Wanklyn, B. M. J. *Phys. C: Solid State Phys.* **1977**, *10*, L459.
28. Gupta, G. P.; Dickson, D. P. E.; Johnson, C. E. J. *Phys. C: Solid State Phys.* **1978**, *11*, 215.
29. Gupta, G. P.; Dickson, D. P. E.; Johnson, C. E. J. *Phys. C: Solid State Phys.* **1979**, *12*, 2411.
30. Vlasse, M.; Matejka, G.; Tressaud, A.; Wanklyn, B. M. *Acta Cryst.* **1977**, *B33*, 3377-3380.
31. Takács, L.; Reiff, W. M. *J. Chem. Phys. Solids* **1989**, *50*, 33.
32. Calage, Y.; Reiff, W. M. *J. Solid State Chem.* **1994**, *111*, 294-299.
33. Kwiecen, M. J.; Takács, L.; Reiff, W. M. *Inorg. Paper No. 170*, **1987**, American Chemical Society: Denver, CO.
34. Reiff, W. M.; Moron, M. C.; Calage, Y. Inorg. Chem. 1996, (in press).
35. Cooper, D. M.; Dickson, D. P. E.; Johnson, C. E. *Hyp. Interact.* **1981**, *10*, 783-788.
36. Prelorendjo, L. A.; Johnson, C. E.; Thomas, M. F. Wanklyn, B. M. *J. Phys. C: Solid State Phys.* **1982**, *15*, 3199.

Dimers, Oligomers, and Clusters

Chapter 9

Design, Synthesis, and Characterization of π-Cross-Conjugated Polycarbenes with High-Spin Ground States

Kenji Matsuda[1,2], Nobuo Nakamura[2], Kazuyuki Takahashi[2], Katsuya Inoue[2], Noboru Koga[2,3], and Hiizu Iwamura[1,2,4]

[1]Institute for Fundamental Research in Organic Chemistry, Kyushu University, 6–10–1 Hakozaki, Higashi-ku, Fukuoka 812–81, Japan
[2]Department of Chemistry, Graduate School of Science, University of Tokyo, Hongo, Tokyo, Japan
[3]Faculty of Pharmaceutical Science, Kyushu University, 3–1–1 Maidashi, Higashi-ku, Fukuoka 812–82, Japan

Six polydiazo compounds **3d-8d** with pseudo-two-dimensional structures were designed and synthesized. The key reactions used for constructing the skeletons consisted of cyclotrimerization of aryl ethynyl ketones to form 1,3,5-triaroylbenzenes and deoxygenation of a calix[6]arene. The diazo compounds were photolyzed in MTHF solid solution at cryogenic temperatures and analyzed *in situ* by means of Faraday magnetometry. From the field dependence of the magnetization, the hexacarbenes **3-5** were concluded to have tridecet (S = 6) ground states, demonstrating the importance of the topological symmetry, i.e., connectivity of the carbenic centers on the π-cross-conjugated frameworks to determine the spin multiplicity of the alternant hydrocarbon molecules. On the other hand, photolyses of highly-branched nona- and dodecadiazo compounds **6d-8d** gave photoproducts exhibiting spin multiplicities less than those expected for the corresponding **6** and **8**. Intramolecular cross-linking between the carbene centers appeared to limit the extension to highly branched dendritic structures.

Synthesis and characterization of organic molecules with very high-spin ground states are current topics of great interest (*1-3*). In typical organic molecules, all the electrons are paired to form singlet ground states, and triplets with two parallel spins are often observed only as lowest excited states. Two singly occupied orbitals in diradicals are often independent or weakly overlapped to make their singlet and triplet states nearly degenerate or stabilize the former stabilized relative to the latter, respectively. When they are orthogonal in space as in diphenylcarbene or in terms of topological symmetry as in *m*-quinodimethane, the triplets become the ground states according to Hund's rule.

By incorporating the last two structural features for orbital orthogonality within one molecule, Itoh and Wasserman prepared the *m*-phenylenedicarbene **1** ($n = 2$) and

[4]Corresponding author

0097–6156/96/0644–0142$15.00/0

demonstrated that it has a quintet ground state ($S = 2$) (*4*). The dicarbene has been extended to linear tri-, tetra, and pentacarbenes **1** ($n = 3$, 4, and 5, respectively). These were established from their EPR fine structures and/or magnetization data to have $S =$ 3, 4, and 5 ground states, respectively (*5*).

The linear structure can in principle be extended to poly(*m*-phenylenecarbene)s **1** ($n \to \infty$). In practice, however, there are a number of drawbacks in the linear structure. Firstly, it becomes more and more difficult to produce all the carbene centers without fail and keep them intact. Once such a chemical defect is formed in the middle of the cross-conjugated main chain, the high spin multiplicity would be halved (*6*). The linear polyketones, key precursors, become less and less soluble in typical organic solvents in which further chemical transformations have to be carried out. One-dimensional alignment of spins is said to be unstable from a statistical mechanics point of view. Any magnetic linear chain including **1** is expected to exhibit no spontaneous magnetization at finite temperature because the magnitude of the required enthalpy for multiply degenerate lowest excited states is only $2J$, where $2J$ is the exchange coupling parameter between the adjacent spins and not much greater than kT (*7*).

To overcome these problems, construction of a high-dimensional network became an objective. A rigid structure would also help to reduce the high reactivity of triplet carbene centers as one-center diradicals toward recombination, etc. Thus we have arrived at network structure **2** as a long-range goal of the strongly magnetic super-high-spin polycarbenes (Figure 1).

Results and Discussion

Molecular Design and Synthesis of the Three Kinds of Hexacarbenes. As a step nearer to the ideal structure **2**, we have employed three unique structures present in **2**: 1,3,5-trisubstituted and partially dendritic structures **3** (*8,9*) and **4** (*9*) for branched pseudo two-dimensional structures, and a macrocycle **5** (*9*) for the two-dimensional cyclic structure. It was of interest to explore the synthetic strategy and to see if all the three hexacarbenes really have $S = 6$ ground states and offset the demerits of the linear structures described above. A 1,3,5-benzenetriyl unit can be deemed as good as and even better than a *m*-phenylene unit in assembling organic free radical centers in higher concentrations within a molecule and aligning those spins in parallel (*10*).

These polycarbenes were obtained by photolysis of the corresponding diazo precursors **3d**, **4d**, and **5d**. The preparation of the diazo precursors was carried out in a usual way from the corresponding ketones **3k**, **4k**, and **5k**. An exploratory study was made on the synthetic route to these ketones (Scheme 1).

The first and second types of hexaketones **3k** and **4k** were synthesized from *m*-bromobenzaldehyde and tribromobenzene, respectively. In the synthetic sequence for **3k**, the cyclotrimerization of aryl ethynyl ketones was performed. This sequence of reactions making use of cyclotrimerization reaction is also applicable to the higher analogs **6**, **7**, and **8**.

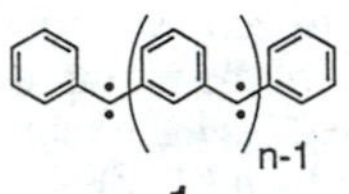

n-1

1

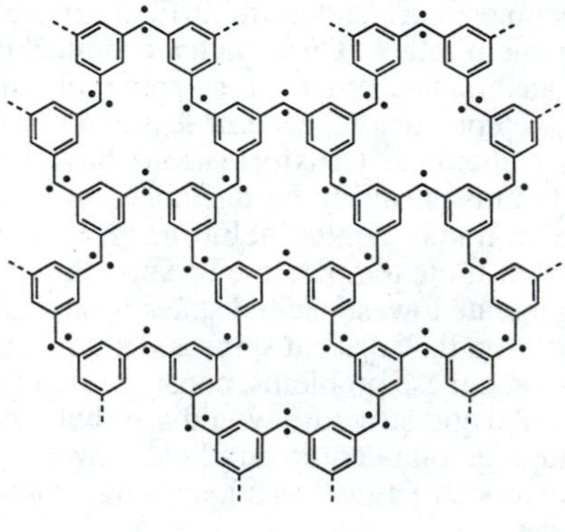

2

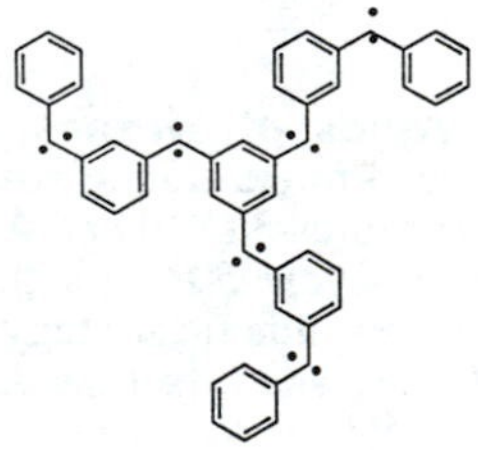

3 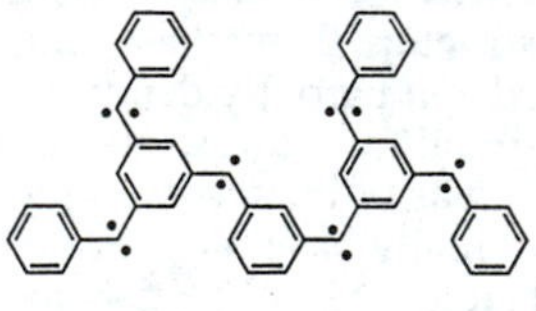 **4** 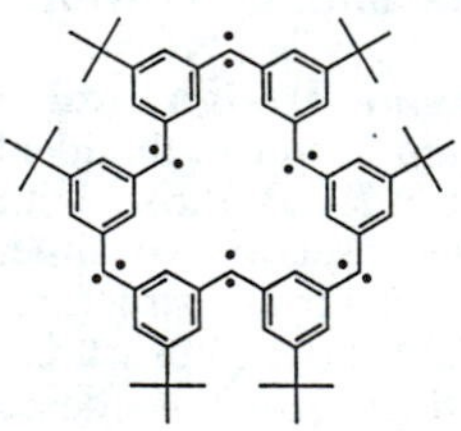 **5**

$$\mathbf{3\text{-}8k}\ (C{=}O) \longrightarrow \mathbf{3\text{-}8d}\ (C{=}N_2) \xrightarrow{h\nu} \mathbf{3\text{-}8}\ (C{:})$$

Scheme 1.

The third type of cyclic hexaketone **5k**, precursor of **5**, was synthesized via a calix[6]arene derivative. The ring structure was constructed by using the established method of synthesis of calix[6]arene. The hydroxyl groups were removed so that their facile insertion or the hydrogen migration reactions may not occur with the carbene centers to be generated (*11*). Hexa-*p-tert*-butyl[1^6]metacyclophane was easily oxidized to hexaketone by a method used in the formation of other types of hexaketones, while [1^4]metacyclophane was not. The synthetic route is summarized in Scheme 2.

These ketones **3k**, **4k**, and **5k** were fully characterized by ^1H, ^{13}C NMR, IR, and mass spectroscopy. HH-COSY NMR spectroscopy was firmly in support of their structures. These ketones were converted to the corresponding hydrazones. The oxidation reactions with yellow mercury oxide were carried out in the dark, monitored by thin layer chromatography (TLC) on alumina. The diazo compounds were purified by column chromatography on alumina (activity IV) in the dark.

The UV-vis spectra of the three diazo compounds **3d**, **4d**, and **5d** had the absorption maxima at 520 nm attributable to the n-π^* absorption. The molar absorptivities (ε) of these absorptions were in the range from 510 to 608, values nearly six times as large as that of diphenyldiazomethane, indicating that these molecules had six cross-conjugated diazo groups.

Molecular Design and Synthesis of the 'Starburst'-Type Nona- and Dodecadiazo Compounds. The 'Starburst'-type structure as given in Figure 2 is popular in chemistry as a dendrimer and in physics as a Bethe lattice (*12*). Ising spins on a Bethe lattice is calculated to show a magnetic phase transition at finite temperature. Thus, the number of spins which is connected to the system becomes drastically larger as the system extends. We have examined two nonacarbene **6** (*13,15*) and **7** (*14*) and one dodecacarbene **8** (*15*).

The synthesis of 'Starburst'-type structures as performed by utilizing the cyclotrimerization of aryl ethynyl ketones. The preparation of the diazo precursor was carried out in an usual way from the corresponding ketone **6k**, **7k**, and **8k.** The nonaketones **6k** and **7k**, the precursor of nonacarbenes **6** and **7**, were synthesized by a simple extension of hexaketone **3k**; the 1,3,5-trisubstituted benzene ring was constructed by cyclotrimerization of 1-[3-[[,3'-(phenylmethyl)phenyl]methyl]phenyl]-2-propyn-1-one and 1-[3,5-bis(phenylmethyl)phenyl]-2-propyn-1-one, respectively. The dodecaketone **8k**, the precursor of dodecacarbene **8**, which has a more highly branched structure, was synthesized by way of Fréchet's convergent approach to constructing dendrimer. For this purpose, two units were required as building blocks. Secondary-amine-catalyzed cyclotrimerization of two kinds of aryl ethynyl ketone (*16*) and deprotection of the trimethylsilyl protecting group were used as key steps in the construction of the structure (Scheme 3 and 4).

The diazo compounds **6d**, **7d**, and **8d** were obtained via hydrazones as described for **3d**, **4d**, and **5d**. The diazo compounds were purified by column chromatography on alumina (activity IV) in the dark.

The IR spectrum of these diazo compounds showed an absorption characteristic of the diazo group at ca. 2040 cm^{-1}. The UV-vis spectra had an absorption maximum at 520 nm attributable to the n-π^* absorption. The molar absorptivities (ε) were 950, 954, and 1660 for **6d**, **7d**, and **8d**, respectively, a value nearly nine and twelve times as large as that of diphenyldiazomethane, indicating that **6d**, **7d**, and **8d** had nine and twelve cross-conjugated diazo groups, respectively.

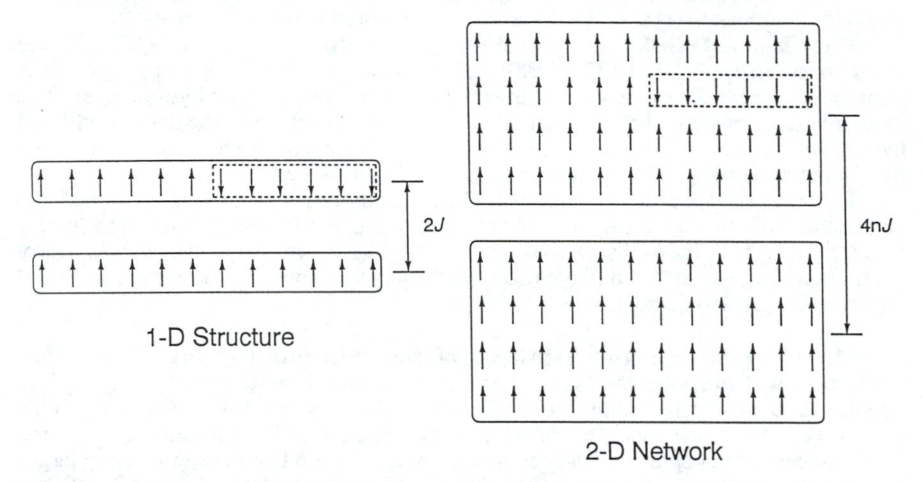

Figure 1. Spin alignment in 1-D chain vs. 2-D network. *n* and *J* are the number and the exchange coupling parameter of the neighboring spins. Intra- and interchain *J* values are presumed to be the same for simplicity.

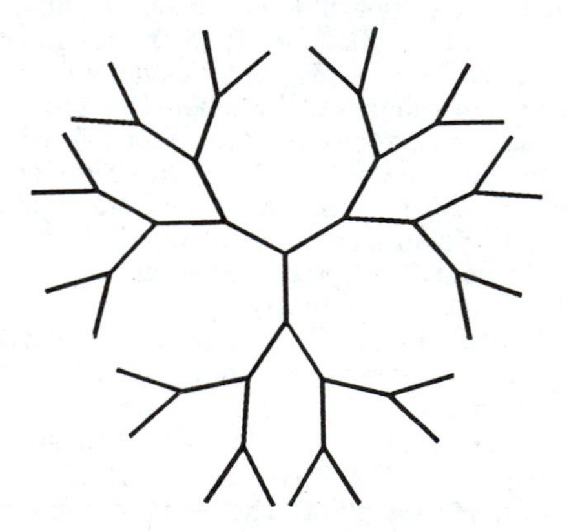

Figure 2. Dendritic Structure or Bethe Lattice.

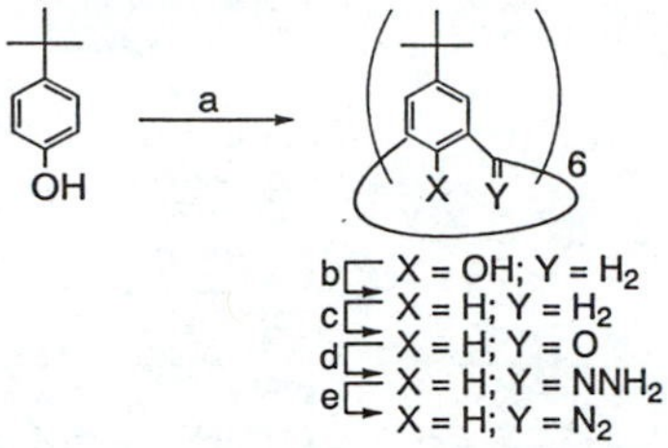

Scheme 2. Reagents and Conditions: (a) 37 % HCHO, KOH, Xylene, 74.3 %; (b) $(EtO)_2POCl$, Bu_4NBr, 50 % NaOH aq., and then K, NH_3; (c) $Na_2Cr_2O_7$, AcOH, 31 %; (d) N_2H_4, N_2H_4•HCl, DMSO, quant.; (e) HgO, EtOK, CH_2Cl_2, benzene, 9 %.

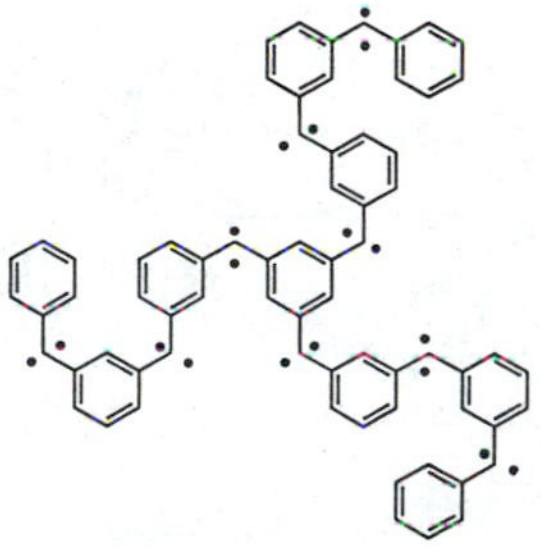

6

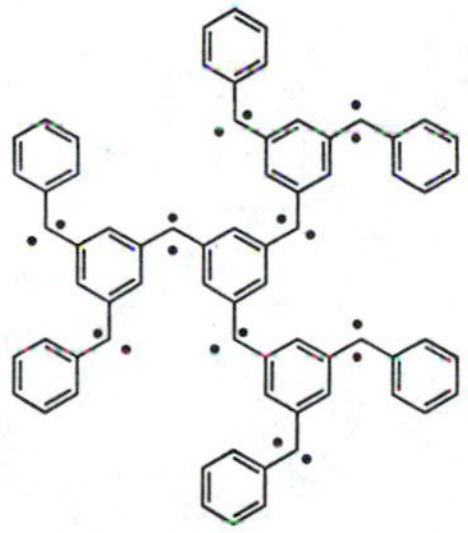

7

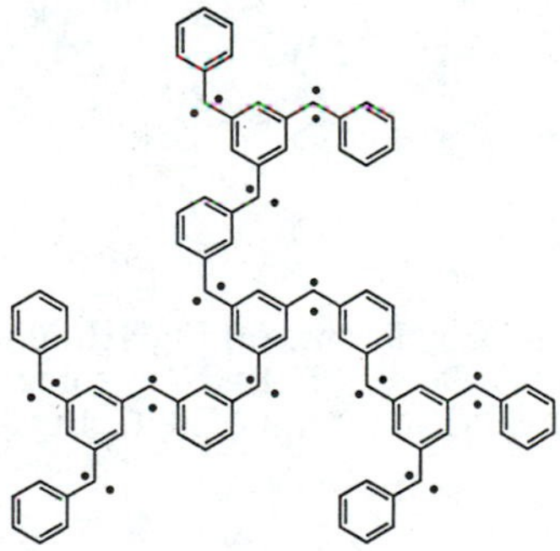

8

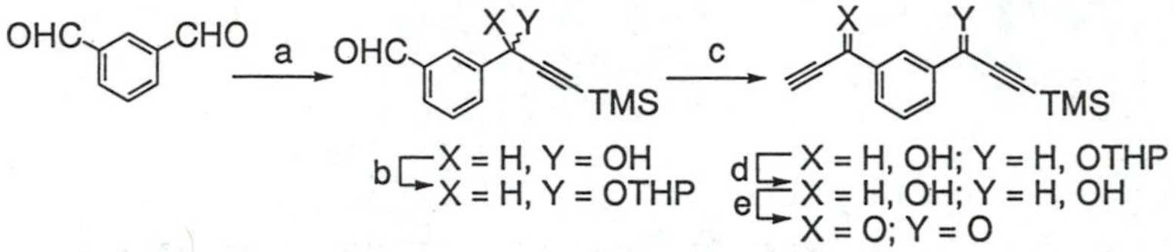

b: X = H, Y = OH → X = H, Y = OTHP

d: X = H, OH; Y = H, OTHP → X = H, OH; Y = H, OH; e: → X = O; Y = O

Scheme 3. Reagents and Conditions: (a) 1 equiv. of TMS-C≡C-Li, TMEDA, THF, -78 °C, 26%; (b) DHP, PPTS, CH_2Cl_2; (c) HC≡C-Li, TMEDA, THF, -78 °C; (d) EtOH, PPTS; (e) 4/3 equiv. of CrO_3-H_2SO_4, acetone, 79%, 4 steps.

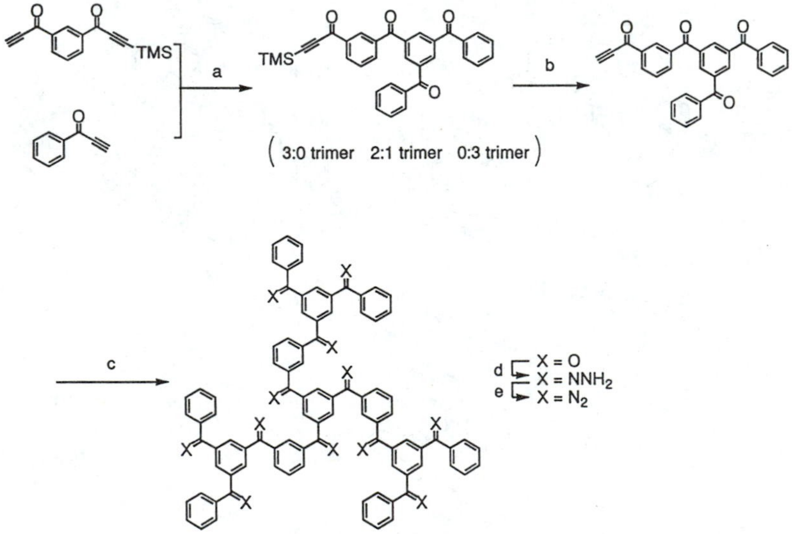

(3:0 trimer 2:1 trimer 0:3 trimer)

d: X = O → X = NNH_2; e: → X = N_2

Scheme 4. Reagents and Conditions: (a) Et_2NH, toluene, then GPC, 31%; (b) KF, MeOH, -20 °C, 96%, (c) Et_2NH, toluene, 35%; (d) N_2H_4, N_2H_4•HCl, DMSO, 90°C, 63%; (e) HgO, EtOK, Et_2O, CH_2Cl_2, 24 %.

Magnetization of Photoproducts. The photolyses were carried out under similar conditions at 2 K in the sample room of a Faraday balance. The light (400 nm < λ < 500 nm) from a Xe lamp was introduced through a quartz light guide installed for irradiation of the precursor. The field dependence of magnetization of the photoproduct was determined *in situ*. The temperature and magnetic field dependences of the magnetization were analyzed in terms of the Brillouin function (eq 1):

$$M = NgJ\mu_B B_J(x) \quad (1)$$

where

$$B_J(x) = \frac{2J+1}{2J}\coth\left(\frac{2J+1}{2J}x\right) - \frac{1}{2J}\coth\left(\frac{x}{2J}\right) \quad (2)$$

$$x = \frac{Jg\mu_B H}{k_B T} \quad (3)$$

N is the number of the molecule, J is the quantum number of the total angular momentum, μ_B is the Bohr magneton, g is the Landé g-factor, and k_B is the Boltzmann constant. Since these carbenes are hydrocarbons and have only light elements, the orbital angular momentum should be negligible and J can be replaced with spin quantum number S in eqs 1~3.

The observed data fitted best with the Brillouin function with $J = S = 6.0$ for **3** as expected for a paramagnet with a tridecet state. The data for **4** and **5** in reference to the theoretical $S = 6$ curve are given in Figure 3. The data were collected at three different temperatures, but no temperature dependence was observed, confirming that the tridecet is a ground state and that there is no antiferromagnetic intermolecular interaction present.

The magnetization data of hexacarbene **4** were obtained at different stages of the photolysis. Theoretical curves with the spin quantum numbers in the range 5-6 were fitted with the observed data irrespective of the progress of the photolysis. These values were larger than those expected for random generation of the carbenes. If the elimination of nitrogen occurred randomly with similar probability at all the diazo groups, the expected value would be derived from a binomial distribution and would be rather small. From this consideration, the diazo groups of the same molecule are likely to be eliminated simultaneously. A one-photon-multicleavage reaction may be evoked (*5e*). However, the possibility of the photolytic reaction occurring from the irradiated surface and progressing inward of the sample solid solutions may not be excluded.

The data of the photoproduct of nonadiazo compound **6** showed the saturation of magnetization at a field less than 1 T and obeyed the theoretical curve eq. 1 with $S = 9$, indicating that nonacarbene **6** with a nonadecet ground state was obtained (Figure 4). The nonacarbene did not show any hysteresis and therefore remained to be a paramagnet with an exceptionally large magnetic moment.

Nonadiazo compound **7d** was photolyzed and the field dependence of magnetization was measured. The observed data collected at two different temperatures fitted best with the same Brillouin function with $S = 7$ confirming that the pentadecet is a ground state. It will be discussed later why $S = 7$ was observed instead of the theoretically expected $S = 9$.

In order to obtain further information, we measured UV-vis spectra before and after the magnetic measurements. Before the magnetic measurement the sample had not yet been irradiated, and after the measurement the sample had been irradiated. The rate

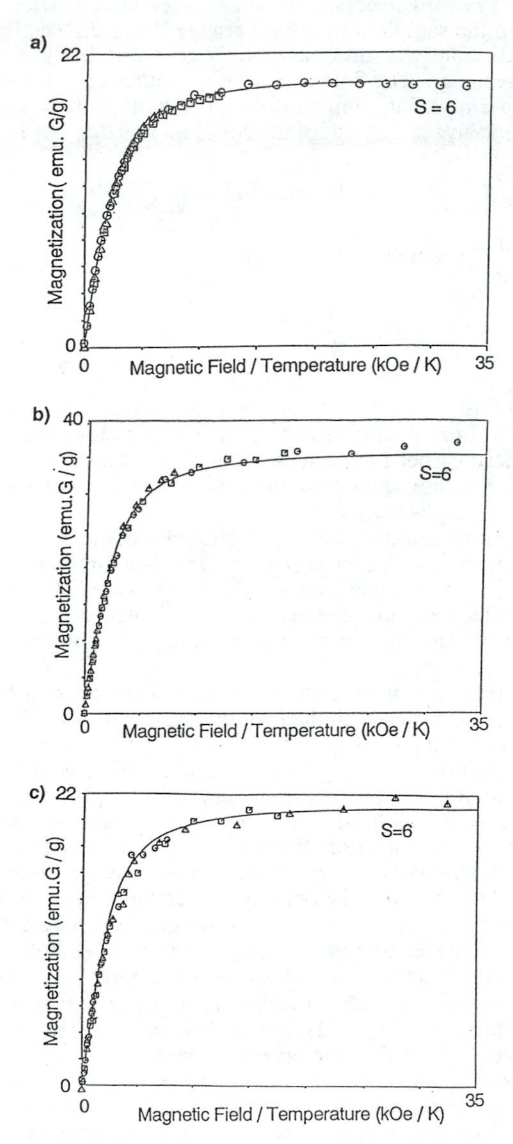

Figure 3. Field dependence of the magnetization of hexacarbenes: (a) **3** in 5 mM MTHF matrix, measured at 2.1 (○), 4.8 (□), and 10.0 (Δ) K; (b) **4** in 0.09 mM MTHF matrix, measured at 2.1 (○), 4.0 (□), and 8.9 (Δ) K; (c) **5** in 0.1 mM MTHF matrix, measured at 2.1 (Δ), 4.2 (□), and 9.0 (○) K. The ordinates are normalized by the amount of the starting diazo compounds and uncorrected for the degree of photolysis.

of the photolysis could be monitored by comparing its n-π* absorptions. From the comparison of the UV-vis data, the degree of the photolysis was concluded to be 90 % complete as a whole (Figure 5). From these data, two theoretical curves were obtained. The $S = 9$ curve in Figure 6 is for the ideal case, i.e. all the carbene centers are assumed to be generated and kept intact. The $S = 7$ curve corresponds to the case where only seven carbene centers are intact and ferromagnetically coupled to one another.

Photolysis of dodecadiazo compound **8d** was carried out under conditions similar to those of the other diazo compounds. Field dependence of the magnetization of the photoproduct of **8d** is shown in Figure 7a. The data obtained at three different temperatures did not agree with one another. The lower the temperature, the lower were the magnetization values. From this observation, an antiferromagnetic interaction among the carbene centers is suggested to be operative. When the sample concentration as low as 0.1 mM is taken into account, the observed antiferromagnetic interaction is most reasonably assigned to an interbranch interaction within a molecule rather than between molecules.

Furthermore the observed data did not obey the theoretical curve with $S = 12$ or any other single spin quantum number. As H/T was increased, the magnetization rose rapidly but leveled off rather slowly. This behavior could be explained as the mixing of low-spin species and high-spin species. Thus the formation of several species of different spin multiplicities is suggested in this photoproduct. As the field is increased in the region of low magnetic field, magnetization grows in more rapidly and approaches to saturation for high-spin species. By observing this region, we can roughly estimate the spin multiplicity of the species obtained. In Figure 7b, the expansion of Figure 7a, the observed data fitted a theoretical curve with $S = 7$. This means that there was high-spin species at least with S = 7 included in this photoproduct.

Feasible Structures for the Nonacarbene 7 Responsible for the Observed $S = 7$. Our analyses showed that, while the nine diazo groups were photolyzed from nonadiazo compound **7d**, only seven carbene centers appear to be aligned in parallel in the photoproduct. It is not likely that the nonacarbene **7** has a pentadecet ground state ($S = 7$). Two terminal carbene centers are required to have strong antiferromagnetic interaction to each other to cancel four spins out . One of the feasible and extreme cases of such an interaction would be a chemical bond formation: the formation of heptacarbene **9**. Diphenyldiazomethane is known to give azine, anthracene and phenathrene as well as tetraphenylethylene when photolyzed in high concentration (*17*). Recombination of two triplet carbene centers to form a singlet ethylene is considered to be a chemical version of triplet-triplet annihilation and spin-allowed. The corresponding azine and phenanthrene structures in which the stilbene unit in **9** is replaced with these chromophore may not be excluded. The limited amount of the photoproduct from our experiments did not allow us to obtain independent structural proof for **9**. The nonacarbene **7** has the most branched Starburst structure, and therefore, after the release of the nitrogen molecules, two carbene centers might be generated in close proximity and undergo facile recombination. After bonding, the molecule would not be flexible enough to undergo further recombination. The spin quantum number was constant at 7 regardless of the degree of the photolysis. An energetically cascade one-photon reaction of the multiple diazo group cleavage and/or surface effect of the solid photolysis may explain the phenomenon and a preferential release of only seven diazo groups from nonadiazo compound **7d** seems to be a remote possibility.

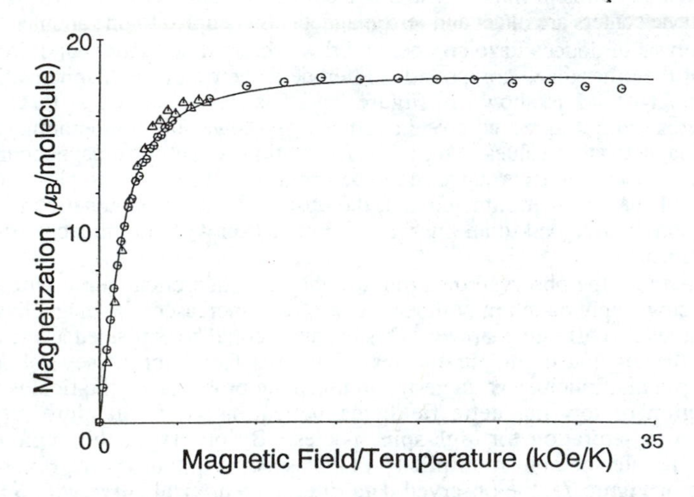

Figure 4. Field dependence of the magnetization of the photoproduct of nonacarbene **6** in 0.10 mM MTHF matrix, measured at 2.1 (O) and 10.0 (Δ) K. The ordinates are normalized at 18 μ_B. Solid line shows theoretical curve with $S = 9$.

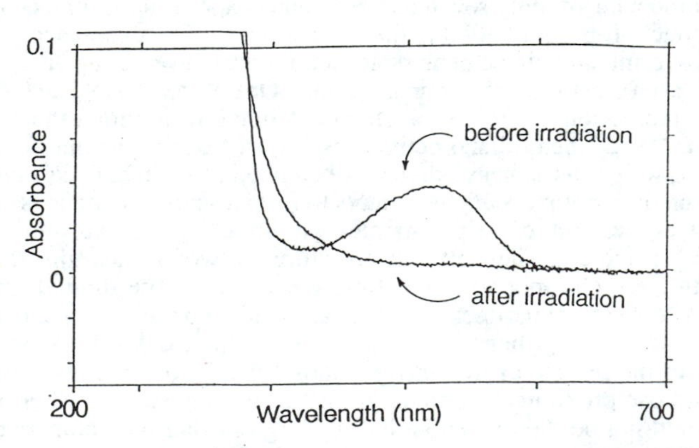

Figure 5. UV-vis spectra measured before and after photolysis of nonadiazo compound **7d**.

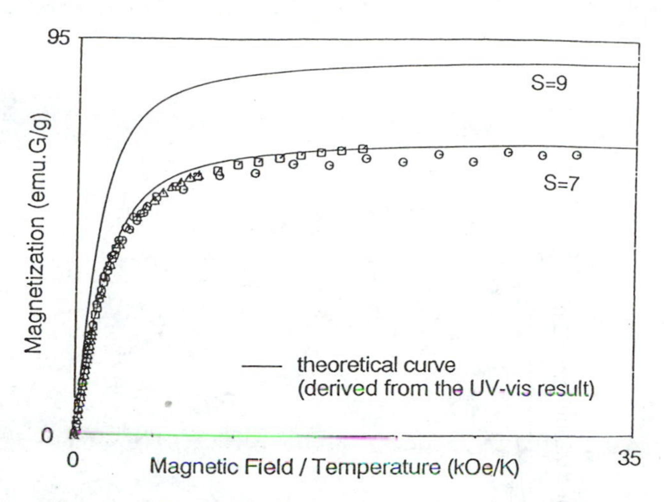

Figure 6. Magnetization curves (measured at 2.2 (O), 3.9 (□), and 9.0 (Δ) K) of photoproduct of **7d** compared to the theoretical curve derived from the result of UV-vis spectra.

Conclusion

In this paper, we have described the synthesis of polycarbenes with pseudo-two-dimensional connectivity and magnetic characterization of them. The photolysis of the diazo precursors proceeded smoothly in the MTHF solid solutions at cryogenic temperature. The magnetic field dependence of magnetization of hexacarbenes **3-5** measured by the Faraday method was in good agreement with the theoretical value formulated by Brillouin function with $J = S = 6$. On the basis of these observations it was demonstrated that all the carbenes posses tridecet ($S = 6$) ground state regardless of the molecular shapes. It has most dramatically been demonstrated that it is not molecular shape or geometrical symmetry of the molecules but the topological symmetry which is most important in determining the spin multiplicity of the alternant hydrocarbon molecules. This strategy could be extended to nonacarbene **6** which was found to have nonadecet ($S = 9$) ground state, the highest spin even reported for an organic molecule.

In the case of nonacarbene **7**, the obtained magnetization curves were consistent with $S = 7$ regardless of the degree of the photolysis of the precursor **7d**. The nonacarbene **7** has the most branched Starburst structure, and therefore, after the release of the nitrogen molecules from the diazo groups, two carbene centers might be generated in close proximity and undergo facile recombination. The side chains of the less branched nonacarbene **6** are regarded as more stretched and may contain some solvent molecules buried among the chains that would protect the cross-linking of the carbene centers than those of **7**. Higher cross-linking among the carbene centers was evidenced in the photoproduct from the most branched dodecadiazo compound **8d** by the observed S values not greater than $S = 7$ and appearance of the antiferromagnetic interaction. Construction of the rigid structure like **2** is concluded to be an important factor in the design of super high spin molecules containing highly reactive moieties such as carbene centers.

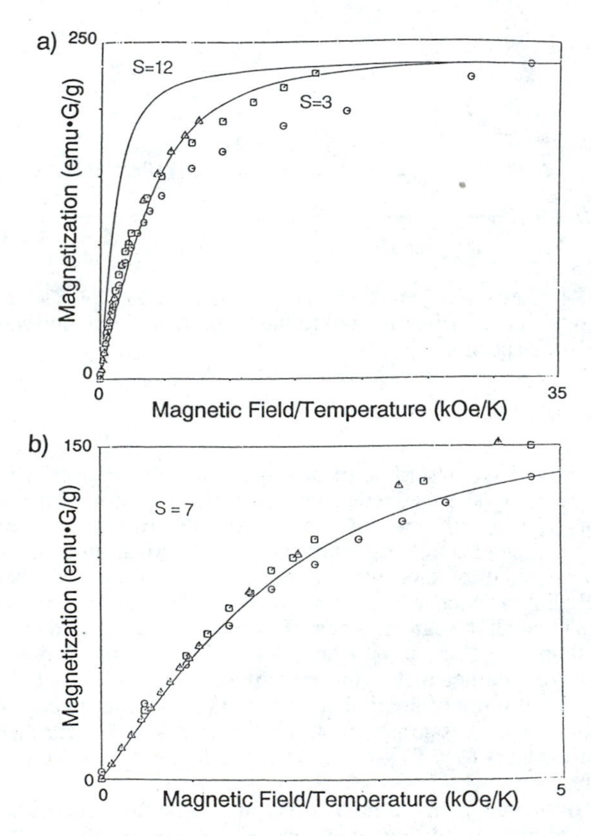

Figure 7. Field dependence of the magnetization of the photoproduct of dodecadiazo compound **8d** in 0.10 mM MTHF matrix, measured at 2.1 (O), 4.2 (□), and 9.1 (Δ) K. The ordinates are normalized by the amount of the starting diazo compounds and uncorrected for the degree of photolysis. a) full region; b) expansion of a) in the region shown in the axis.

9

References

1. For reviews see: (a) Iwamura, H. *Adv. Phys. Org. Chem.* **1990**, *26*, 179. (b) Dougherty, D. A. *Acc. Chem. Res.*, **1991**, *24*, 88. (c) Iwamura, H.; Koga, N. *Acc. Chem. Res.*, **1993**, *26*, 346. (d) Kurreck, H. *Angew. Chem., Int. Ed. Engl.* **1993**, *32*, 1409. (e) Miller, J. S.; Epstein A. J. *Angew. Chem., Int. Ed. Engl.* **1994**, *33*, 385. (f) Rajca, A. *Chem. Rev.*, **1994**, *94*, 871. (g) Miller, J. S.; Epstein, A. J. *Chem. Eng. News*, **1995**, Oct. 2, 30.
2. Perchloropolyarylmethyl: (a) Veciana, J.; Rovira, C.; Crespo, M. I.; Armet, O.; Domingo, V. M.; Palacio, F. *J. Am. Chem. Soc.* **1991**, *113*, 2552. (b) Veciana, J.; Rovira, C.; Ventosa, N.; Crespo, M. I.; Palacio, F. *J. Am. Chem. Soc.* **1993**, *115*, 57.
3. Polyarylmethyl: (a) Rajca, A. *J. Am. Chem. Soc.* **1990**, *112*, 5890. (b) Rajca, A.; Utamapanya, S.; Xu, J. *J. Am. Chem. Soc.* **1991**, *113*, 9235. (c) Rajca, A.; Utamapanya, S.; Thayumanavan, S. *J. Am. Chem. Soc.* **1992**, *114*, 1884. (d) Rajca, A.; Utamapanya, S. *J. Am. Chem. Soc.* **1993**, *115*, 2396. (e) Rajca, A.; Rajca, S.; Padmakumar, R. *Angew. Chem., Int. Ed. Engl.* **1994**, *33*, 2091. (f) Rajca, A.; Rajca, S.; Desai, S. R. *J. Am. Chem. Soc.* **1995**, *117*, 806.
4. (a) Itoh, K. *Chem. Phys. Lett.* **1967**, *1*, 235. (b) Wasserman, E.; Murray, R. W.; Yager, W. A.; Trozzolo, A. M.; Smolinski, G. *J. Am. Chem. Soc.* **1967**, *89*, 5076.
5. (a) Teki, Y.; Takui, T.; Itoh, K.; Iwamura, H.; Kobayashi, K. *J. Am. Chem. Soc.* **1983**, *105*, 3722. (b) Sugawara, T.; Bandow, S.; Kimura, K.; Iwamura, H.; Itoh, K. *J. Am. Chem. Soc.* **1984**, *106*, 6449. (c) Teki, Y.; Takui, T.; Yagi, H.; Itoh, K. *J. Chem. Phys.* **1985**, *83*, 539. (d) Sugawara, T.; Bandow, S.; Kimura, K.; Iwamura, H.; K., I. *J. Am. Chem. Soc.* **1986**, *108*, 368. (e) Teki, Y.; Takui, T.; Itoh, K.; Iwamura, H.; Kobayashi, K. *J. Am. Chem. Soc.* **1986**, *108*, 2147. (f) Fujita, I.; Teki, Y.; Takui, T.; Kinoshita, T.; Itoh, K.; Miko, F.; Sawaki, Y.; Iwamura, H.; Izuoka, A.; Sugawara, T. *J. Am. Chem. Soc* **1990**, *112*, 4074.
6 (a) Rajca, A.; Utamapanya, S. *J. Am Chem. Soc.* **1993**, *115*, 10688-10694; (b) Tyutyulkov, N. N.; Karabunarliev, S. C. *Int. J. Quantum Chem.* **1986**, *29*, 1325-1337; (c) Mataga, N. *Theor. Chim. Acta.* **1968**, *10*, 372-376.
7. Carlin, R. L. *Magnetochemistry;* Springer-Verlag: Berlin, 1986.
8. (a) Nakamura, N.; Inoue, K.; Iwamura, H.; Fujioka, T.; Sawaki, Y. *J. Am. Chem. Soc.* **1992**, *114*, 1484-1485; (b) Furukawa, K.; Matsumura, T.; Teki, Y.; Kinoshita, T.; Takui, T.; Itoh, K. *Mol. Cryst. Liq. Cryst.* **1993**, *232*, 251-260
9. Matsuda, K.; Nakamura, N.; Takahashi, K.; Inoue, K.; Koga, N.; Iwamura, H. *J. Am. Chem. Soc.* **1995**, *117*, 5550-5560.

10.(a) Kanno, F.; Inoue, K.; Koga, N.; Iwamura, H. *J. Chem. Phys.* **1993**, *97*, 13267. (b) Yoshizawa, K.; Hatanaka, M.; Matsuzaki, Y; Tanaka, K.; Yamabe, T. *J. Chem. Phys.* **1994**, *100*, 4453. (c) Wasserman, E.; Schueller, K.; Yager, W. A. *Chem. Phys. Lett.* **1968**, *2*, 259. (d) Takui, T.; Itoh, K. *Chem. Phys. Lett.* **1973**, *19*, 20.
11.(a) Goren, Z.; Biali, S. E. *J. Chem. Soc., Perkin Trans.* I **1990**, 1484. (b) Grynszpan, F.; Biali, S. E. *J. Phys. Org. Chem.* **1992**, *5*, 155. (c) de Vains, J-B.; Pellet-Rostaing, S.; Lamartine, R. *Tetrahedron Lett.* **1994**, *35*, 8147.
12.Domb, C. In *Phase Transitions and Critical Phenomena;* Domb, C., Green, M. S., Eds.; Academic Press: London, 1974; vol. 3, pp 357-484 and references therein.
13.Nakamura, N.; Inoue, K.; Iwamura, H. *Angew. Chem., Int. Ed. Engl.* **1993**, *32*, 872.
14.Matsuda, K.; Nakamura, N.; Inoue, K.; Koga, N.; Iwamura, H. *Chem. Eur. J.* **1996**, *2*, 259-264.
15.Matsuda, K.; Nakamura, N.; Inoue, K.; Koga, N.; Iwamura, H. *Bull. Chem. Soc. Jpn.* **1996**, *69* , in press.
16.(a) Balasubramanian, K.; Selvaraj, P. S.; Venkataramani, P. S. *Synthesis* **1980**, 29-30; (b) Matsuda, K.; Nakamura, N.; Iwamura, H. *Chem. Lett.* **1994**, 1765-1768.
17.Turro, N. J.; Aikawa, M.; Butcher, J. A., Jr. *J. Am. Chem. Soc.* **1980**, *102*, 5127-5128.

Chapter 10

Origin of Superparamagnetic-Like Behavior in Large Molecular Clusters

D. Gatteschi and R. Sessoli

Department of Chemistry, University of Florence, via Maragliano 77, Florence, Italy

A short tutorial is provided for the magnetic properties to be expected in large molecular clusters. The results reached so far in the understanding this new class of materials are exemplified with manganese and iron compounds. The central role of magnetic anisotropy in determining the superparamagnetic-like behavior of these materials is stressed

The synthesis and investigation of the magnetic properties of large molecular clusters is one of the current challenges in molecular magnetism [1-6]. In fact in this way it is conceivable to obtain new types of magnetic particles of nanometric size which, compared with those which can be obtained with other approaches, have the advantage of being absolutely monodispersed. Further they can be kept separated one from the other by dissolving them in appropriate matrices and/or polymers, in such a way to minimize the magnetic interactions among them.

The reasons why nanometric size magnetic particles are interesting are numerous [7-8]. From the theoretical point of view they can provide information on the detailed mechanism according to which the transition from paramagnetic to bulk magnetic behavior occurs. Further, as will be clarified below, they can show the coexistence of quantum and classic behavior, a hot topic nowadays. From the experimental point of view they can show many appealing properties, ranging from magnetic bistability of single molecules to large magnetocaloric effects.

In the following we will first briefly outline the magnetism to be expected in large clusters, placing particular emphasis on superparamagnetism, and then we will show the most successfull examples so far reported of manganese and iron clusters in which this behavior has been experimentally observed.

0097–6156/96/0644–0157$15.00/0

Magnetism of Clusters

The simplest approach to describe the magnetic properties of paramagnetic systems is that of using a spin hamiltonian [9-10]. For a system which has no orbital degeneracy this can be conveniently written as:

$$H = \mu_B \, \mathbf{H.g.S} + D\,[S_z^2 - S(S+1)/3] + E\,(S_x^2 - S_y^2) \qquad (1)$$

where the first term is the Zeeman, and the second and third are the zero-field splitting terms. E is zero for axial and D for cubic symmetry. For transition metal ions they both are determined by low symmetry components of the crystal field and spin-orbit coupling. The hamiltonian is applied to a basis of (2S+1) $|S\ M_S\rangle$ functions, the matrix is diagonalized, and the magnetic susceptibility is calculated using standard techniques.

When more than one magnetic center is present in the cluster, the hamiltonian (1) is written for all the individual spins S_i and summed. The interaction between the spins is taken into account by a sum of terms associated with all the possible pairs. A convenient expression is:

$$H = J_{ij}\, \mathbf{S_i.S_j} + \mathbf{S_i.D_{ij}.S_j} + \mathbf{d_{ij}.\ S_i x S_j} \qquad (2)$$

The first term, the isotropic one, is often the most important. In this notation a positive value of the J_{ij} parameter corresponds to antiferromagnetic coupling of the two spins, while a negative value corresponds to ferromagnetic coupling. The second term, the anisotropic one, introduces an additional zero field splitting of the levels. It has its origin both in the through space interaction between the spins (magnetic dipolar in nature) and in the exchange interaction. The third term can be neglected in most cases.

The procedure for calculating the magnetic properties of the cluster is an extention of that outlined for the single ions. The basis of functions to be used for the calculation of the hamiltonian matrix however grows rapidly with the number of coupled spins. For N equivalent spins S there are $(2S+1)^N$ states. The matrices can be block diagonalized exploiting the symmetry of the system. The most obvious one is that of the total spin S, which is obtained from the individual spins with vector addition rules, $|S_1-S_2| \leq S \leq S_1+S_2$. The total spin states for clusters of six S= 5/2 spins are given in Table I.

Table I. Total spin states and degeneracies for a cluster of six S= 5/2 spins

S=15	1	**S=14**	5	**S=13**	15	**S=12**	35
S=11	70	**S=10**	126	**S=9**	204	**S=8**	300
S=7	405	**S=6**	505	**S=5**	581	**S=4**	609
S=3	575	**S=2**	475	**S=1**	315	**S=0**	111

The most efficient procedure to calculate the energy levels, in terms of memory storage in a computer, is provided by irreducible tensor operator approaches (ITO) [11-13]. As an example we show in Figure 1 the energy levels calculated for a ring of six S= 5/2 spins. These calculations have been used in order to interpret the magnetic properties of $[MFe_6(OCH_3)_{12}dbm]Cl$ [14], where dbm is dibenzoyl-methanide, and M can be Li, Na, which comprises a ring of six iron(III) ions. Once the energy levels are available it is possible to calculate the magnetic susceptibility as a function of temperature. The best fit to the experimental susceptibility gave a value for the nearest-neighbor coupling constant J= 19 cm^{-1}. This implies a ground S= 0 level, with an S= 1 level at 12.7 cm^{-1}, an S= 2 at 38 cm^{-1}, an S= 3 at 76 cm^{-1}, and an S= 4 at 126.7 cm^{-1}. The quality of the calculated levels was confirmed by strong field magnetization studies, which provided the experimental energies of the excited levels of increasing multiplicity.

These data, and the calculated levels of Figure 1 show that the spin levels merge into a quasi continuum for intermediate energy, but they are still discrete close to the ground state. Therefore we may conclude that the magnetic properties of clusters of increasing size are essentially those of complex paramagnetic species, which do not follow the Curie law, but a simple extention of that taking into account the temperature dependent population of the levels. However if we imagine having available clusters with increasing numbers of spins, we immediately realize that at some stage the size becomes so large that the connected network of interacting spins gives rise to a magnetically ordered state, be it ferro-, antiferro- or ferri-magnetic. In other terms the size of the particles must become practically infinite, and the magnetic properties become those of an infinite array. The problem which is open is that of the critical size for observing such phenomena, and which must be expected to be the first manifestations of this behavior. In order to do this we may use the knowledge already available for bulk magnets.

Superparamagnetism

A magnetically ordered state corresponds to a system in which the spin correlation function diverges. Let us focus on a ferromagnet first, which corresponds to the simplest possible spin arrangement, all parallel to each other. It is well known that in a ferromagnet below the critical temperature the total energy of a particle is minimized if domains are formed [15]. The domains are spatial regions within which all the spins are parallel to each other; however the prevalent orientations of the spins in different domains are different. As a result the particle is not magnetized in the absence of an external field. The different domains are separated by border regions in which the prevalent orientation of the spins is rapidly changing to connect the two domains, as shown in Figure 2. The magnetic exchange interaction and magnetic anisotropy of the compound influence in opposite ways the width of the border region, the so-called Bloch wall. In fact from the example of Figure 2 we see that in order to pass from the preferred orientation with spin up on the left to the preferred orientation of spin down on the right the spins are no longer parallel to each other. The effort will be shared by all the spins in the wall, and, if the exchange interaction is high, more spins will be needed in order to bear the effort. At the same time, however, more spins are not oriented along the preferred direction and therefore the higher the anisotropy the smaller is the Bloch wall.

When the size of the Bloch walls becomes larger than the size of the particle, only one domain will be present. This is a single domain particle. For iron, spherical particles smaller than 60 nm become a single domain. On further reducing the size of the particles the anisotropy barrier, which can be written as:

$$A = KV \tag{3}$$

where K is the anisotropy per volume and V is the volume of the particle, finally becomes comparable to thermal energy. The magnetization of the particle flips freely, therefore its time average in the absence of an external field is zero. On introducing a field, a large

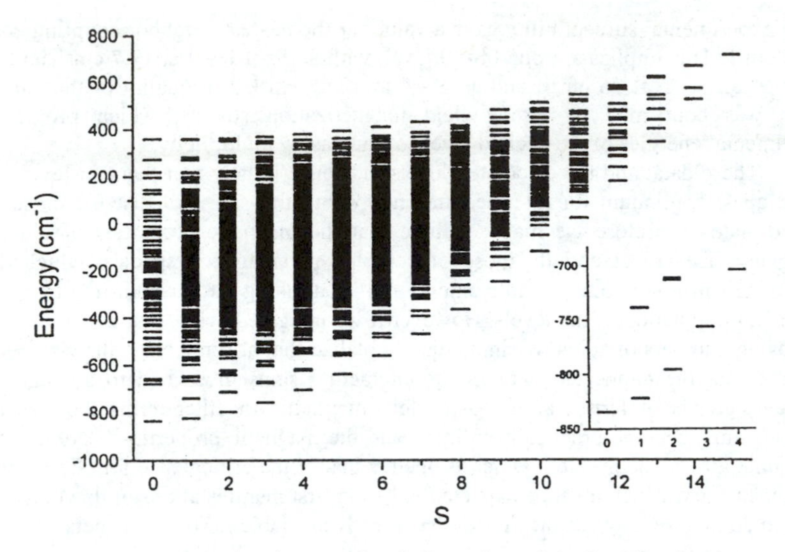

Figure 1. Calculated energy levels for a ring of six S= 5/2 spins antiferromagnetically coupled with J= 19 cm^{-1}.

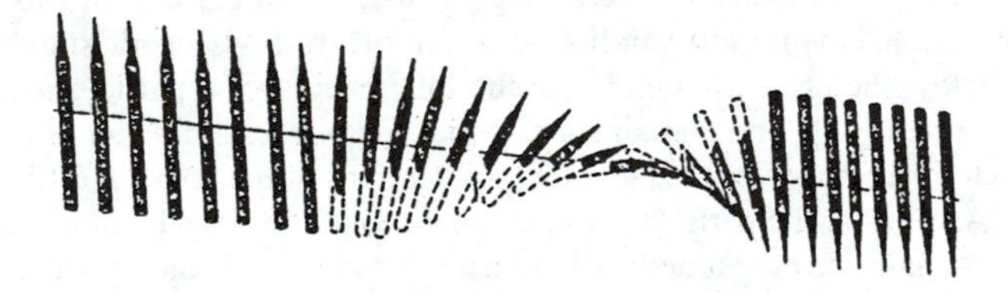

Figure 2. Scheme of two domains separated by a Bloch wall.

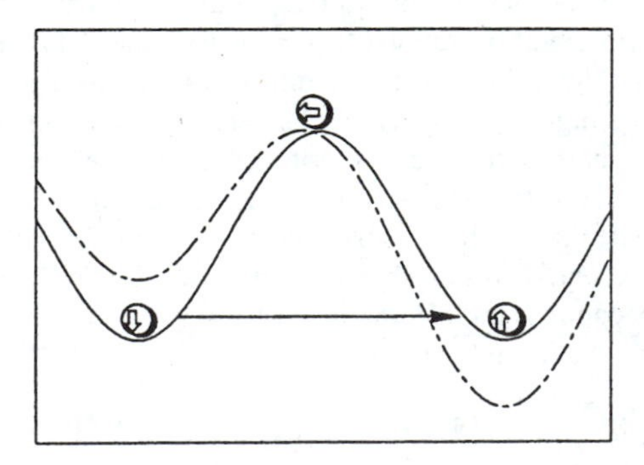

Figure 3. Anisotropy barrier for the reorientation of the magnetization of a particle. ___ with no applied field; _ _ _ with applied field.

magnetic moment is observed and the magnetization follows the Brillouin function as expected for a paramagnet with S tending to infinity. This is superparamagnetic behavior. Since the process of reverting the magnetization is thermally activated, the magnetization relaxes according to the following equation:

$$\tau = \tau_0 \exp(KV/kT) \quad (4)$$

The pre-exponential factor τ_0 corresponds to the relaxation time of the magnetization in the absence of the barrier. Its value is $\tau_0 = 10^{-9} - 10^{-11}$ s for classical ferromagnets. A pictorial representation of the anisotropy barrier is shown in Figure 3.

From (4) it is apparent that the magnetization of a particle at low temperature will be essentially blocked in one of the two minima, and it will show regular ferromagnetic behavior. The temperature at which the magnetization is blocked in one of the walls is called the blocking temperature, T_b. It can be defined as the temperature at which τ becomes equal to the characteristic time τ_{exp} of the experimental technique used for monitoring the relaxation of the magnetization. τ_{exp} is in the range of $10^{-4} - 10^{2}$ s for ac susceptibility measurements, while it is of the order of 10^{-8} s for Mössbauer spectroscopy. The latter has been widely used due to the large number of inorganic magnets based on iron.

The behavior of ferrimagnets is also similar, but at the beginning it may be surprising to realize that also antiferromagnets can show superparamagnetic behavior. The simplest antiferromagnets are those formed by two ferromagnetic lattices which have reversed orientations. Since the two lattices are identical the total magnetization is zero. However this is true only for the spins of the bulk. The surface spins cannot keep the same orientation as in the bulk, and a non-zero component of the magntization arises due to canting of these spins. The smaller the particle the larger the surface-to-volume ratio is and the larger are the surface components. Therefore small antiferromagnetic particles behave like weak ferromagnets, and at high temperatures they can behave like superparamagnets. The resulting mangetic moment is represented by a vector, called the Néel vector [16].

The experimental measurement of superparamagnetic behavior can be done using ac susceptibility and Mössbauer spectroscopy, as indicated above. In an ac susceptibility measurement the sample is placed between two coils in which a current of a given frequency is flowing. The sample magnetization responds to the oscillating magnetic field of the coils and a signal proportional to it is recorded. If the relaxation time of the magnetization is fast only an in-phase response is detected, while if the relaxation is long an out-of-phase component is detected. The blocking of the magnetization of a superparamagnet therefore in an ac susceptibility experiment is detected by the observation of an out-of-phase component, χ''. In experiments performed at different frequencies χ'' has maxima at different temperatures.

In Mössbauer spectra of iron compounds recorded without an applied magnetic field in general two signals are observed, associated with the quadrupole splitting. The position and the extent of the splitting provide information on the chemical nature of the iron ion. When the system is placed in an external magnetic field the doublet splits into six components, therefore the blocking of the magnetization is monitored by the transition from two to six signals.

The above treatment is a classic one, and it assumes that at low temperature the particle is blocked in one of the two minima, without any possibility to reorient. However if we take into account the quantum nature of the particles we cannot exclude the possibility of tunneling between the two minima of Figure 3. The problem is very challenging both for basic understanding of the properties of matter at mesoscopic scale and also for possible applications. In fact the presence of quantum tunneling reduces the possibility of exploiting small particles as memory storage elements, because the information stored in one of these bits could not remain stored forever, but only for the time associated with the tunneling process.

The theoretical interest of quantum tunneling in superparamagnets is associated with the verification of the correspondence principle which states that the classic behavior can be obtained as the limit of quantum behavior when the size of the particles tends to infinity. Up to now it was essentially only superconductivity which provided the opportunity to test the effects of quantum phenomena in mesoscopic particles, but now magnetism is providing additional testing grounds [8,17].

One interesting aspect of quantum effects is that there are several theoretical predictions which wait for suitable materials on which to be tested, such as that quantum tunneling effects should be better observed in antiferromagnets than in ferromagnets, and that they would be forbidden in systems with odd number of unpaired electrons. Molecular clusters in principle can provide answers to these requirements.

Be it classic or quantum in nature, central to the understanding of the properties of superparamagnets is the origin of the anisotropy barrier which the particle must overcome in order to reorient its magnetization. In a bulk particle the anisotropy has in principle at least three origins: shape, single ion, and exchange contribution. The first is associated with the dipolar fields generated by the correlated spins. It can be calculated once the shape of the particle is known. For a sphere it is zero, as it depends on the demagnetization factors, while it is of the Ising type for an infinite cylinder, and of the XY type for a thin ring.

The second component to the anisotropy comes from the magnetic anisotropy of the individual ions. The magnetic anisotropy of the individual ion comes from unquenched orbital contributions compatible with the crystal field felt by the metal ions. Information on it may come from the knowledge of the coordination of the individual ions, and of their electron configuration. For instance high spin iron(III), which has a d^5 electron configuration has a very small Zeeman anisotropy. However the ground $^6A_{1g}$ of octahedral symmetry may be largely split in lower symmetry due to the admixture of excited states determined by spin orbit coupling. Manganese(III) which in octahedral symmetry has a ground ^5E_g state is subject to Jahn-Teller distortions and therefore it has generally large anisotropy components, essentially associated with the zero field splitting.

The third contribution to magnetic anisotropy for the particles comes from the second and third terms of the hamiltonian (2), which are determined by both through space dipolar interactions between the individual spins, and by exchange interactions involving one ion in its ground state and the other ion in an excited state. In general these terms decrease on increasing the distance between the two interacting ions and on decreasing the exchange interaction between them [18].

All these terms can be taken into account in clusters, with standard spin projection techniques which provide the values of the parameters for clusters as a function of those of the individual ions. However in clusters in general the contribution from shape anisotropy is small, and the other two dominate.

From the experimental point of view the determination of the magnetic anisotropy of clusters can be performed by measuring the susceptibility of single crystals, provided that the clusters are all iso-oriented in the unit cell. Another technique which can provide important information is EPR, which can give explicitely the **g** anisotropy and the zero field splitting of the ground state, and in some cases also of the excited states.

Manganese Clusters

Manganese carboxylates form a large series of compounds of general formula $[Mn_{12}O_{12}(RCOO)_{16}]$, which can have a varying number of water molecules [6,19,20]. They contain eight manganese(III) and four manganese(IV) ions. Structural evidence suggested trapped valences, as shown by the Jahn-Teller distortions of the manganese(III) ions. Recently Christou and Hendrickson reported also partially reduced clusters, where one manganese(III) ion is substituted by a manganese(II), and also compounds in which four manganese(III) are susbstituted by iron(III). A sketch of the core of the cluster is shown in Figure 4. The acetate derivative, Mn12ac, was shown to have a ground S= 10 state, with the first excited state, S= 9, separated by 35 K [20]. The nature of the ground state was determined by magnetic susceptibility measurements and confirmed by EPR and inelastic neutron scattering experiments [21]. The ground state can be justified by the simple spin scheme shown in Figure 4, which has all the manganese(III) spins (S=2) up and the manganese(IV) spins (S= 3/2) down. Assuming a dominant antiferromagnetic interaction between the manganese(III) and the manganese(IV) ions bridged by two oxo groups the low lying states can be computed taking into account four S= ½ states (corresponding to the ground state of the manganese(III)-manganese(IV) pair, and four S= 2 states of the remaining manganese(III) ions. Sample calculations showed that indeed lowest state can be S= 10 for reasonable values of the exchange parameters. [20]

The ground S= 10 state is largely split in zero field, as shown by High Frequency EPR spectroscopy and single crystal magnetic anisotropy measurements. The former technique, in which the frequency of the exciting radiation can be extended up to 500 Ghz provides unequivocal evidence of the value of D because the spectra can be interpreted in the strong field approximation (Zeeman term larger than the zero field splitting) [22]. It provided clear evidence that the lowest component of the S= 10 multiplet is M=± 10, with M=± 9 separated by about 14 K. If this separation is attributed to a zero field splitting described by the spin hamiltonian (1), this corresponds to D= 0.74 K. This means that the separation between the lowest M=± 10 and the highest M= 0 level is calculated as 100 D= 74 K. This value can be considered as an estimation of the anisotropy barrier which a spin must undergo at low temperature, as it will be more clearly stated below. The anisotropy is of the Ising type, which means that the preferred spin orientation is parallel to the tetragonal axis of the cluster.

An extention of the method used for relating the zero field splitting parameters for pairs to those of the individual ions [18] suggests for the ground S= 10 state:

$$D_{10} = a\,D_{Mn(III)} + b\,D_{Mn(IV)} \quad (5)$$

where a and b are numerical coefficients, which can be calculated by spin projection techniques. Assuming that the ground state is as schematized in Figure 4 and that $D_{Mn(III)}$ is much larger than $D_{Mn(IV)}$ the former is calculated to be -4 cm^{-1}, a value in line with those usually observed in simple manganese(III) complexes and with the distortion of the coordination environment for these ions. Therefore the observed large zero field splitting of the S= 10 state appears to be justified by the single ion contribution.

If the barrier is 74 K the magnetic anisotropy is expected to be very high at low temperature. In fact, in low field, only the lowest M= ± 10 levels are populated. Their splitting is $\pm 10\ g\ \mu_B H_{\parallel}$ when the field is parallel to the symmetry axis of the cluster, and zero when the field is perpendicular to the axis. This prediction was confirmed by single crystal measurements which showed that the susceptibility is very high parallel to the tetragonal axis and very close to zero orthogonal to that. The magnetization must overcome a high barrier and its relaxation becomes very slow at low temperature. The temperature dependence of the relaxation time between 2 and 10 K, follows the exponential behavior expected from eq. (3), with A/k= 61 K. The preexponential factor $\tau_0 = 2.1 \times 10^{-7}$ s, is much longer than observed for normal ferromagnets. In the range 4-10 K the relaxation time was measured by ac susceptibility measurement techniques, while at lower temperatures it was measured by saturating the magnetization in a field, switching it off and measuring the magnetization at different times. At 2 K the relaxation time corresponds to two months!

The very slow relaxation of the magnetization gives rise to hysteresis effects [23], as shown in Figure 5. Hysteresis effects are usually observed in bulk magnets as a consequence of the restricted motion of the domain walls. In this case there is no evidence of magnetic order above 50 mK. Further evidence from this comes from the possibility of observing slow relaxation by dissolving the clusters in solvents and measuring the susceptibility in glassy matrices. Similar results were reported by Hendrickson by dissolving the clusters in polymer matrices [6]. In this case the hysteresis is not due to movements of domain walls, which are not present, but to the very slow relaxation of the magnetization associated with the magnetic anisotropy

On the basis of the EPR estimate the barrier is 74 K, while the fit of the relaxation time suggests 61 K. More accurate HF-EPR spectra now show that the fine pattern is not constant, suggesting that eq (1) is not completely accurate for describing the energies of the M components of S= 10. In fact it is known that the spin hamiltonian approach for a spin S must in principle be extended up to terms of degree 2S in the spin operators. Therefore for S= 10 one might include terms up to twentieth degree in the spin operators. In order to be realistic and avoid an awful overparameterization one can include only quartic terms, which have been used for the interpretation of the EPR spectra of large spins. The simplest possible form is:

$$H= DS_z^2+ A_4 S_z^4+ B_4 (S_x^4+S_y^4) \qquad (6)$$

The fit of the spectra is in progress. Here we only refer to these data to show that the inclusion of quartic parameters may in principle give a better agreement between the barrier calculated with the EPR data and that coming from the experimental fit of the relaxation time.

The relaxation time of the magnetization has been justified [24] with a theoretical treatment which can be considered as an extention of the Orbach process used for simple paramagnets. In the presence of a weak field the spins will be in the lowest M= -10 state, and in order to relax to the M= + 10 state they will have to climb all the eigenstates M, up to that of highest energy M= 0, as shown in Figure 6. The theoretical treatment suggests that at low temperature the pre-exponential factor τ_0 of eq. (4) is inversely proportional to S^6, and directly proportional to the third power of the energy barrier, suggesting that the slow relaxation in Mn12ac is certainly determined by the large spin value, and by the Ising type anisotropy.

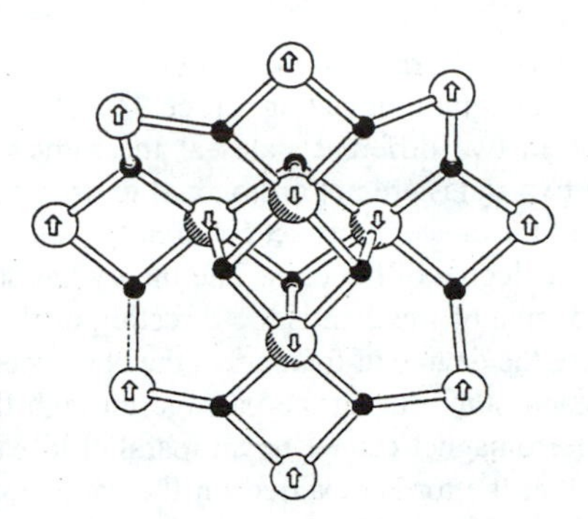

Figure 4. Sketch of the structure of Mn_{12} with preferred spin orientation in the ground state.

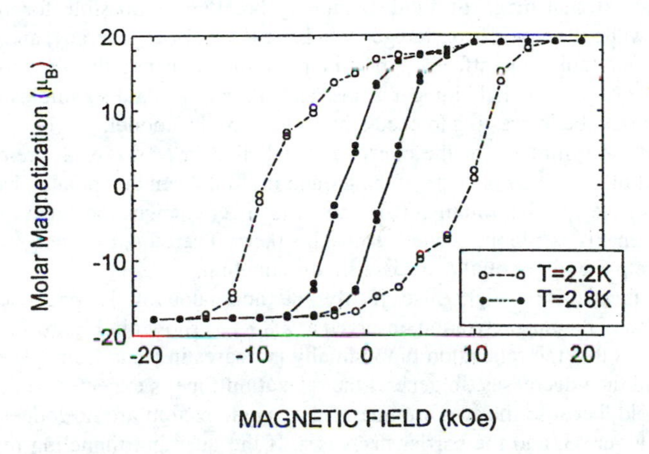

Figure 5. Magnetic hysteresis of Mn_{12}ac clusters.

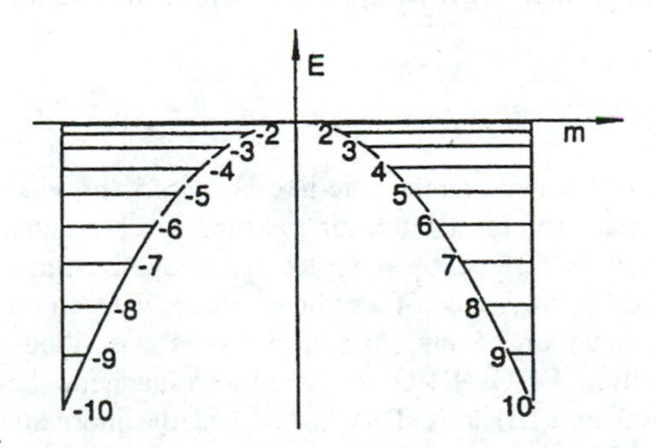

Figure 6. Scheme of the ladder of spin levels belonging to the zero field split ground S= 10 multiplet.

Below 2 K the relaxation rate of the magnetization becomes independent of temperature, suggesting a quantum tunneling process [25]. It is well known that a classical object cannot be in two different states at the same time. However quantum objects can, provided the two states are not orthogonal to each other. In order to observe such a behavior the spins of the cluster must act coherently, in such a way that the particle can be described by a single degree of freedom. The theory has so far been developed for ferromagnets, where the degree of freedom is the direction of the magnetic moment, and for antiferromagnets, where the degree of freedom is the Néel vector. In a ferromagnet the spins remain parallel to each other during the passage through the barrier. On the other hand the spins in an antiferromagnet cannot be antiparallel to each other while crossing the barrier. The result is that the torque exerted on the canted spins is stronger and the tunneling time is much shorter for antiferromagnets than for ferromagnets. thus making the observation of quantum tunneling easier for the former than for the latter. A ferrimagnet is expected to be more similar to an antiferromagnet.

In order to observe quantum tunneling a symmetry breaking is needed in the absence of an external magnetic field. Symmetry breaking is possible for systems with integer spin, while those with non-integer spin have Kramers degeneracy, and cannot give rise to quantum tunneling effects. Since in principle it is not difficult to synthesize clusters with integer and half integer spins, and already several examples of both are available, it would be interesting to check the validity of this model.

When the symmetry of the cluster is axial, like in Mn12ac, the removal of the degeneracy of the $\pm$ 10 components of the ground multiplet can be operated by the quartic components of the spin hamiltonian (6). The term in A_4 changes the separation between the $\pm$M components, while the second term splits them. Therefore the analysis of the EPR spectra can help shed light on the mechanism of tunneling.

This mechanism might also justify the anomalous field dependence of the relaxation time of the magnetization observed at 2.8 K, as shown in Figure 7 .
The anomaly is that the relaxation time initially increases in weak fields, goes through a maximum and then decreases. In general the relaxation time is expected to decrease with an applied field, because the two minima of the magnetization are no longer equivalent, as shown in Figure 3, and the barrier decreases. If the quantum tunneling mechanism is assumed to be persistent at 2.8 K between the +2 and -2 components of the ground multiplet a weak applied field would be expected to quench it, thus decreasing the magnetic relaxation time. In a larger field the thermally activated mechanism would take over.

Iron-Oxo Clusters

The other major class of superparamagnetic-like clusters is provided at the moment by iron oxo clusters. Perhaps the most interesting cluster is represented by ferritin the iron storage protein [26]. It is formed by a proteic part, apoferritin, which consists of a segmented protein shell in the shape of a hollow sphere, with an outer diameter of 12.5 nm and an inner diameter of 7.5 nm. *In vivo* the sphere is filled with ferrihydrite, of approximate composition $Fe_2O_3.9H_2O$. It is antiferromagnetic below 240 K, and it comprises up to 4,500 iron(III) ions. Due to the size the inorganic core behaves as a superparamagnet, and Mössbauer spectroscopy has long been used in order to measure the size of the inorganic core, taking advantage of the relation between the anisotropy barrier, experimentally determined, and the volume of the particle. Two iron(III) clusters

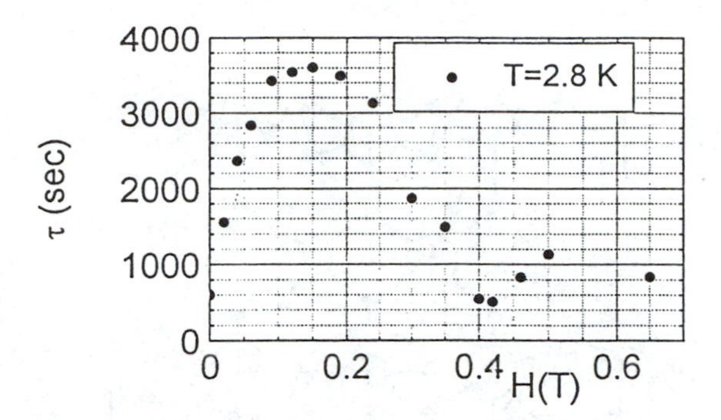

Figure 7. Relaxation time of the magnetization of $Mn_{12}ac$ at 2.8 K in an external magnetic field.

have also been shown by Mössbauer spectroscopy to give rise to slow relaxation of the magnetization at low temperature. They are $[Fe_{11}O_6(OH)_6(O_2CPh)_{15}]$.6THF, Fe11, [5], and $[Fe_{16}MO_{10}(OH)_{10}(O_2CPh)_{20}]$, Fe16M, where M= Mn^{2+}, Co^{2+}, Fe^{2+}, [27]. Both clusters are rather three-dimensional in structure and they show antiferromagnetic coupling, with a low ground spin state, presumably S= 1/2 for Fe11 and S= 5/2 for Fe16Mn. At 4.2 K Fe11 shows significant broadening in the Mössbauer spectra, an indication of slow relaxation of the magnetization, but no magnetic hyperfine lines are observed even at 1.8 K. On the other side Fe16Mn exhibits relaxation phenomena at about 7 K, with magnetic hyperfine lines developing below 6 K, superimposed on a broad absorption envelope at the center of the spectrum. Detailed analysis of the spectra with and without an applied magnetic field lead the authors to suggest [27-29] that the onset of superparamagnetic behavior occurs for iron clusters Fe_n in the vicinity of $n \geq 10$, and it is more readily observed in pseudo three-dimensional than in pseudo two- and one-dimensional clusters. However, given the limited range of temperatures in which they could perform experiments they were not able to experimentally test the exponential behavior of the relaxation time, nor had they an estimation of the anisotropy barrier. This was only assumed to be determined by the shape effects of the clusters.

$[(tacn)_6Fe_8(\mu_3\text{-}O)_2(\mu_3\text{-}OH)_{12}]^{7-}$, Fe8, has the structure [30] shown in Figure 8. The cluster is flat, with the iron ions lying approximately on a plane. Therefore a cluster like this must be considered as an example of a two-dimensional rather than of a threee-dimensional lattice. Magnetic susceptibility measurements provided evidence [31] of a ground S= 10 state, which can be justified by the spin arrangment depicted in Figure 8 HF-EPR spectra provided evidence of the zero field splitting of the ground state D/k= -0.27 K [32]. Therefore Fe8 has important similarities with Mn12ac, with the same ground state, and a negative zero field splitting. The latter however is much smaller than in the manganese cluster. The barrier for the orientation of the magnetization is calculated to be 27 K, which should start to produce effects in ac susceptibility measurements below 3 K, if eq. 1 works with a similar τ_0 parameter as observed in Mn12. Further the system should show evidence of slow relaxation in Mössbauer spectroscopy at much higher temperatures, and therefore provide for the first time the opportunity to detect the superparamagnetic behavior with both the techniques.

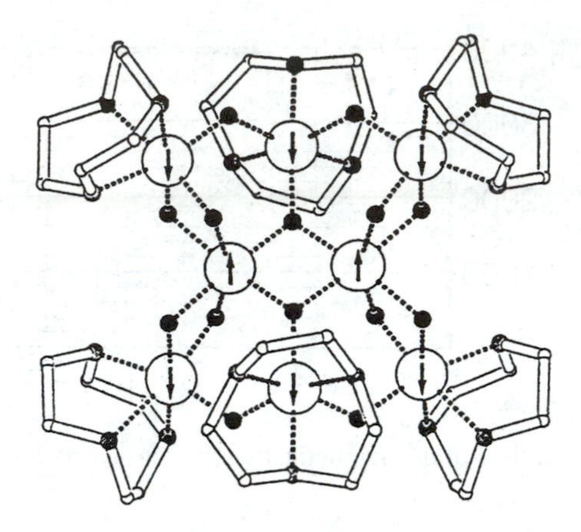

Figure 8. Sketch of the structure of Fe8 with preferred spin orientation in the ground state.

These predictions have been confirmed by experiments which showed that for both ac susceptibility and Mössbauer relaxation times eq 1 is closely obeyed [32]. The calculated value of the barrier corresponds well to that obtained from the EPR spectra. Since these are only sensitive to molecular properties the anisotropy of the cluster cannot be due to shape effects, but it has most probably a single ion origin. In fact proceding as outlined above for the manganese cluster it is possible to calculate, at the order of magnitude accuracy, a single ion contribution of -0.75 K, in agreement with the values often found in simple iron(III) complexes.

References

1. Powell, A.K.; Heath, S.L.; Gatteschi, D.; Pardi, L.; Sessoli, R.; Spina, G.; Del Giallo, F.; Pieralli, F. *J. Am. Chem. Soc.*, **1995,** *117*, 2491
2. Nakamura, N.; Inoue, K.; Iwamura, H. *Angew. Chem. Int. Ed. Engl.*,**1993,** *105*, 900
3. Fujita, I.; Teki, Y.; Takui, T.; Kinoshita, T.; Itoh, K.; Miko, F.; Sawaki, Y.; Iwamura, H.; Izuoka, A.; Sugawara, T.*J. Am. Chem. Soc.*, **1990,** *112*, 4074
4. Utamparaya, S.; Rajca, A.; Xu, J. *J. Am. Chem. Soc.*, **1991,** *113*, 9233
5. Gorun, S.M.; Papaefthymiou, G.C.; Frankel, R.B.; Lippard, S.J. *J. Am. Chem. Soc.*, **1987,** *109*, 3337
6. Eppley, H.J.; Tsai, H.-L.; de Vries, N.; Folting, K.; Christou, G.; Hendrickson, D.N. *J. Am. Chem. Soc.*, **1995,** *117*, 301
7. Awschalom, D.D.; Di Vincenzo, D.P. *Phys. Today*, **1995,** 40
8. Leggett, A.J. *Phys. Rev. B,* **1984**, *30*, 1208.
9. Carlin, R.L. *Magnetochemistry*, Springer Verlag, Berlin, 1986.
10. Kahn, O. *Molecular magnetism*, VCH, Weinheim, 1993.
11. Gatteschi, D.; Pardi, L. *Gazz. Chim. It.,* **1993,** *123*, 23
12. Silver, B.L. *Irreducible Tensor Methods*, Academic Press, New York, 1976.
13. Gatteschi, D.; Pardi, L. in *Magnetochemistry*, O'Connor, C.J. (Ed), World Scientfic, Singapore, 1993.
14. Caneschi, A.; Cornia, A.; Lippard, S.J. *Angew. Chem. Int. Ed. Engl.*, **1995,** *107*, 511
15. Morrish, A.H., *The Physical Principles of Magnetism*, (1966), J. Wiley & Sons, New York

16. Néel, L. *Ann. Geophys.*, **1949,** *5*, 99
17. Gider, S.; Awschalom, D.D.; Douglas, T.; Mann, S.; Chaparala, M. *Science*, **1995**, *268*, 77
18. Bencini, A.; Gatteschi, D. *EPR of Exchange Coupled Systems*, Springer, Berlin, 1990
19. Lis, T. *Acta Crystallogr. Sect. B.*, **1980**, *36*, 2042
20. Sessoli, R.; Tsai, H.L.; Schake, A.R.; Wang, S.; Vincent, J.B.; Folting, K.; Gatteschi, D.; Christou, G.; Hendrickson, D.N. *J. Am. Chem. Soc.*, **1993**, *115*, 1804
21. Hennion, M.; Pardi, L.; Mirabeau, I.; Suard, E.; Sessoli, R.; Caneschi, A. work in progress.
22. Barra, A.-L-; Caneschi, A.; Gatteschi, D.; Sessoli, R. *J. Am. Chem. Soc.*, **1995**, *117*, 8855.
23. Sessoli, R.; Gatteschi, D.; Caneschi, A.; Novak, M.A. *Nature*, **1993**, *365*, 141
24. Villain, J.; Hartmann-Boutron, F.; Sessoli, R.; Rettori, A. (), *Europhys. Lett.*, **1994**, *27*, 159.
25. Paulsen, C.; Park, J.G.; Barbara, B.; Sessoli, R.; Caneschi, A. *J. Magn. Magn. Mater.*, **1995**, *140-144*, 379
26. Mann, S.; Frankel, R.B. (1989), in Mann, S.; Webb, J., Williams, R.J.P. (eds), *Biomineralization. Chemical and Biochemical Perspectives*, VCH, Weinheim, 389.
27. Papaefthymiou, G.C. *Phys. Rev. B*, **1992**, *46*, 10366
28. Micklitz, W.; Lippard, S.J. *J. Am. Chem. Soc.*, **1989**, *111*, 6856
29. Papaefthymiou, G.C. (1993) *Nanophase and Nanocomposite Materials*, Kormarmeni, S.; Parker, I.C.; Thomas, G.J. (eds), Mat. Res. Soc. Symposium Proceedings, 286, 67.
30. Wieghardt, K.; Pohl, K.; Jibril, I.; Huttner, G. *Angew. Chem., Int. Ed. Engl.* **1984**, *24*, 77.
31. Delfs, C.; Gatteschi, D.; Pardi, L.; Sessoli, R.; Wieghardt, K.; Hanke, D. *Inorg. Chem.*, **1993**, *32*, 3099
32. Barra, A.-L.; Debrunner, P.; Gatteschi, D.; Schulz, C.E.; Sessoli, R. submitted for publication

Chapter 11

Magnetostructural Correlations in Dinuclear Cu(II) and Ni(II) Complexes Bridged by μ_2-1,1-Azide and μ_2-Phenoxide

L. K. Thompson[1], S. S. Tandon[1], M. E. Manuel[1], M. K. Park[1], and M. Handa[2]

[1]Department of Chemistry, Memorial University of Newfoundland, St. John's, Newfoundland, A1B 3X7, Canada
[2]Department of Materials Science, Interdisciplinary Faculty of Science and Engineering, Shimane University, Nishikawatsu, Matsue 690, Japan

Magnetostructural correlations for dinuclear copper(II) complexes bridged by a combination of μ_2-1,1-azide and a diazine, and macrocyclic complexes bridged by phenoxide are discussed. For the azide bridged complexes the magnetic properties change from ferromagnetic to antiferromagnetic at $\approx 108°$ ($\Delta 2J \approx 46$ cm^{-1}/deg.), with evidence to suggest a temperature variable exchange integral. The phenoxide bridge angle/exchange integral correlation does not extrapolate to a normal crossover from antiferromagnetic to ferromagnetic behavior, and π conjugation routes within the macrocyclic ligand are implicated in the total exchange process. The nickel/phenoxide case conforms to the expected oxygen bridged correlation with $-2J=0$ cm^{-1} at $\approx 97°$.

Intramolecular magnetic exchange interactions in dinuclear transition metal complexes bridged by simple ligands have fascinated coordination chemists and physicists, and theorists, since the pioneering work of Bleaney and Bowers in the early 1950's on the dimeric copper acetate molecule (*1*). The variable temperature magnetic properties of simple dinuclear copper(II) complexes can be described by the Bleaney-Bowers equation (equation 1; terms have their usual meaning with χ expressed per mole of copper, and based on the exchange Hamiltonian $H = -2JS_1.S_2$) (1). Single atom bridging ligands e.g. hydroxide, have been studied widely,

$$\chi_m = \frac{N\beta^2 g^2}{3k(T-\theta)}[1+1/3\exp(-2J/kT)]^{-1}(1-\rho) + \frac{[N\beta^2 g^2]\rho}{4kT} + N\alpha \text{-----} (1)$$

particularly for copper(II) complexes, because of the ease with which such

0097–6156/96/0644–0170$15.00/0

complexes form in aqueous solvent media, even in the absence of base. The now classical study by Hatfield *et al* (*2*) on simple dihydroxo-bridged dicopper(II) complexes, with $d_{x^2-y^2}$ copper ground states, revealed the importance of bridge angle as a fundamental factor controlling spin exchange, with a demonstrated linear relationship between Cu-OH-Cu bridge angle and exchange integral. The bridge angles ranged from 95.6° to 104.1°, which fortuitously spanned the critical angle at which the magnetic properties changed from antiferromagnetic to ferromagnetic (97.5°). These experimental observations were consistent with extended Hückel MO calculations, which predicted that accidental orthogonality (the situation in which the symmetric and antisymmetric MO combinations involving the metal magnetic orbitals and appropriate symmetry bridge orbitals, in the triplet state, are degenerate) would occur at 92° (*3,4*). Subsequent studies on alkoxide bridged dicopper(II) complexes revealed a similar situation (*5,6*). While numerous examples of open chain complexes with hydroxide or alkoxide bridges were known, with a reasonable range of bridge angles, this was not true for phenoxide bridged systems, and in particular those involving macrocyclic ligands. Macrocyclic phenoxide bridged dinickel(II) and dicopper(II) complexes will be discussed in this report, along with consideration of inductive contributions to exchange.

The step from simple oxygen bridged dicopper(II) complexes to μ_2-1,1-azide bridged analogues followed logically, and a few examples of systems with small Cu-N_3-Cu angles (< 106°) were reported in the early 1980's (*3,7,8*). These were found to be exclusively ferromagnetic, and the ability of the μ_2-1,1-azide to propagate ferromagnetic coupling was attributed to a *spin polarization* effect, involving an interaction between the two copper d_{xy} orbitals and the π_g MO on the azide (*8*). The theoretical argument to support this explanation arose from the fact that in this angle range the energy splitting between the two molecular orbitals constructed from the d_{xy} magnetic orbitals in the triplet state, and appropriate symmetry ligand orbitals (Δ), according to extended Hückel calculations (*3*), was considered to be very small, and so any antiferromagnetic term (J_{AF}) would be small and not compensate for or exceed any inherent ferromagnetic term (J_F)(equation 2; S is the overlap integral and C the two-electron exchange integral for a system with two thermally populated spin levels).

$$J = J_{AF} + J_F \qquad J_{AF} = -2\Delta S;\ J_F = 2C \tag{2}$$

This lead to the suggestion that perhaps all 1,1-azide bridged complexes would be ferromagnetic, regardless of angle. This chapter will explore the previously uncharted antiferromagnetic realm of the μ_2-1,1-azide bridge, and also the fundamental question of the temperature dependence of the exchange integral itself, or more appropriately the result of Δ being temperature dependent.

Phenoxide Bridged Macrocyclic Dinickel(II) Complexes

Phenoxide bridged macrocyclic complexes derived by template condensation of diformylphenols and diamines have been produced in large numbers since the original report by Robson in 1970 (*9*) describing the dinuclear species $[M(L)]^{2+}$

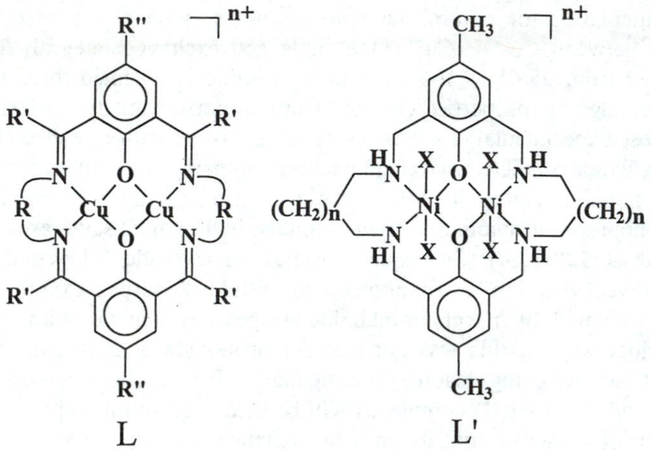

L L'

R=$(CH_2)_n$ (n=2-4), C_6H_4, C_6F_4,$C_2(CN)_2$
R'=H, CH_3
R"=CH_3,t-Bu, C_3F_7

n= 2,3

Figure 1. Macrocyclic dicopper and dinickel complexes.

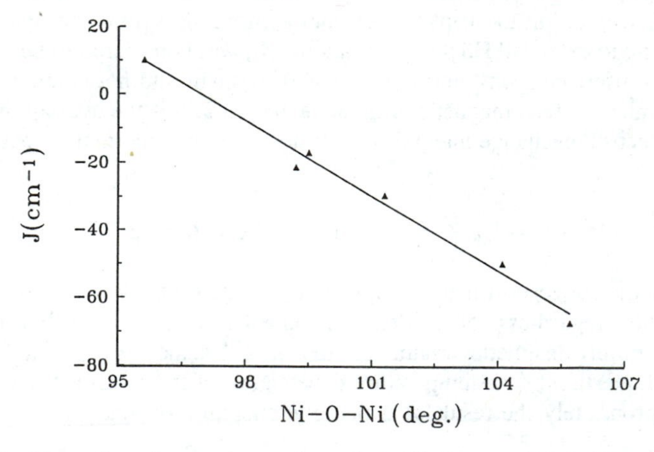

Figure 2. Plot of exchange integral against Ni-$O_{phenoxide}$-Ni bridge angle for macrocyclic dinickel(II) complexes.

(M=Cu,Ni,Co,Fe,Mn) (Figure 1). However, despite many studies describing both structures and magnetic properties of such systems, no meaningful magnetostructural correlations relating exchange interaction with M-OPh-M angle were reported until recently. In order to develop such a correlation a series of structurally related complexes is required, within which factors contributing to the exchange process other than the phenoxide bridge angle are kept essentially constant. A series of dinickel(II) complexes of the related saturated macrocyclic ligand L', with phenoxide bridge angles in the range 99-106° (Figure 1, n=3), which were reported recently (*10*), showed a linear relationship between Ni-OPh-Ni angle and exchange integral and a crossover from antiferromagnetic to ferromagnetic behavior at ≈97°, consistent with the dihydroxy-bridged copper systems (Figure 2). Further support for this correlation comes from the complex $[Ni_2(L^{2,2})(CH_3COO)_2].10H_2O$ (Ni-OPh-Ni 95.6°, J=10.1 cm^{-1}; $L^{2,2}$=L'(Figure 1; n=2)) (*11*), which fits exactly on the line, and extends the correlation over a range of angles of 10°. The similarity between dicopper(II) and dinickel(II) complexes, where exchange depends only on the bridging oxygen, is considered to be reasonable despite the fact that for nickel(II) both the $d_{x^2-y^2}$ and d_{z^2} orbitals are magnetic orbitals. However the involvement of symmetric d_{z^2} orbitals in exchange coupling will be minimal in complexes of this sort where the exchange process is dominated by equatorial interactions.

Phenoxide Bridged Macrocyclic Dicopper(II) Complexes

Numerous dicopper(II) complexes of unsaturated ligands L (e.g. n=2-4, X=H_2O, Cl, Br, I, ClO_4) (Figure 1) have been synthesized and studied, mostly with square-pyramidal copper centers. Control over the dinuclear center dimensions, e.g. Cu-Cu separation and Cu-O-Cu angle, is somewhat limited, but varying the linker group size (R) in general leads to increased angles and metal-metal separations as the chain length increases (*12-18*) . The presence of electron withdrawing ligands (e.g. X=Cl, Br, I) were demonstrated to exert an inductive effect, which modulated the exchange process, leading to reduced antiferromagnetic coupling in comparison with structurally related complexes, and electron withdrawing groups on the macrocyclic ligand itself also had a similar effect (*13,17,18*). By choosing related complexes with no inductive perturbations, and consistent copper ion magnetic ground state ($d_{x^2-y^2}$), and with minimal distortions of e.g. the copper ion coordination sphere and the Cu_2O_2 dinuclear center itself, a series of suitable complexes for a magnetostructural correlation can be selected. A typical example, $[Cu_2(L)(H_2O)_2](BF_4)_2$, is shown in Figure 3 (L(R=$(CH_2)_2$, R'=CH_3, R"=CH_3), Cu-O-Cu 98.8(4)°, Cu-Cu 2.997(3) Å; -2J = 689 cm^{-1}). Figure 4 illustrates a plot of -2J against averaged Cu-OPh-Cu angle for this series (*12,14,17,19,20,21*) (Thompson, L.K.; Mandal S.K.; Tandon, S.S.; Bridson, J.N.; Park, M.K., *Inorg. Chem.*, in press.), with the best fit line. Although the correlation is not as precise as in the nickel case the trend is clear, and assuming that a linear relationship is appropriate the data extrapolate to an angle of ≈77° for -2J=0 cm^{-1}. This is clearly inconsistent with the dinickel case, and also the copper/hydroxide case, and begs the question as to why -2J values are so high at angles close to the normal experimental angle of accidental orthogonality for a dioxygen bridged system.

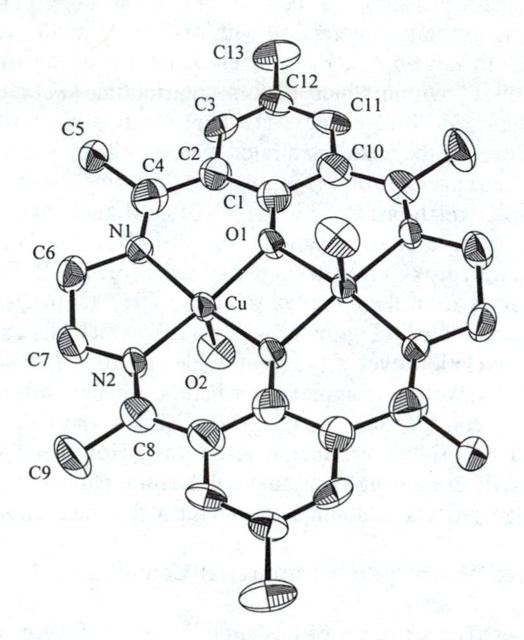

Figure 3. Structural representation for $[Cu_2(L)(H_2O)_2](BF_4)_2$.

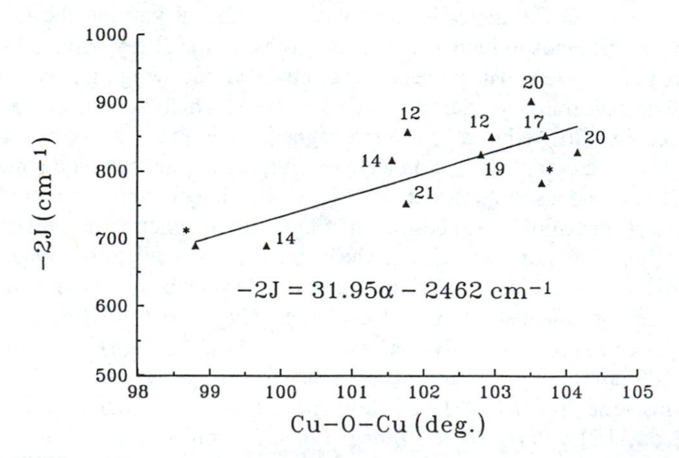

Figure 4. Plot of exchange integral (-2J) against Cu-$O_{phenoxide}$-Cu bridge angle for macrocyclic dicopper(II) complexes (data from refs. 12,14,17,19,20,21 (▲); Thompson, L.K. et al, Inorg. Chem., in press (★)).

A test of the validity of the relationship is obtained by considering comparable complexes with electron withdrawing ligand groups. The square-pyramidal complexes $[Cu_2(L)X_2]$ (L(R=$(CH_2)_3$, R'=H, R"=CH_3); X=Cl, Br) in which the halogens are axially bonded to the copper centers have much lower -2J values than predicted based on their Cu-O-Cu angles (*17*). The pseudo-octahedral complexes $[Cu_2(L)(ClO_4)_2]$ (L (R=C_6F_4, R'=H, R"=tBu) and $[Cu_2(L)(ClO_4)_2]$ (L (R=C_6F_4, R'=H, R"=C_3F_7), which have perfluoro-phenyl linker groups (R), both have small Cu-O-Cu angles (99.2(1)°, 100.8(1)° respectively)(*18*), and are more weakly antiferromagnetically coupled (-2J=581 cm^{-1}, 526 cm^{-1} respectively) than would be expected based on their bridge angles, consistent with the presence of electron withdrawing fluorine atoms on the macrocyclic ligand. The complex $[Cu_2(L)](ClO_4)_2.3H_2O.CH_3OH$ (L(R=$C_2(CN)_2$, R'=H, R"=CH_3) has peripheral CN groups bound much more closely to the copper centers (Figure 5), and although it is still strongly antiferromagnetically coupled, the coupling is weak in comparison with all the other complexes in this class. The five-membered chelate rings subtended at the dicyano-alkene bridges would dictate small phenoxide bridge angles (≈ 99°), consistent with the perfluoro-phenyl complexes. That this complex is much more weakly coupled is, no doubt, the result of a cyanide inductive effect at the copper centers, attenuating the exchange. While no x-ray structure is available for this, or related compounds, a neutral derivative produced by recrystallization from acetone (Figure 5) in which a most unusual conjugated addition of acetone has occurred across two imine groups, is still antiferromagnetically coupled despite having Cu-OPh-Cu angles of 92.0(2)° and 92.8(2)°. The complex is severely bent along the phenoxide O-O axis with an angle between the CuN_2O_2 mean planes of 151.1°, and pronounced pyramidal distortion at the oxygen bridges themselves (MacLachlan, M.J.; Park, M.K.; Thompson, L.K., unpublished results). These factors, of necessity, would diminish antiferromagnetic coupling, strengthening the argument that an unusual exchange situation prevails with these complexes. A common feature in all cases (Figures 1,5) involves π-conjugated macrocyclic fragments within each aromatic half of the molecule. These provide additional six-bond pathways through which exchange coupling could occur between the copper centers. This would involve a long exchange distance (<8 Å), but this is not unreasonable given reports of significant antiferromagnetic exchange between square-pyramidal copper(II) centers separated by conjugated, aromatic ligand groups with six bonds (7.6 Å; -2J = 21-26 cm^{-1})(*22*), eight bonds (≈11.8 Å; -J = 36-210 cm^{-1})(*23*), and nine bonds (11.25 Å; -2J= 140 cm^{-1})(*24*). If such a through ligand exchange route prevailed in all the macrocyclic dicopper(II) complexes with unsaturated ligands, through a broad range of phenoxide angles, then it would add a roughly constant component to the overall exchange process, and lead to a situation where, for low phenoxide bridge angles, the actual exchange would be more negative (i.e. more antiferromagnetic) than expected.

The inevitable comparison with the dihydroxide case raises the question of whether replacing hydrogen on hydroxide by slightly more electronegative carbon would affect the exchange process. Kahn has expressed the view that this would mainly affect the oxygen $2p_y$ orbital and make e.g. the $3d_{xy}/2p_y$ molecular orbital less antibonding, with the result that a larger energy gap (Δ) between the two singly

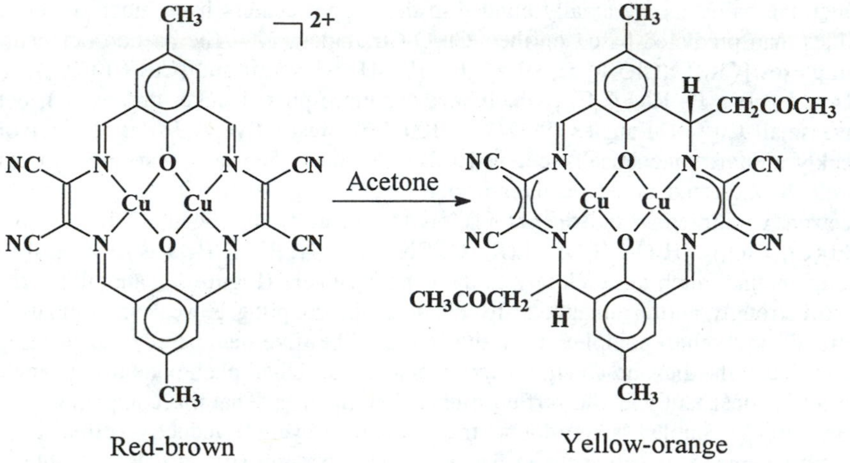

Acetone

Red-brown

Yellow-orange

$g = 2.00(2)$, $-2J = 465(4)\ cm^{-1}$, $\rho = 0.012$, $TIP = 45 \times 10^{-6}$ emu.

$g = 2.024(3)$, $-2J = 25.2(3)\ cm^{-1}$, $\rho = 0.034$, $TIP = 57 \times 10^{-6}$ emu.

Figure 5. Macrocyclic dicopper complex based on diaminomaleonitrile and its di-acetone adduct (data from equation 1).

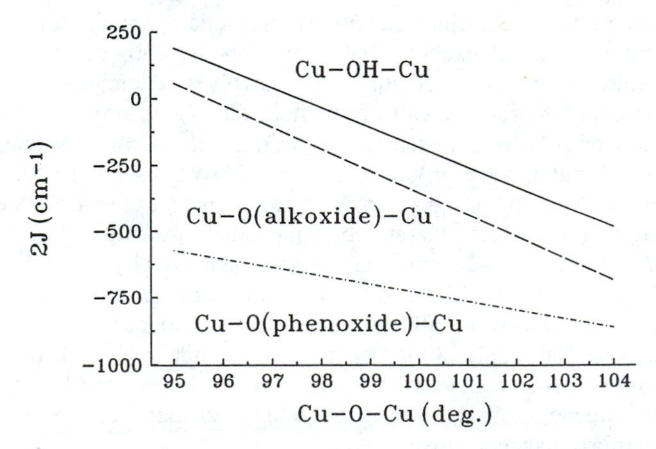

Figure 6. Plots of exchange integral (-2J) against bridge angle for hydroxide, alkoxide and phenoxide bridged dicopper complexes.

occupied molecular orbitals in the triplet state would result, leading to an enhanced antiferromagnetic contribution to overall exchange (*25*). However, while dialkoxide bridged dicopper(II) systems are consistent with this argument, and have slightly more negative exchange integrals for comparable bridge angles, the phenoxide case is clearly anomalous. Figure 6 illustrates idealized plots for the dihydroxide (*1*) and dialkoxide (*5,6*) cases, and also the diphenoxide system, where it is clear that absolute -2J values for the phenoxide bridged complexes are very much larger. A test of this proposed two path exchange model would rest with the study of a series of dicopper complexes with only saturated macrocyclic ligand groups. Such a study is currently underway in our laboratory.

μ_2-1,1-Azide Bridged Dicopper(II) Complexes

The dinuclear center dimensions in a single atom bridged dinuclear complex cannot effectively be controlled if the primary ligand groups have no bridging interaction themselves. The early studies on dihydroxide bridged dicopper(II) complexes (*2*) and di μ_2 1,1-azide bridged dicopper(II) complexes (*3*) did not involve any such constraint, and the dinuclear center dimensions were largely left to the whims of nature. The design of a suitable class of dinucleating ligand, which will allow the group of interest to act as a bridge, and generate a large range of bridge angles is not a trivial matter. Ideally it should have a minimal effect other than to provide the right geometrical template. A series of tetradentate diazine ligands based on thiadiazole, pyridazine and phthalazine (Figure 7) have proved to be very successful in this regard, and have specifically generated a series of μ_2-1,1-azide bridged complexes with a very large range of angles (98.3-124.1°)(*26,27*). The general structural representation (e.g. for 6,6,6 ligands) includes an equatorial bridge ($C=N_3$), and possible axial anionic bridge (D). The key geometric factors that are responsible for controlling the dinuclear center dimensions are the size of the diazine ring and the size of the chelate rings involving the peripheral donors (the numerical designation defines the sequence chelate ring, diazine ring, chelate ring). A comparison of hydroxide bridged complexes of this class of ligands (C=OH, D= anion) showed that for complexes of 5,6,5 ligands Cu-OH-Cu angles were in the range 115-126°, while for 6,6,6 ligands angles fell in the range 100-115° (bridge groups D influenced geometry as well)(*28*). A similar situation prevails for comparable azide complexes with angles in the range 107.9-124.1°(*26,27*), while smaller angles are achieved using the 7,5,7 ligand BMPTD. Despite the five-membered diazine ring, which perhaps would have created a large Cu-Cu separation (e.g. 3.4-3.5 Å in the case of typical 5,6,5 ligands), small Cu-N_3-Cu angles (98.3-105.9°) and Cu-Cu separations (3.07-3.14 Å) are found, which are considered to be a function of the large peripheral chelate ring size. Examples of two typical complexes ($[Cu_2(BMPTD)(\mu_2\text{-}N_3)(\mu_2\text{-}Cl)Cl_2]$ (Cu-N_3-Cu 105.9°) and $[Cu_2(PPD35Me)(\mu_2\text{-}N_3)Br_3(CH_3OH)]$ (Cu-N_3-Cu 122.5°) are illustrated in Figures 8 and 9 respectively.

The magnetic properties of these two complexes are completely different. ($[Cu_2(BMPTD)(\mu_2\text{-}N_3)(\mu_2\text{-}Cl)Cl_2]$ is very strongly ferromagnetically coupled (g=2.214(10), 2J=168(3) cm^{-1}), in keeping with the earlier work of Kahn, while

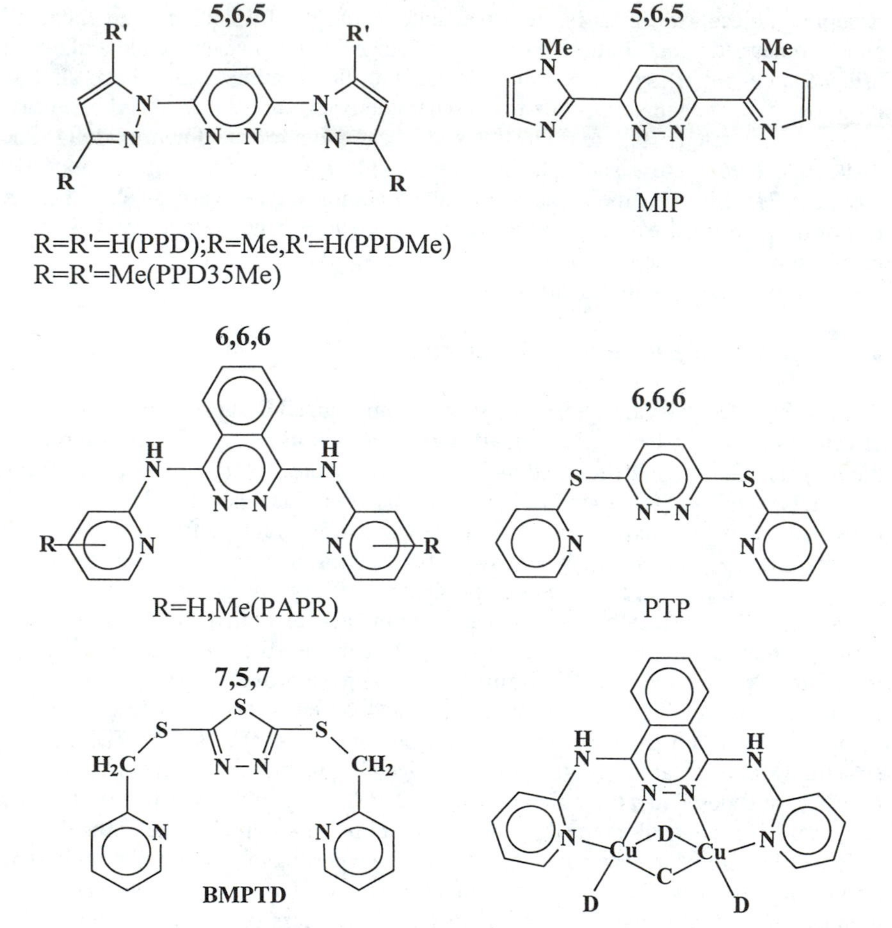

5,6,5

R=R'=H(PPD);R=Me,R'=H(PPDMe)
R=R'=Me(PPD35Me)

5,6,5

MIP

6,6,6

R=H,Me(PAPR)

6,6,6

PTP

7,5,7

BMPTD

Figure 7. Tetradentate (N_4) diazine ligands based on pyridazine, phthalazine and thiadiazole.

$[Cu_2(PPD35Me)(\mu_2\text{-}N_3)Br_3(CH_3OH)]$ is very strongly antiferromagnetically coupled (g=2.21(2), -2J=921(9) cm^{-1}), demonstrating for the first time the existence of an antiferromagnetic realm for the μ_2-1,1-azide bridge (*26,27*). A series of dicopper(II) complexes of various N_4 diazine ligands (Figure 7) was examined, all with the same magnetic ground state ($d_{x^2-y^2}$), and a linear correlation was established for the angle range 100-125° (Figure 10)(*26,27*). The linear equation (Figure 10) indicates a bridge angle of ≈106° for the crossover from antiferromagnetic to ferromagnetic behavior. However the correlation also includes the diazine bridge, and various estimates indicate that a reasonable exchange contribution for such a bridge at the point where $2J_{azide}$ = ZERO cm^{-1} would be of the order of $2J_{diazine} \approx -100\ cm^{-1}$. The complex $[Cu_2(BMPTD)(\mu_2\text{-}Cl)_2Cl_2]$ (Figure 11) contains two $d_{x^2-y^2}$ ground state copper centers bridged equatorially by the thiadiazole diazine, and so this is a

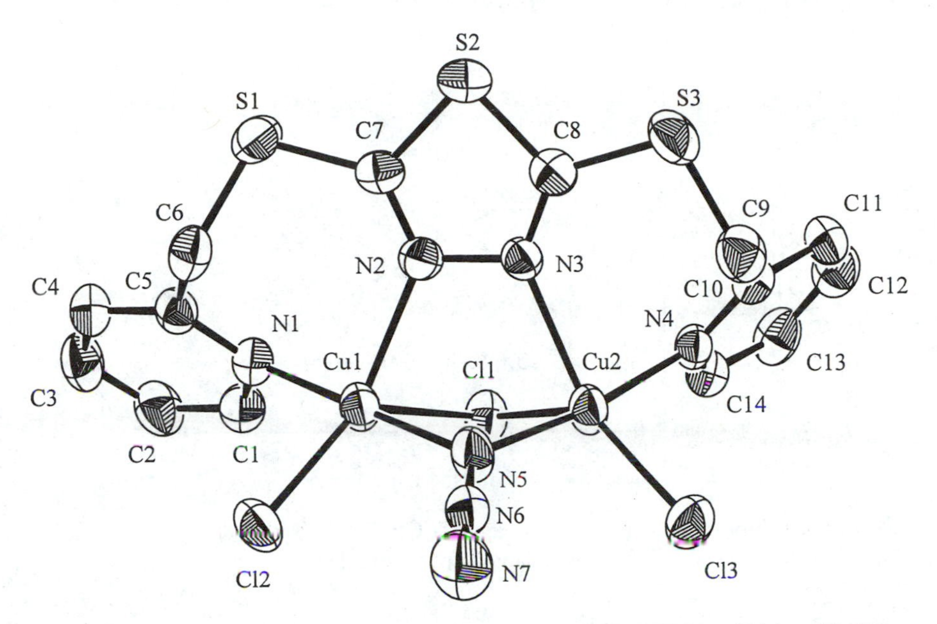

Figure 8. Structural representation for $[Cu_2(BMPTD)(\mu_2\text{-}N_3)(\mu_2\text{-}Cl)Cl_2]$ (Reproduced with permission for ref. 26. Copyright 1994, American Chemical Society).

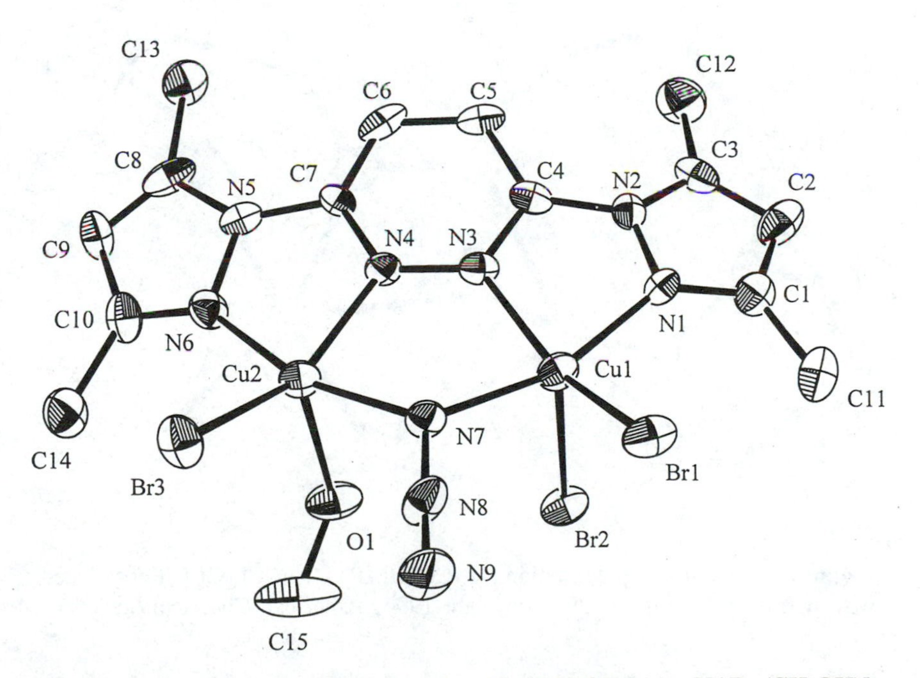

Figure 9. Structural representation for $[Cu_2(PPD35Me)(\mu_2\text{-}N_3)Br_3(CH_3OH)]$ (Reproduced with permission from ref. 27. Copyright 1995, American Chemical Society).

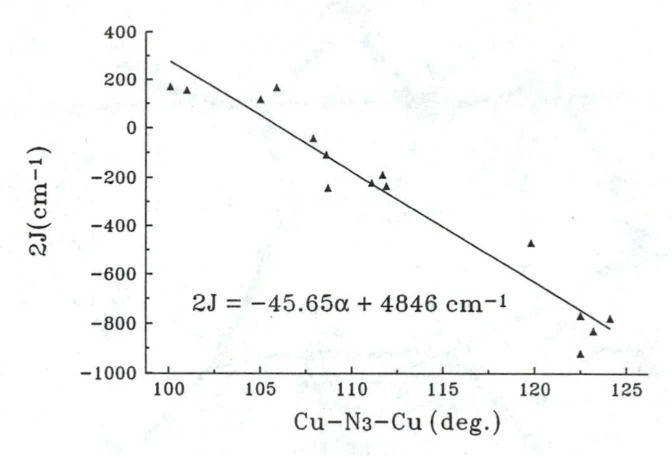

Figure 10. Plot of exchange integral against Cu-N_3-Cu bridge angle for a series of dicopper-diazine complexes (adapted from ref. 27).

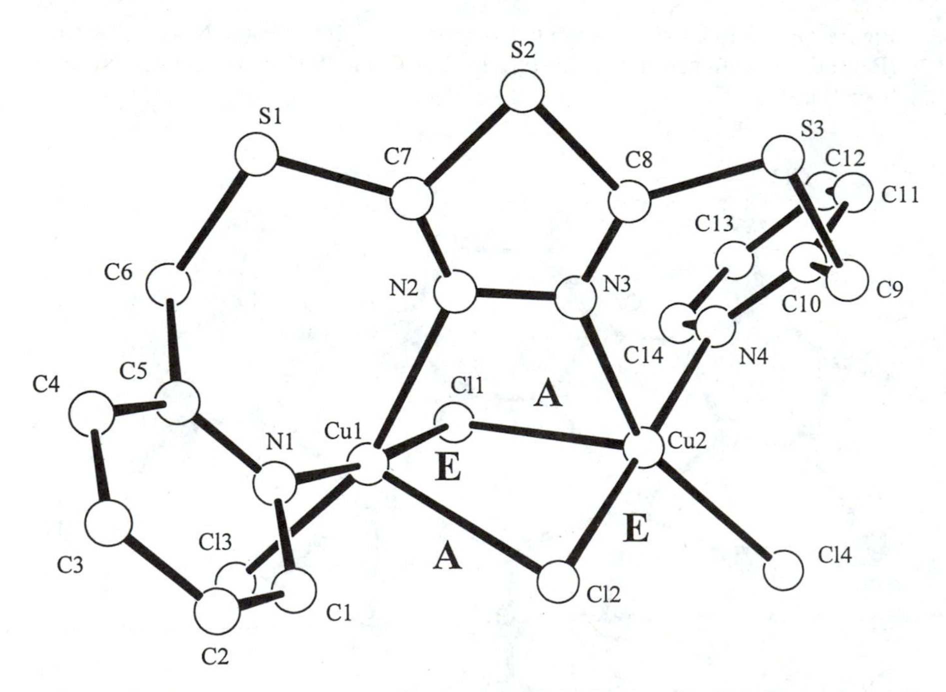

Figure 11. Structural representation for $[Cu_2(BMPTD)(\mu_2\text{-}Cl)_2Cl_2]$ (Reproduced with permission from ref. 29. Copyright 1994, American Chemical Society).

magnetically active bridge. However the axial/equatorial arrangement of the two chlorine bridges creates an orthogonal connection in both cases, thus eliminating these bridges as active superexchange bridges in an antiferromagnetic sense (the copper magnetic, unpaired electron lies in the $d_{x^2-y^2}$ orbital). Therefore, given the fact that any likely ferromagnetic exchange through these bridges will be insignificant in the overall exchange process, one can obtain an estimate of the exchange capacity of the thiadiazole bridge in an environment similar to that in the related azide complexes. This complex is weakly antiferromagnetically coupled ($g=2.075(7)$, $-2J=59.3(7)\ cm^{-1}$)(*29*). Related tetrachloro-complexes with exactly the same chlorine and diazine bridging arrangement also involve pyridazine ($-2J=131\ cm^{-1}$)(*30*) and phthalazine ($-2J=55.2\ cm^{-1}$(*28*), $124\ cm^{-1}$(*31*)) bridges. Therefore, assuming an average residual exchange due to the diazine of $-2J \approx 100\ cm^{-1}$, the angle at which the azide bridge changes from antiferromagnetic to ferromagnetic behavior is closer to 108°.

The slope of the azide correlation line (Figure 10) indicates that for every one degree change in azide bridge angle, assuming other factors are essentially constant, 2J changes by about 46 cm^{-1}. Complexes with J values close to the point at which the azide changes its magnetic behavior, which would have bridge angles close to 108°, might therefore exhibit net antiferromagnetism despite the fact that the azide was responsible for ferromagnetic coupling (equation 2). This should theoretically create a simple situation in which the net exchange effect would be the resultant of two antagonistic terms, and the system should obey the Bleaney-Bowers equation (*1,32*). In practice for the compounds that have been studied with $Cu-N_3-Cu$ angles $\approx 108°$ this turns out not to be the case. The complex $[Cu_2(PAPH)(\mu_2-N_3)Cl_3]$ (Figure 12) has a dinuclear center with two $d_{x^2-y^2}$ copper atoms bridged equatorially by the phthalazine and the 1,1-azide. The $Cu-N_3-Cu$ angle (107.9(2)°)(*26*) is close to the limiting azide angle for antiferromagnetic coupling, and so one might expect only a small -2J value if the sample exhibits net antiferromagnetism. The χ_{Cu} vs. temperature profile for this compound has a $\chi_{max.}$ at $\approx 27.5K$ (▲; Figure 13), but the susceptibility data do not fit the Bleaney-Bowers equation for a simple dinuclear copper complex. Nominally -2J is roughly proportional to $T_{\chi max.}$ ($\approx 1.6kT_{\chi max}\ cm^{-1}$) for a normal dinuclear complex, and a comparison of the susceptibility data for a system which obeys the Bleaney-Bowers equation (solid line in Figure 13) and has a comparable $T_{\chi max.}$, shows that the absolute χ values for $[Cu_2(PAPH)(\mu_2-N_3)Cl_3]$ are very much lower throughout the whole temperature range. Such a situation is also apparent with several other related complexes with azide angles close to 108° (Thompson, L.K.; Tandon, S.S. *Comments on Inorg. Chem.*, in press, *27,33*). There are no structural features present in these systems, e.g. intermolecular associations, which would require a different treatment of the magnetic data, and so it is difficult to explain why these complexes have such unusual magnetic properties. Given the fact that for more strongly antiferromagnetically coupled 1,1-azide bridged complexes (*26,27*) reasonable fits to the Bleaney-Bowers equation can be obtained, which become better as -2J increases, and that identical measurement techniques were employed using the same equipment for all complexes studied, there is likely to be a more fundamental rationale to the problem.

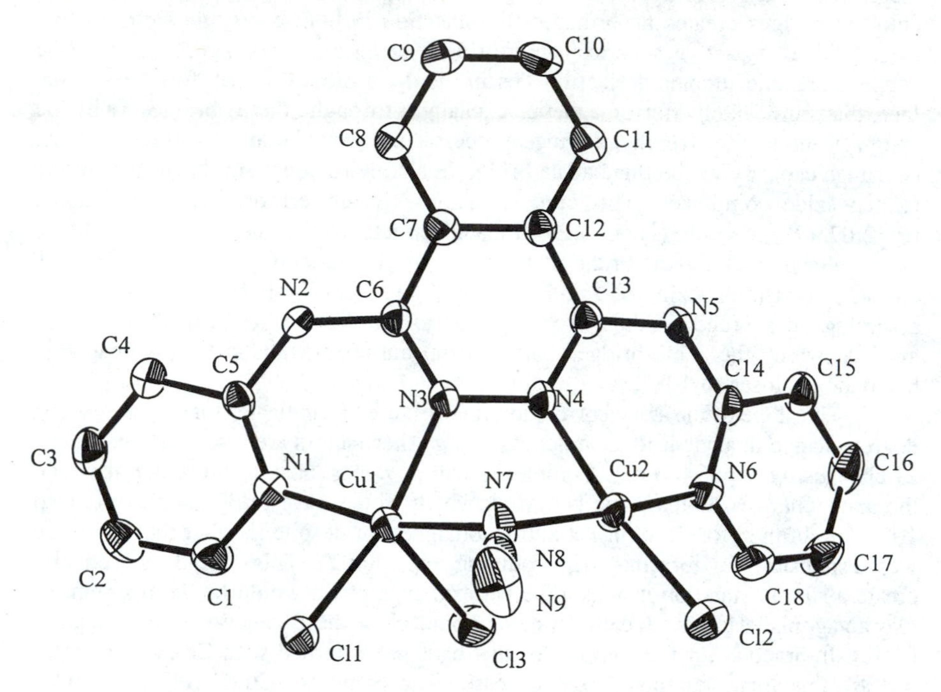

Figure 12. Structural representation for $[Cu_2(PAPH)(\mu_2\text{-}N_3)Cl_3]$ (Reproduced with permission from ref. 26. Copyright 1994, American Chemical Society).

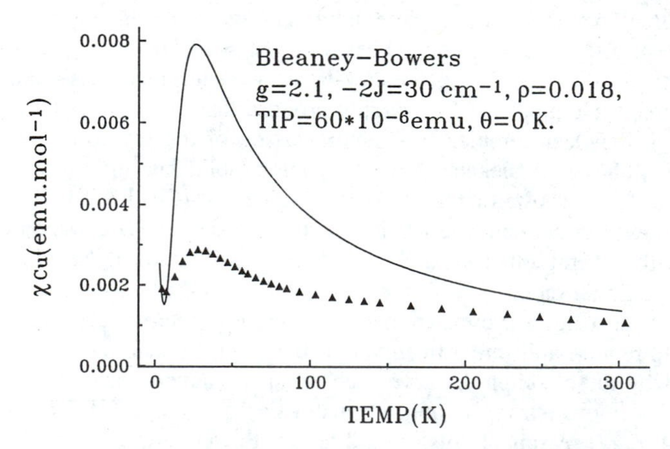

Figure 13. Plot of χ_{Cu} versus temperature for $[Cu_2(PAPH)(\mu_2\text{-}N_3)Cl_3]$ compared with data for a typical exchange coupled dinuclear copper(II) complex.

Should the Exchange Integral be Temperature Independent?

The assumption is made, almost exclusively, when using any exchange equation that the J value does not vary as a function of temperature. Since it is clear from the magnetostructural studies on hydroxide (*2*) and azide (*26,27*) bridged dinuclear copper complexes that even small changes in dinuclear center dimensions, e.g. bridge angle, could lead to a significant change in J value and overall magnetic properties, if such a geometric change occurred in the experimental temperature range it would be unrealistic to expect the Bleaney-Bowers equation to apply. A change in bridge angle of **one** degree in the temperature range 4.2-300K, which would nominally correspond to a change in 2J of $\approx 75\ cm^{-1}$ for the di-hydroxide complexes (*2*), and $\approx 46\ cm^{-1}$ for the di-azide complexes (*27*), could lead to quite different overall magnetic properties, and difficulty in fitting data to the Bleaney-Bowers equation. Such a problem would be most noticeable for weakly coupled systems, where absolute susceptibility values are relatively high, while for strongly coupled systems, where susceptibilities are small, and measurements are generally less accurate, such a problem might not be noticed. The study described by Hatfield (*2*) does not report any such anomalies, but the lower limit of antiferromagnetic coupling is 130 cm^{-1}.

The azide/diazine complexes are complicated by the presence of a second bridge, the diazine itself, whose geometry is not likely to change significantly as a function of temperature. However the overall geometry of the dinuclear center, particularly the Cu-Cu separation and the Cu-N_3-Cu angle, might change quite significantly with temperature. Although few studies have attempted to address this question, one recent report on a bis-μ_2-1,1-azide complex of nickel shows that over a 100K temperature range the Ni-N_3-Ni angle changes by 0.3°(104.0° (290K), 103.7° (190K)) (*34*). Extrapolation of such a temperature effect to 4K could lead to a change in azide bridge angle of $\approx 1°$! For the complex $[Cu_2(PAPH)(\mu_2\text{-}N_3)Cl_3]$ it is reasonable to assume that the azide is acting as a ferromagnetic bridge group (Cu-N_3-Cu 107.9°; $T_{\chi max.} \approx 27.5$ K), but the complex displays overall antiferromagnetic behavior because of the antiferromagnetic nature of the phthalazine bridge. However this ignores the possible effect of a temperature dependent J upon the overall shape of the χ vs. T profile. Attempts to fit the magnetic data for this compound were made by using a modification of the Bleaney-Bowers equation in which -2J was substituted by a temperature dependent function of J. In the absence of a functional probe of the actual geometric changes that would occur over the temperature range 4-300K an arbitrary temperature function was chosen (equation 3). A fit using the Bleaney-Bowers equation, incorporating this term is shown in Figure 14. Although this is not a good fit, it is very much closer than is possible using the normal equation.

$$-2J = -2J_{299K} - C(299-T)^2 \qquad (3)$$

This fit would correspond to a value of $-2J_{299K}=246\ cm^{-1}$, and $-2J_{9K}=37\ cm^{-1}$. It should be emphasized that this is an empirical relationship only, and serves to illustrate the point that the experimental data can be roughly modelled using such an expression. Another example with an azide bridge angle of 107.9° is illustrated

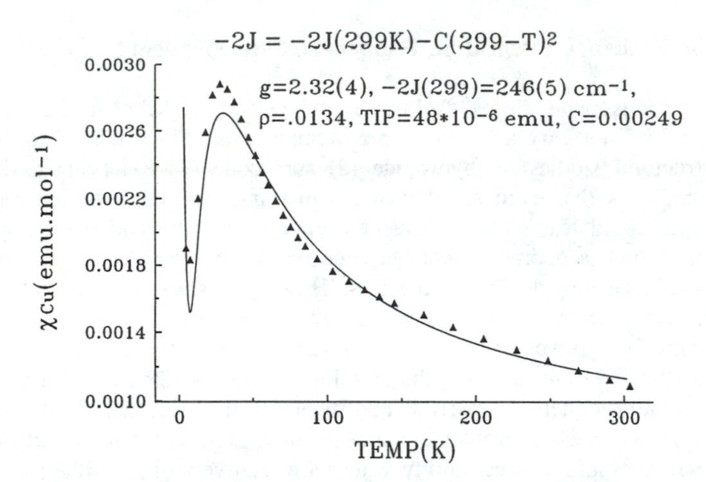

Figure 14. Curve fitting of the variable temperature susceptibility data for $[Cu_2(PAPH)(\mu_2\text{-}N_3)Cl_3]$ using the Bleaney-Bowers equation (equation 1) and a temperature variable exchange integral (equation 3).

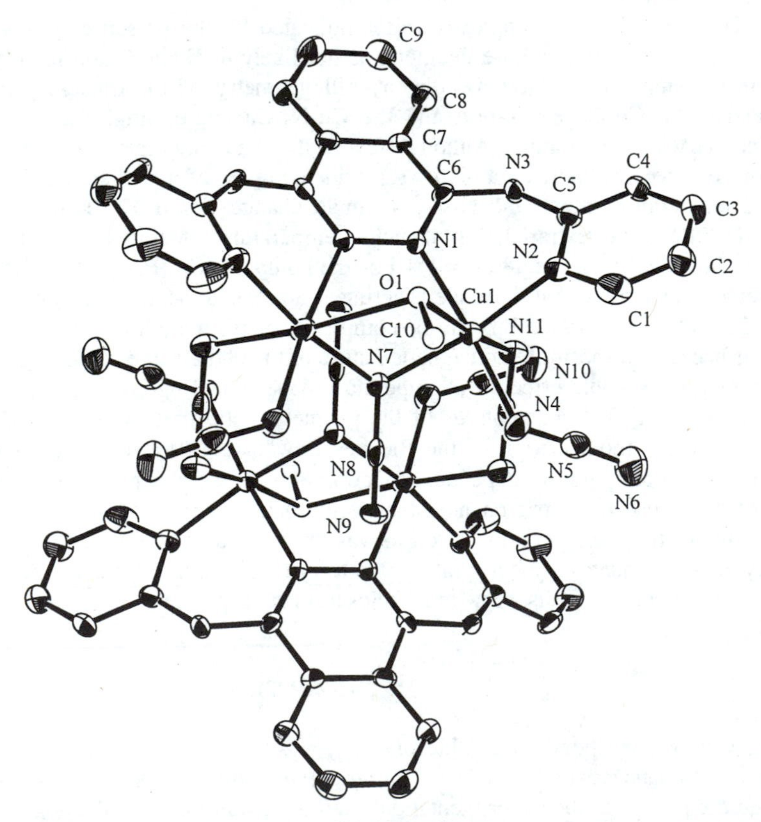

Figure 15. Structural representation for $[Cu_4(PAPH)_2(\mu_2\text{-}1,1\text{-}N_3)_2(\mu_2\text{-}1,3\text{-}N_3)_2(\mu_2\text{-}CH_3OH)_2]$ (Reproduced with permission from ref. 33. Copyright 1995, Royal Society of Chemistry).

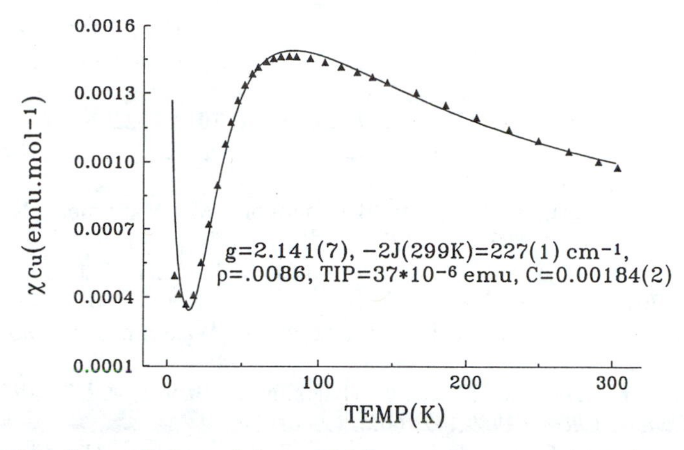

Figure 16. Curve fitting of the variable temperature susceptibility data for $[Cu_4(PAPH)_2(\mu_2\text{-}1,1\text{-}N_3)_2(\mu_2\text{-}1,3\text{-}N_3)_2(\mu_2\text{-}CH_3OH)_2]$ using the Bleaney-Bowers equation (equation 1) and a temperature variable exchange integral (equation 3)

in Figure 15. This rather unusual complex is a tetranuclear dimer in which two dinuclear complex fragments, which contain equatorial 1,1-azide and phthalazine bridges, are joined by two 1,3-azides which bridge via axial interactions. An axial methanol also bridges the two copper centers within each dinuclear half (*33*). From a magnetic perspective the only antiferromagnetic bridges are the diazine and the μ_2-1,1-azide, since the axial bridging interactions are orthogonal. The variable temperature magnetic data for this compound cannot be fitted to the Bleaney-Bowers equation unless a Curie-Weiss like θ correction of -177K is used, which is meaningless in the context of the structure (*33*). However using equation 3 in combination with the Bleaney-Bowers equation gives a reasonable data fit (Figure 16), with $-2J_{299K}=227$ cm^{-1} and $-2J_{9K}=72$ cm^{-1}. While the magnetic data for these unusual compounds can be simulated by using a temperature dependent J, the structures of key compounds must be examined as a function of temperature before an appropriate magnetic model can be established. Such studies are currently underway.

Acknowledgements

We thank the Natural Sciences and Engineering Research Council of Canada for financial support for these studies.

Literature Cited

1. Bleaney, B.; Bowers, K.D. Proc. R. Soc., London, **1952**, *A 214*, 451.
2. Crawford, V.H.; Richardson, H.W.; Wasson, J.R.; Hodgson, D.J.; Hatfield, W.E. *Inorg. Chem.*, **1976**, *15*, 2107.
3. Comarmond, J.; Plumeré, P.; Lehn, J.-M.; Agnus, Y.; Louis, R.; Weiss, R.; Kahn, O.; Morgenstern-Badarau, I. *J. Am. Chem. Soc.* **1982**, *104*, 6330.
4. Hay, P.J.; Thibeault, J.C.; Hoffmann, R. *J. Am. Chem. Soc.*, **1975**, *97*, 4884.
5. Merz, L; Haase, W. *J. Chem. Soc., Dalton,* **1980**, 875.
6. Handa, M.; Koga, N.; Kida, S. *Bull. Chem. Soc., Jpn.*, **1988**, *61*, 3853.

7. Sikorav, S.; Bkouche-Waksman, I.; Kahn, O. Inorg. Chem. **1984**, *23*, 490.
8. Kahn, O.; Sikorav, S.; Gouteron, J.; Jeannin, S.; Jeannin, Y. Inorg. Chem. **1983**, *22*, 2877.
9. Pilkington, N.H.; Robson, R. *Aust. J. Chem.*, **1970**, *23*, 2225.
10. Nanda, K.K.; Thompson, L.K.; Bridson, J.N.; Nag. K. *J. Chem., Soc., Chem. Commun.*, **1994**, 1337.
11. Aratake, Y.; Ohba, M.; Sakiyama, H.; Tadokoro, M.; Matsumoto, N.; Ōkawa, H. *Inorg. Chim. Acta,* **1993**, *212*, 183.
12. Mandal, S.K.; Thompson, L.K.; Newlands, M.J.; Gabe, E.J. *Inorg. Chem.*, **1989**, *28*, 3707.
13. Lacroix, P.; Kahn, O.; Theobald, F.; Leroy, J.; Wakselman, C. *Inorg. Chim. Acta*, **1988**, *142*, 129.
14. Mandal, S.K.; Thompson, L.K.; Newlands, M.J.; Biswas, A.K.; Adhikary, B.; Nag, K. *Can. J. Chem.*, **1989**, *67*, 662.
15. Carlisle, W.D.; Fenton, D.E.; Roberts, P.B.; Casellato, U.; Vigato, P.A.; Graziani R. *Trans. Met. Chem.*, **1986**, *11*, 292.
16. Hoskins, B.F.; McLeod, N.J.; Schaap, H.A. *Aust. J. Chem.*, **1976**, *29*, 515.
17. Mandal, S.K.; Thompson, L.K.; Newlands, M.J.; Gabe, E.J. *Inorg. Chem.*, **1990**, *29*, 1324.
18. Brychcy, K.; Dräger, K.; Jens, K-J.; Tilset, M.; Behrens, U. *Chem. Ber.*, **1994**, *127*, 465.
19. Mandal, S.K.; Thompson, L.K.; Nag. K; Charland, J-P.; Gabe, E.J. Can. J. Chem., **1987**, *65*, 2815.
20. Tandon, S.S.; Thompson, L.K.; Bridson, J.N.; McKee, V.; Downard, A.J. *Inorg. Chem.*, **1992**, *31*, 4635.
21. Tandon, S.S; Thompson, L.K.; Bridson, J.N. *Inorg. Chem.*, **1993**, *32,* 32.
22. Tinti, F; Verdaguer, M.; Kahn, O.;Savariault, J-M. *Inorg. Chem.*, **1987**, *26*, 2380.
23. Tandon, S.S.; Mandal, S.K.; Thompson, L.K. Hynes, R.C. *Inorg. Chem.*, **1992**, *31*, 2215.
24. Chaudhuri, P.; Oder, K.; Wieghardt, K.; Gehring, S.; Haase, W.; Nuber, B.; Weiss, J. *J. Am. Chem., Soc.*, **1988**, *110*, 3657.
25. Galy, J.; Jaud, J,; Kahn, O.; Tola, P. *Inorg. Chim. Acta,* **1979**, *36*, 229.
26. Tandon, S.S.; Thompson, L.K.; Manuel, M.E.; Bridson, J.N. *Inorg. Chem.*, **1994**, *33*, 5555-5570.
27. Thompson, L.K., Tandon, S.S.; Manuel, M.E.; *Inorg. Chem.*, **1995**, *34*, 2356-2366.
28. Thompson, L.K.; Lee, F.L.; Gabe, E., J. *Inorg. Chem.*, **1988**, *27*, 39-46.
29. Tandon, S.S.; Chen, L.; Thompson, L.K.; Bridson, J.N. *Inorg. Chem.*, **1994**, *33*, 490-497.
30. Thompson, L.K.; Mandal, S.K.; Charland, J-P.; Gabe,E.J. *Can. J. Chem.*, **1988**, *66*, 348-354.
31. Chen, L.; Thompson, L.K.; Bridson, J.N. *Inorg. Chem.*, **1993**, *32*, 2938-2943.
32. Mallah, T.; Boillot, M-L.; Kahn, O.; Gouteron, J.; Jeannin, S.; Jeannin, Y. *Inorg. Chem.*, **1986**, *25*, 3058.
33. Tandon, S.S.; Thompson, L.K.; Miller, D.O.; *J. Chem. Soc., Chem. Commun.*, **1995**, 1907-1908.
34. Escuer, A.; Vicente, R.; Ribas, J. *J. Magnetism and Mol. Materials*, **1992,** *110,* 181.

Chapter 12

The Azido Ligand: A Useful Bridge for Designing High-Dimensional Magnetic Systems

R. Cortés[1], L. Lezama[2], F. A. Mautner[3], and T. Rojo[2]

[1]Departamento de Quimica Inorgánica, Facultad de Farmacia, Universidad del Pais Vasco, Apartado 450, E–01080 Vitoria-Gasteiz, Spain
[2]Departamento de Quimica Inorgánica, Facultad de Ciencias, Universidad del Pais Vasco, Apartado 644, E–48080 Bilbao, Spain
[3]Institut für Physikalische und Theoretische Chemie, Technische Universität Graz, A–8010 Graz, Austria

Designing new high-dimensional magnetic molecular systems built from coordination compounds has recently been a point of attention for inorganic chemists. The variety of coordination chemistry provides the synthesizers with a useful tool to build magnetic molecular architectures interesting for their properties, which arise from the interaction among their subunits. Strategies based on the reaction of appropriate terminal and bridging ligands with paramagnetic metal ions allow the preparation of oligomeric species whose nuclearity and magnetic properties may, in any sense, be tailored (*1*).

By using good superexchange bridging groups such as oxalate or cyanide, extended lattices of antiferromagnetic or ferrimagnetic systems which show magnetic order at low temperatures have been achieved (*2,3*). In this way, the azido ligand represents a good choice for the design of new magnetic systems. This ligand is able to give ferromagnetic exchange interactions (through its end-on (EO) coordination mode) and antiferromagnetic ones (through the end-to-end (EE) mode). It is also important to note the great versatility in their coordination modes, giving rise to high nuclearity systems. The Scheme shows the different bridging modes actually observed for this ligand.

Former studies carried out for the azido ligand led to dinuclear entities, generally doubly bridged in the end-to-end form (*4,5*). More unusual end-on mode was obtained in some copper(II) dinuclear systems (*6*). Our preliminary work in this field was done with the aim of obtaining compounds exhibiting the unusual end-on bridging mode. From those previous studies, we have been searching for a continuous increase in the dimensionality of the prepared compounds in order to study their magneto-structural correlations. In this chapter we report the most significant structural and magnetic results obtained by our group for dinuclear, one-, two-, and three-dimensional systems.

0097–6156/96/0644–0187$15.00/0

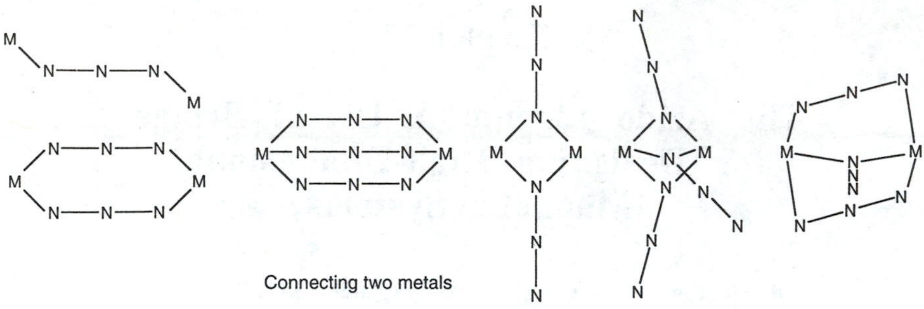

Connecting two metals

Scheme 1a

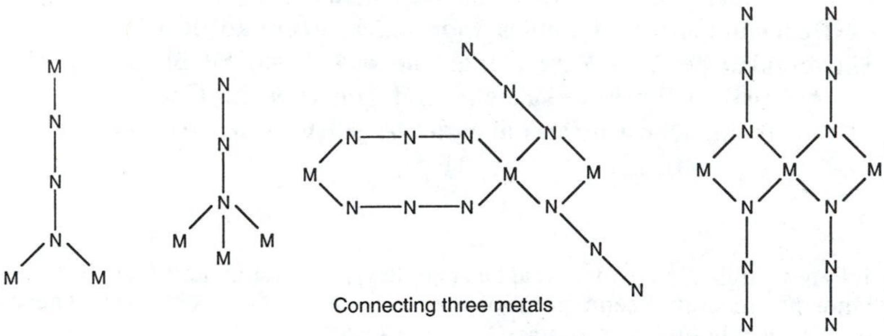

Connecting three metals

Scheme 1b

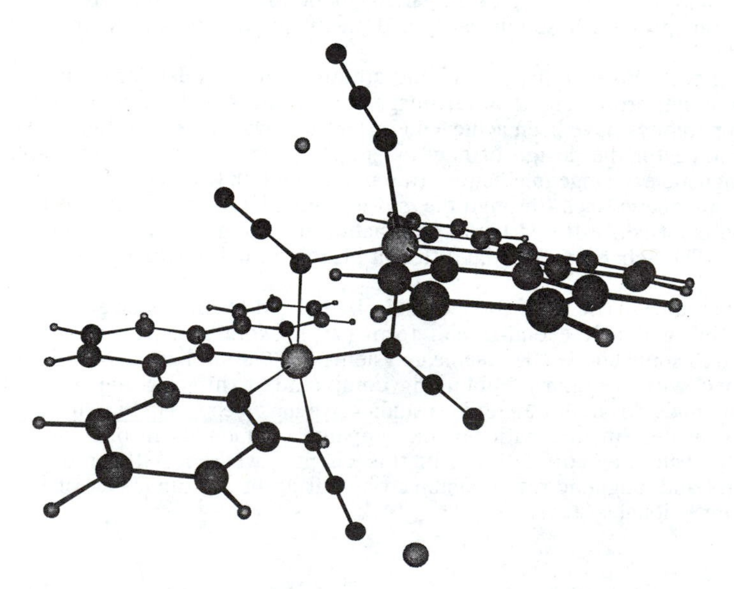

Figure 1. Crystal structure of $[M(N_3)_2(terpy)]\cdot H_2O$ (M= Mn, Ni).

Dinuclear Compounds.

Work centered on copper(II)-halide dinuclear systems led us to observe that the use of rigid tridentate aromatic amines favors a *cis*-coordination of the bridging groups around the metal ion. Using the halide compounds as precursors, we have developed a synthetic strategy to obtain *cis*-pseudohalide-bridged dinuclear compounds of general formula [M(pseudohalide)$_2$(L_{III})]·H_2O [M= Ni, Mn; L_{III}= 2,2':6',2"-terpyridine (terpy), N'-(2-pyridin-2-ylethyl)-pyridine-2-carbaldimine (pepci)] (*7-9*). As a result, several nickel(II) (*7,8*) and manganese(II) (*9*) dinuclear azido compounds, always exhibiting the EO bridging mode, have been obtained. A molecular structure of this kind of dinuclear system is shown in Figure 1.

The general structure for all these complexes consists of centrosymmetric dinuclear units where the metal ions are doubly bridged by EO-N_3 ligands, with the tridentate ligand occupying three of the four equatorial sites of the coordination polyhedron of each metal and a terminal azido group to complete the hexacoordination. In all cases, the values of the bridging angles are lower than 105°.

Ferromagnetic interactions between the metal centers are favored with this kind of bridging (*6-9*). In this way, the magnetic results for the nickel(II) and manganese(II) compounds are illustrated in Figure 2 as an example.

As can be observed, the $\chi_m T$ value increases upon cooling, reaching a maximum after which it rapidly decreases in the case of the nickel compound, as a consequence of the zero-field splitting, while for manganese the $\chi_m T$ value remains stable at lower temperatures. This clearly indicates intradimer ferromagnetic interactions in both complexes. In fact, ferromagnetic exchange constants in the range 5-40 cm^{-1} , which are proportional to the value of the bridging angle, have been obtained for the nickel compounds. However, the only manganese(II) compound known, with a bridging angle of 101°, exhibits a J value of 2.5 cm^{-1}.

Ferromagnetic behavior associated with this kind of bridging has usually been explained by using the spin polarization theory (*10*), which predicts ferromagnetism for the range of bridging angle values. However, in the case of copper dimers it has recently been demonstrated (*11*) that ferromagnetic behavior prevails at small bridging angles (<108°), and at larger angles antiferromagnetic exchange predominates. A recent report (*12*) of a polarized neutron diffraction study on [Cu_2(t-bupy)$_4$(μ_2-1,1-N_3)](ClO_4)$_2$ revealed spin density calculations that contradict the theoretical interpretation using the spin polarization theory. So, this theory must be used with caution and, in fact, it may not be an appropriate description.

Where the tridentate ligand is substituted by two bidentate ones, such as 2,2'-bipyridine, an EE *cis* - dinuclear system is obtained: [Mn(N_3)(bipy)$_2$](ClO_4)·1/2bipy (see Figure 3) (Cortés, R.; Rojo, T. unpublished data).

The crystal structure of this compound shows that the disposition of the azido anion is in an EE bridging mode, where the bridging angles Mn-N-N and N-N-Mn are 128.6° and 131.3°, respectively. The Mn-(N_3)$_2$-Mn unit exhibits a chair configuration, with a torsion angle of 18°. The existence of a non-coordinated bipyridine ligand, which is sited parallel to the bipy ligands of the dimer, is a peculiarity of this structure (see Figure 3). Preliminary magnetic results for this dinuclear compound are shown in Figure 4.

A continuous decrease of the $\chi_m T$ value is observed from room temperature to 4.2 K. This shows the existence of significant antiferromagnetic coupling associated with the EE azido bridges in the complex.

One-Dimensional Magnetic Systems.

One-dimensional azido-compounds of divalent transition metal ions have not been studied extensively. In the case of the Ni(II) compounds, the interest arises from

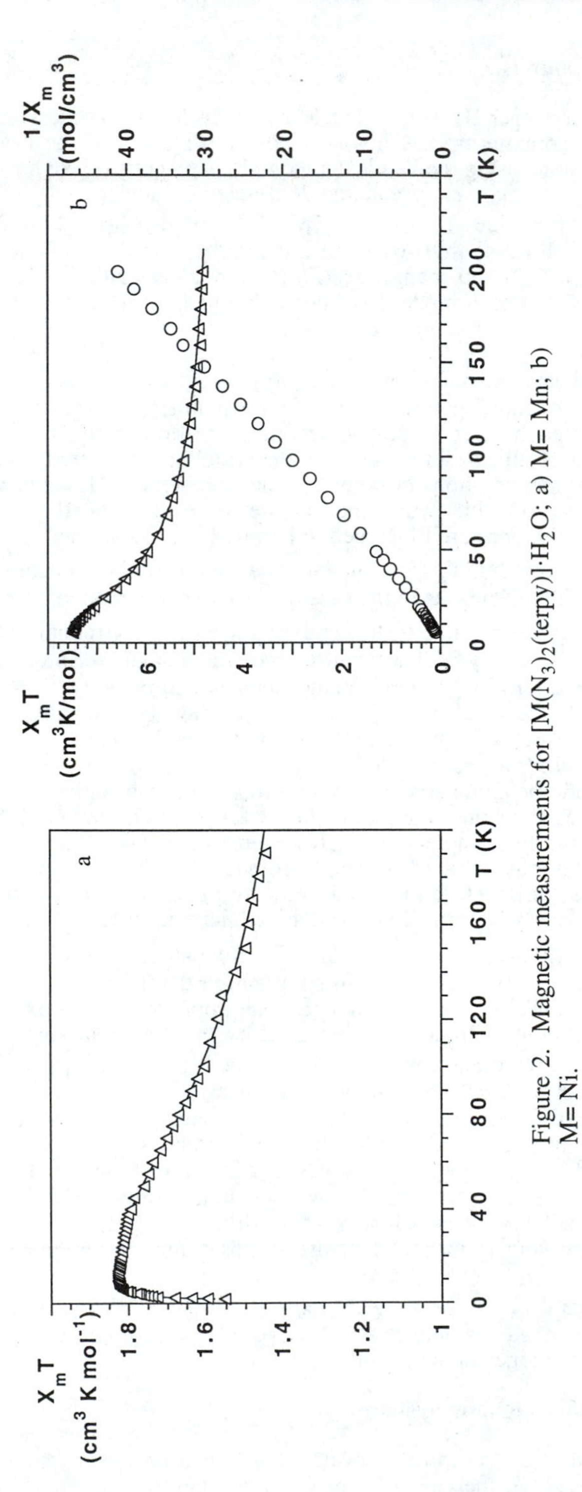

Figure 2. Magnetic measurements for $[M(N_3)_2(terpy)]\cdot H_2O$; a) M= Mn; b) M= Ni.

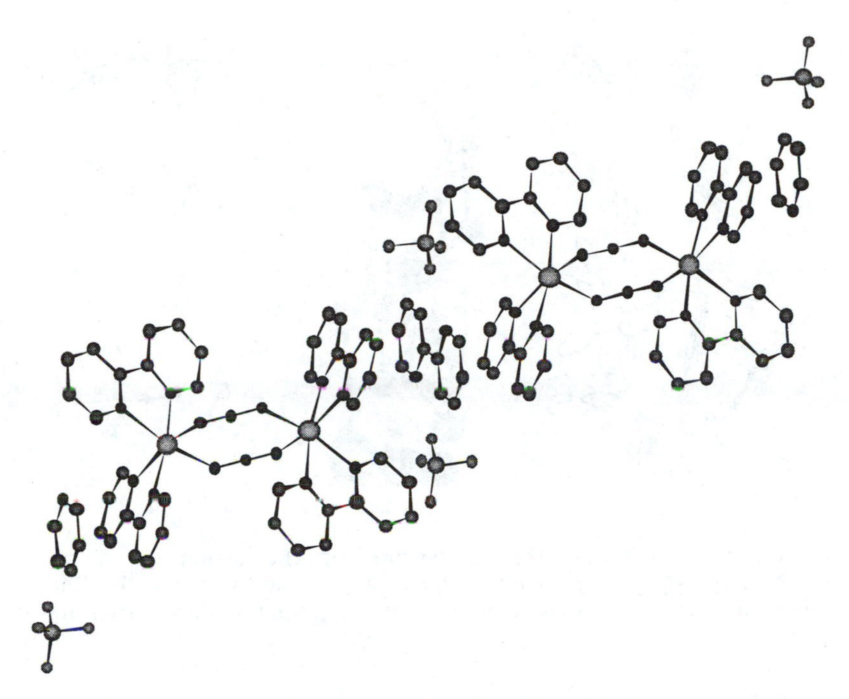

Figure 3. Crystal structure of $[Mn(N_3)(bipy)_2](ClO_4)\cdot 1/2bipy$.

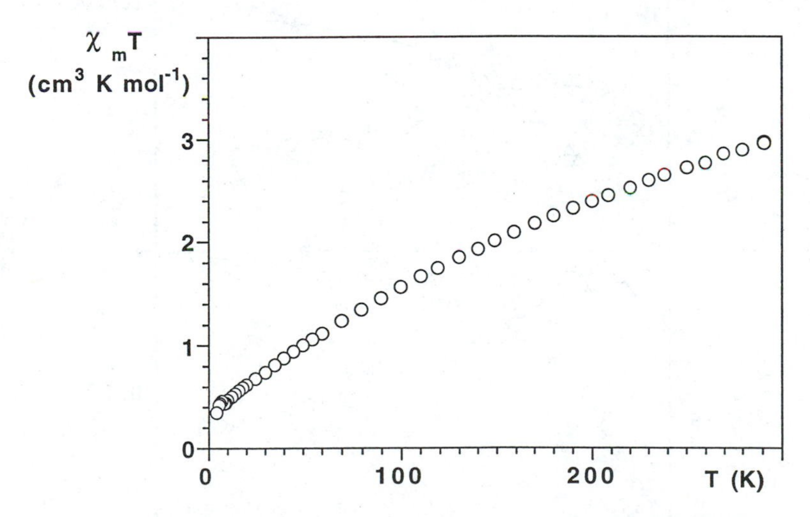

Figure 4. Magnetic measurements for $[Mn(N_3)(bipy)_2](ClO_4)\cdot 1/2bipy$.

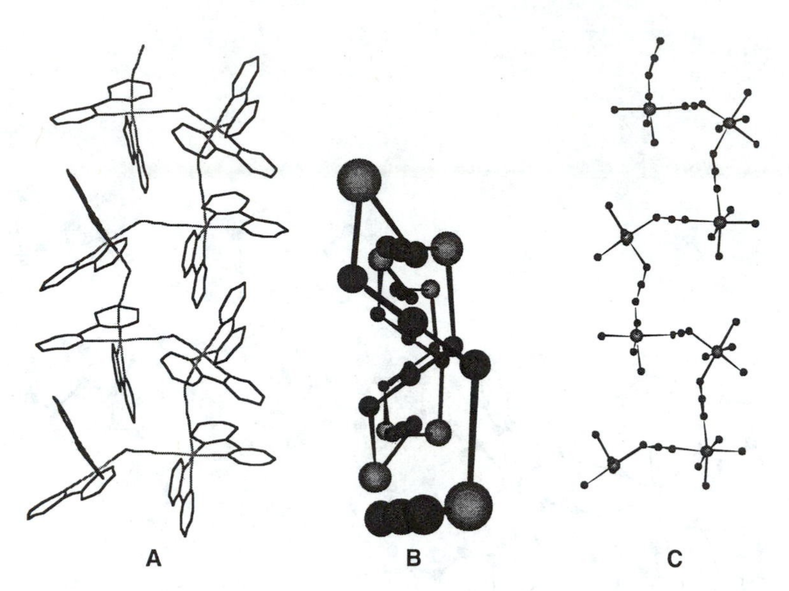

Figure 5. Different representations of the structure of the $[Ni(N_3)(bipy)_2]_n(Y)_n$ (Y= ClO_4, PF_6) chains: A and C along the [001] direction; B from the propagation direction viewpoint. (Adapted from ref. 15.).

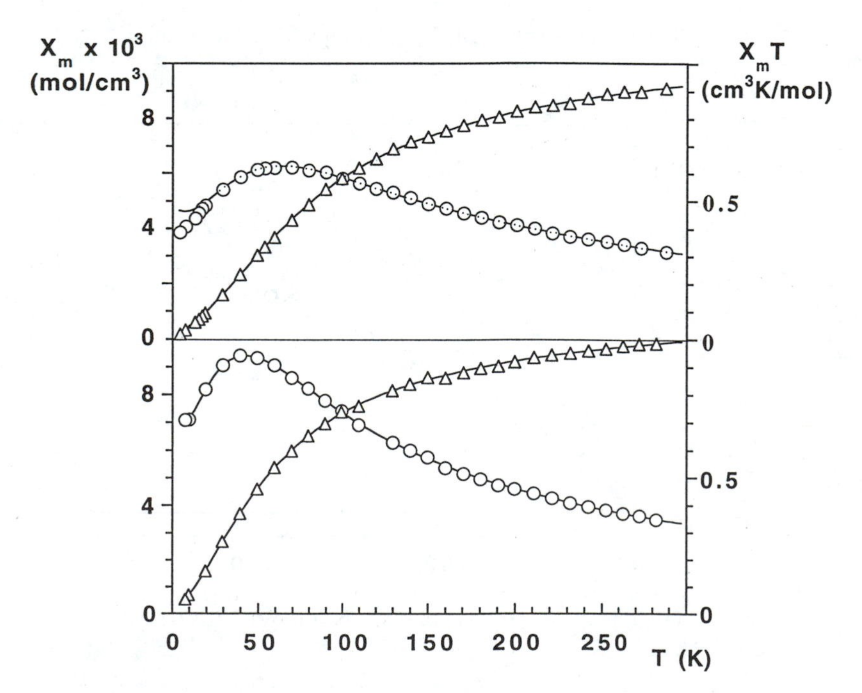

Figure 6. Magnetic measurements for the $[Ni(N_3)(bipy)_2]_n(Y)_n$ (Y= ClO_4, PF_6) chains. (Adapted from ref. 15.).

Haldane's theory which predicts that for the integer spin values S a one-dimensional Heisenberg antiferromagnet (1D-HAF) must exhibit a singlet ground state separated from the first triplet excited state by an energy gap, contrary to the half-integer spin values which have a gapless spectrum. In the case of Mn(II) and other magnetic ions, the existence of 1-D systems is very rare and all new contributions are of interest to increase the knowledge about the structure-magnetism correlations.

Several nickel(II) compounds having a *trans* conformation of the azido ligands in the EE bridging mode have recently been shown (*13,14*). Such a disposition was obtained by using tetradentate planar aliphatic amines to block the four equatorial sites of the metal and connecting these units by means of azido ligands. All of these systems are 1D-antiferromagnetic and exhibit Haldane's gap.

The *cis* arrangement of the bridging azides is favored by the aromatic amines. To date, it has not been possible to obtain an one-dimensional stacking of the metal ions in compounds with rigid tridentate ligands. However, the use of two bidentate aromatic ligands such as 2,2'-bipyridine gave good results. In this way, the first Ni(II)-bipy *cis*-azido chain structures were obtained in the $[Ni(N_3)(bipy)_2]Y$ (Y= ClO_4, PF_6) compounds (*15*). Such a conformation was strongly favored by the *cis* positioning of the bipy ligands, owing to their steric crowding. Figure 5 shows different perspective views of these compounds.

In spite of both compounds being isostructural, the difference in the counter-anions causes significant variations in the bridging and torsion angles values which are responsible for the modifications in the corresponding exchange coupling constants. Figure 6 shows the thermal variation of both χ_m and $\chi_m T$ for the $[Ni(N_3)(bipy)_2]_n(Y)_n$ (Y= ClO_4, PF_6) compounds.

In both cases the molar susceptibility value increases with decreasing temperature, reaches a maximum and then rapidly decreases. The $\chi_m T$ value decreases continuously upon cooling, going towards zero. Both the continuous decrease in the $\chi_m T$ value and the maximum observed in the susceptibility clearly indicate the existence of strong antiferromagnetic intrachain interactions in the compounds. Fitting of the experimental susceptibility values to the empirical expression led to the values of the exchange coupling constant of $J= -33.0\ cm^{-1}$ and $J= -22.4\ cm^{-1}$ for the perchlorate and hexafluorophosphate compounds, respectively.

In the case of one-dimensional systems connected by EO azido bridges, only one compound is actually known (*16*). This compound shows ferromagnetic interactions characteristic of this bridging mode, in which the bridging angle is 96.8°.

Concerning alternating chains, a 1-D complex of Ni(II) with both end-to-end and end-on bridging modes has been found (*17*). This compound shows alternating single EE and triple EO azido bridges. It shows antiferromagnetic interactions explained in the case of the three EO bridges by the existence of very short bridging angles (85°). We have recently obtained a 1-D copper system with the formula $[N(CH_3)_4]_n[Cu(N_3)_3]_n$ in which the Cu(II) ions are connected by a triple bridge, two with EE and one with EO conformations (Figure 7) (Mautner, F. A. unpublished results). The steric crowding caused by the simultaneous existence of both types of bridging between two metal ions can be diminished by the semicoordination ability of the Cu(II) ion. This compound represents the first example of such connectivity.

The magnetic behavior associated with this complex is quite interesting and is shown in Figure 8. A continuous descent in the $\chi_m T$ value is observed upon cooling, reaching a minimum at 10 K after which it rapidly increases. This features seems to be associated with the existence of a "ferrimagnetic" behavior in the complex. Magnetization measurements are in progress to confirm this hypothesis.

An interesting manganese(II) chain exhibiting alternation in the sign of the magnetic coupling has been prepared by using a metal:bidentate ligand stoichiometry of 1:1, which enables four coordination vacancies around the metal to be occupied by the azido anions (*18*). The structure of this compound, $[Mn(N_3)_2(bipy)]_n$, contains alternating double end-on and double end-to-end azido bridging groups (Figure 9).

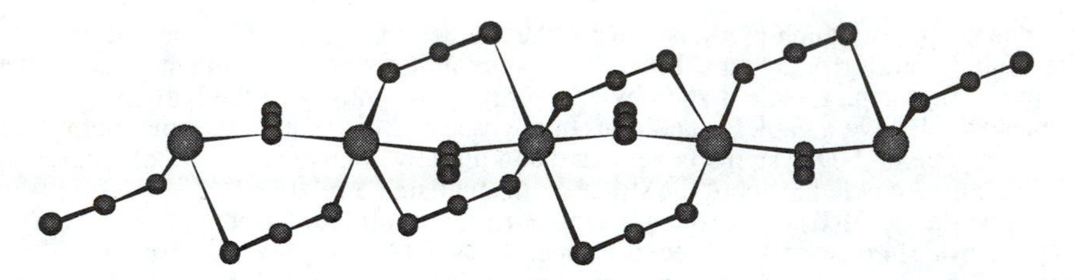

Figure 7. Crystal structure of the $[N(CH_3)_4]_n[Cu(N_3)_3]_n$ chain.

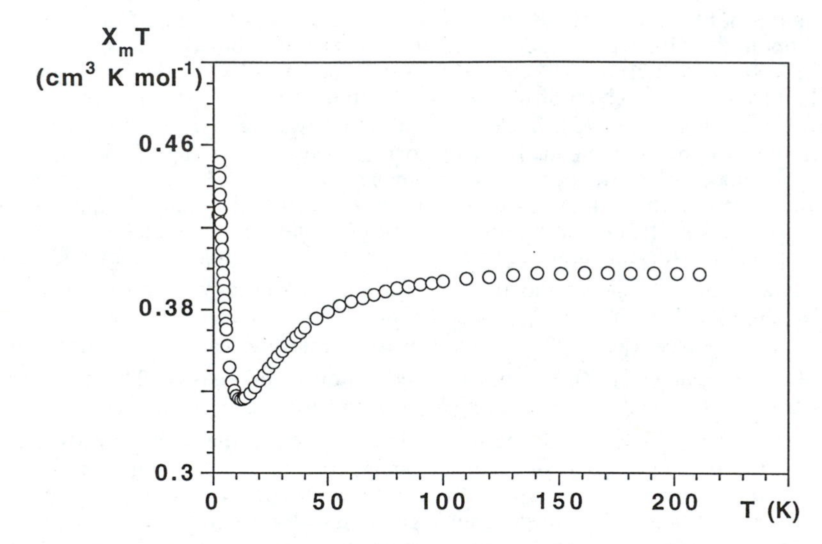

Figure 8. Thermal variation of $\chi_m T$ for $[N(CH_3)_4]_n[Cu(N_3)_3]_n$.

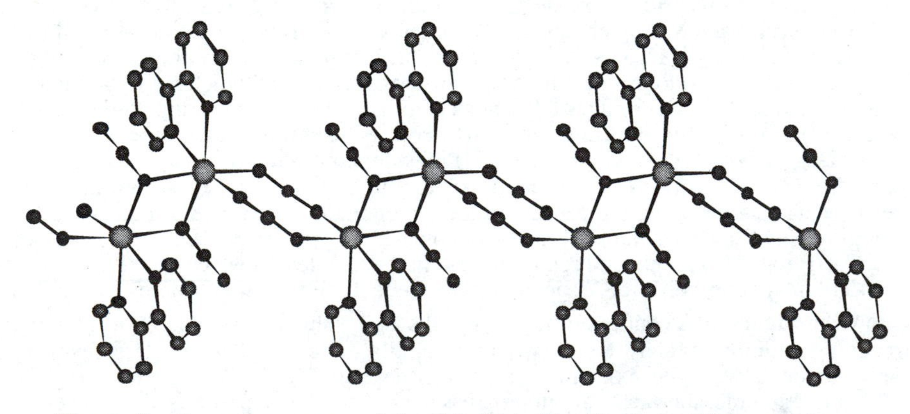

Figure 9. Crystal structure of the $[Mn(N_3)_2(bipy)]_n$ chain. (Adapted from ref. 18.).

The azido bridges are arranged in *cis* positions and are perpendicular to each other. The Mn-N-Mn' angle in the EO bridge is 101.0°. For the EE bridges the Mn-N-N angles are 131.1(5)° and 127.3(5)° for a chair conformation of the Mn-$(N_3)_2$-Mn unit and a torsion angle of 41.9°. Due to these structural features, alternating ferromagnetic (through EO bridges) and antiferromagnetic interactions (through EE bridges) should be expected. The thermal variation of both χ_m and $\chi_m T$ values are shown in Figure 10.

The χ_m value increases upon cooling, reaching a rounded maximum at about 55 K and then drops to zero at lower temperature. The $\chi_m T$ value at room temperature is 4.0 cm^3 K mol^{-1}, which is smaller than that expected for an uncoupled Mn(II) ion. This value decreases continuously with decreasing temperature going towards to zero. The magnetic behavior of this compound can only be explained assuming two alternating exchange constants: a positive J_1 for EO and a negative J_2 for EE pathways. However, no formula is available in the literature for an S=5/2 ferro-antiferromagnetic alternating chain and for this reason a model has been elaborated.

The model is based on the spin Hamiltonian: $\mathcal{H} = -J_1 \Sigma S_{2i}S_{2i+1} - J_2 \Sigma S_{2i+1}S_{2i+2}$, which leads to the expression of the bulk susceptibility shown in equation 1:

$$\chi = \frac{Ng^2\mu_B{}^2}{3kT}\left(\frac{1 + u_1 + u_2 + u_1u_2}{1 - u_1u_2}\right) \quad (1)$$

where $u_1 = \coth(J_1/kT) - kT/J_1$ and $u_2 = \coth(J_2/kT) - kT/J_2$

The best agreement obtained by least squares refinement corresponds to the following set of parameters: J_1 = +21.7 K, J_2= -16.0 K, g= 1.92. The model proposed appears to provide a very good description of the susceptibility down around 20 K (see Figure 10). In the range 20-300 K, the discrepancy does not exceed the experimental uncertainty. At lower temperature, the agreement is not so satisfactory, since instead of showing a non-zero value at low temperature, the experimental susceptibility drops to zero, due probably to the influence of small interchain interactions.

2-D Magnetic Systems.

No 2-D azido bridged magnetic compounds have been described in the literature. They could be obtained by promoting vacancies in the four equatorial sites of the coordination sphere of each metallic ion. This planar conformation could extend into two directions through the azido bridging groups and completing the coordination sphere of the metallic ion to give the desired layered systems. Where bidentate ligands are utilized the results are not satisfactory. In order to favor that conformation it would be helpful to use monodentate ligands, such as pyridine or substituted pyridines, which could occupy the corresponding two axial coordination vacancies and stabilize the global structure. As a result, the [Mn$(N_3)_2$(4-ac-py)$_2$] (4acpy= 4-acetyl pyridine) compound has been obtained (Figure 11) (Goher, M.; Mautner, F. A. unpublished results).

The structure of this compound shows the manganese(II) ions octahedrally coordinated to four azido ligands (filling the equatorial plane), while two molecules of 4acpy are in a trans arrangement at the axial sites of the coordination sphere. All the azido ligands act as EE bridges between two manganese(II) ions to give the global two-dimensional structure. Magnetic measurements are now being carried out in order to show the expected antiferromagnetic interaction due to the type of azido bridging present.

Another alternative way to obtain 2-D compounds could be the connection between 1-D systems through bis-bidentate ligands. After a variety of experiments we have successfully obtained a 2-D complex starting from the alternating ferro-

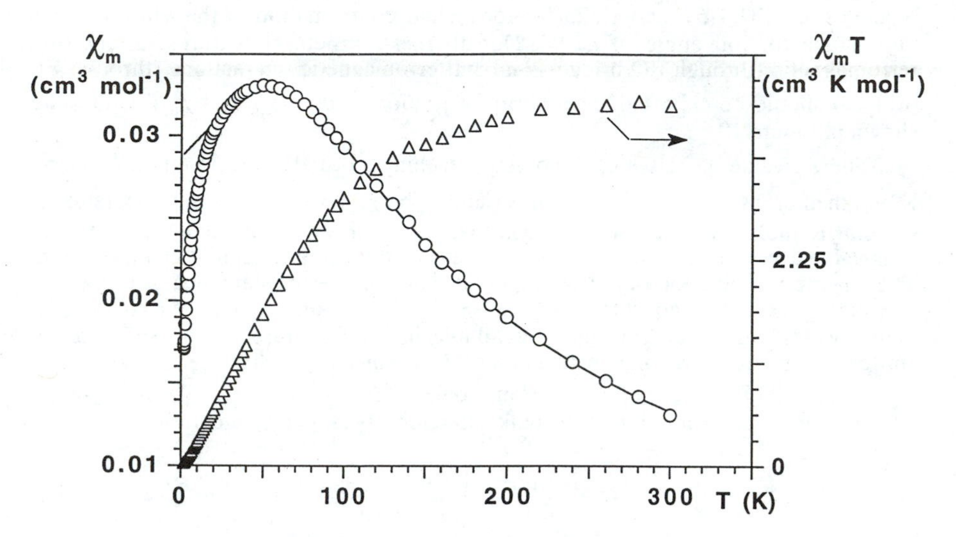

Figure 10. Magnetic measurements for the $[Mn(N_3)_2(bipy)]_n$ chain. (Adapted from ref. 18.).

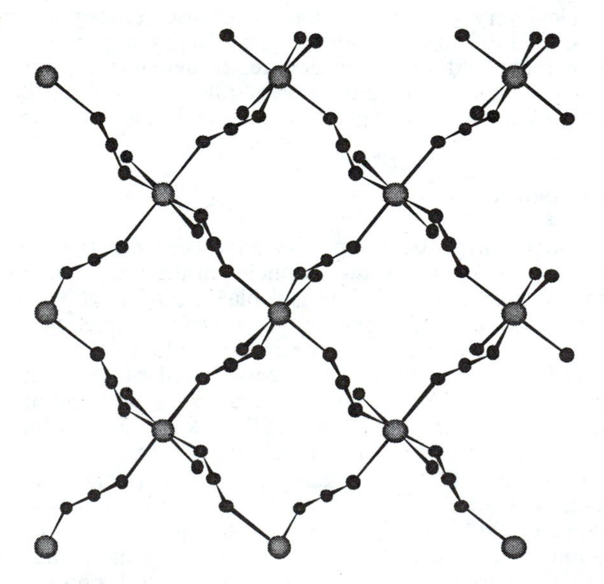

Figure 11. Crystal structure of the $[Mn(N_3)_2(4acpy)_2]_n$ compound.

antiferromagnetic 1-D compound (see Fig. 9). The isolated compound has the formula $[Mn_2(N_3)_4(bipym)]$ (Cortés, R.; Lezama, L.; Pizarro, J. L.; Arriortua, M. I.; Rojo, T. *Science*, submitted) and may be described as manganese(II)-azido-chains connected by bipyrimidine ligands (Figure 12) which can give rise to alternating interactions.

The crystal structure shows chains of manganese(II) ions connected by double EO azido bridges. These chains are alternatively connected by bis-bidentate bipyrimidine ligands. The whole structure can also be described as a honeycomb sheet, consisting of an infinite hexagonal array of manganese(II) ions bridged by bis-bidentate bipyrimidine ligands and double end-on azido bridges. This 2-D polymer extends in the *xy* plane by the repetition of almost circular rings.

The magnetic measurements obtained for the $[Mn_2(N_3)_4(bipym)]_n$ compound are shown in Figure 13. As can be seen, the $\chi_m T$ value increases with decreasing temperature reaching a maximum at about 20 K, after which this value decreases upon cooling down to 4 K. This behavior indicates ferromagnetic interactions in the first temperature range, followed by antiferromagnetic ones in the low temperature range.

3-D Magnetic Systems.

Three-dimensional compounds represent a challenge for researchers in the field of magnetic molecular systems. This is because of the need to both understand their magneto-structural correlations and to obtain 3-D antiferro- or ferromagnets with possible technical applications. 3-D systems with the azido bridging group could probably be obtained by either using metal:monodentate ligands in a 1:1 ratio, connecting 2-D compounds through *trans*-bidentate ligands or preparation without the use of ancillary blocking ligands.

One three-dimensional compound, $[N(CH_3)_4]_n[Mn(N_3)_3]_n$ (*19*), has been obtained by using the last approach cited. The proportion of three azido ligands per divalent ion was accomplished by using an appropriate counter-cation. The crystal structure of the compound is shown in Figure 14.

The structure may be described as a distorted Perovskite where the Mn^{II} is shifted from the origin of the unit cell by approximately 0.25 along the b-axis. The azido groups act as end-to-end bridging ligands between metals to form the three-dimensional network. Each manganese ion shows a slightly distorted octahedral coordination sphere. The $[N(CH_3)_4]^+$ cations are located within the holes formed by the manganese(II)-azido-sublattice (Figure 14). The EE azido bridges form two asymmetric angles with the metal: Mn-N-N around 165° and N-N-Mn' around 135°. The Mn···Mn distances are at 6.4 Å. The compound undergoes a reversible phase transformation at approximately 303 K. The cell length of the pseudo-cubic high temperature (HT) phase at 310 K is 6.446(4) Å, with one formula per unit cell.

The $\chi_m T$ value strongly decreases when the temperature decreases, going towards zero (see Figure 15). Considering this fact, the existence of a sharp maximum on the χ_m curve, and that the extrapolated susceptibility at 0 K is found to be approximately 2/3 of the value of the maximum susceptiblity, a three-dimensional antiferromagnetic lattice can be deduced. The low-high temperature (305 K) structural phase transition has been successfully observed (inset in the $\chi_m T$ vs. T curve). The magnetic behavior of this compound can be explained assuming a regular three-dimensional network as indicated by the crystal structure. The best least-square fit of the experimental data for $[N(CH_3)_4]_n[Mn(N_3)_3]_n$ to the expression of the magnetic susceptibility for a simple cubic Heisenberg antiferromagnetic system (*20*) is obtained with the parameters J/k= -2.5 K and g= 2.01.

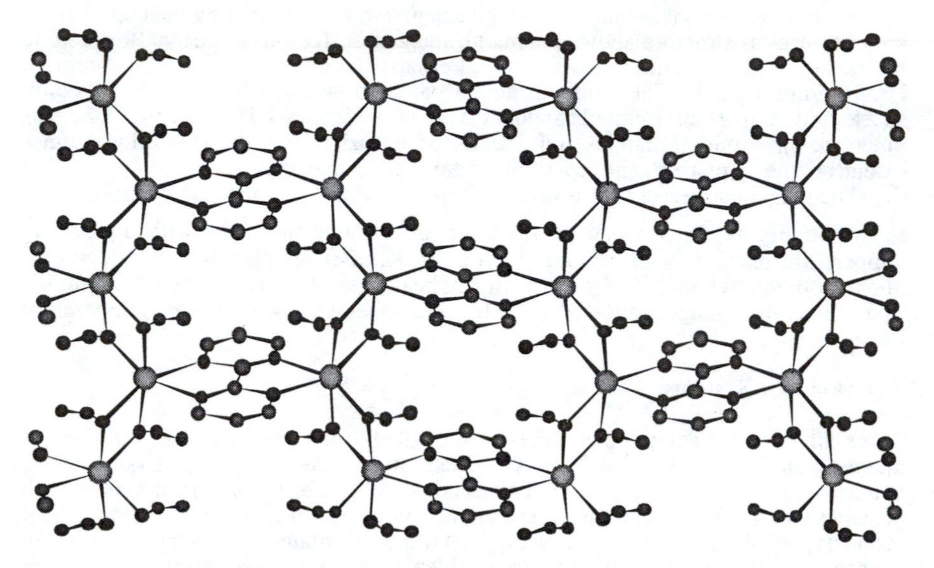

Figure 12. Crystal structure of the $[Mn_2(N_3)_4(bipym)]_n$ compound.

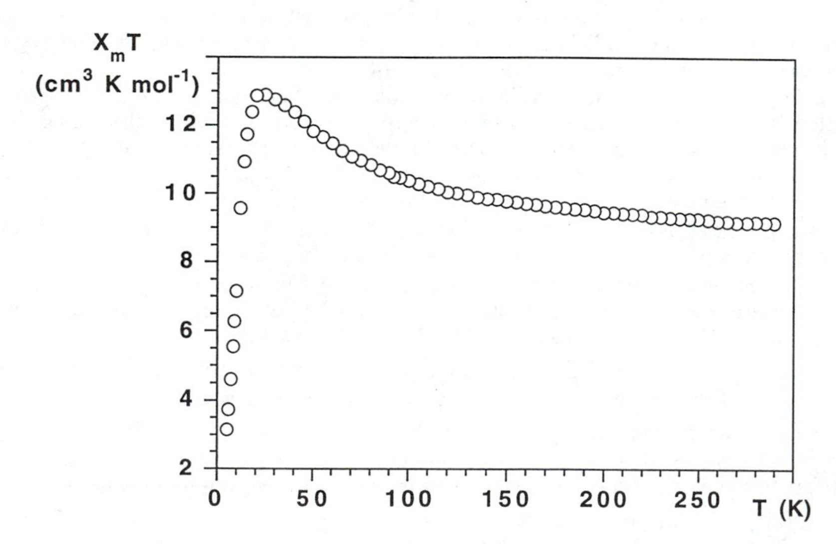

Figure 13. Magnetic measurements for the $[Mn_2(N_3)_4(bipym)]_n$ compound.

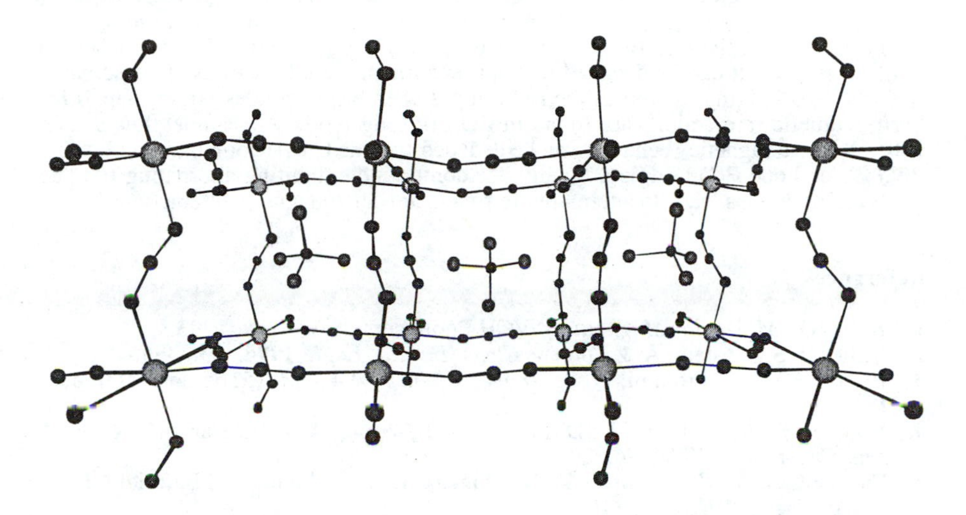

Figure 14. Crystal structure of the $[N(CH_3)_4]_n[Mn(N_3)_3]_n$ compound.

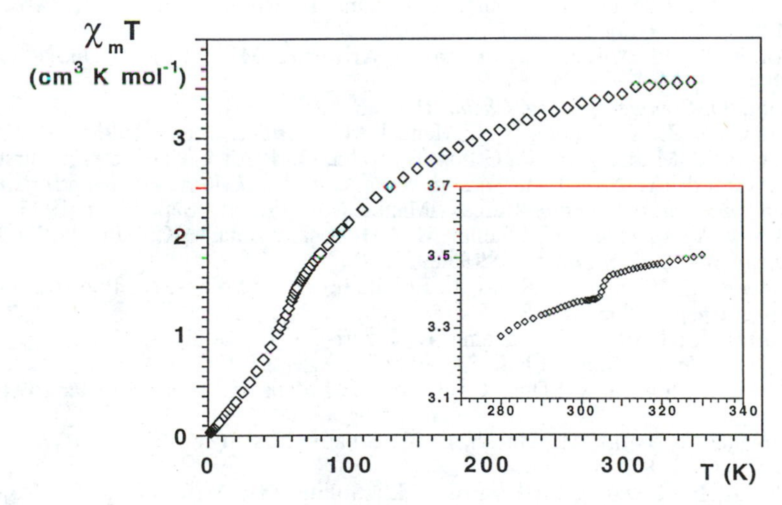

Figure 15. Magnetic measurements for the $[N(CH_3)_4]_n[Mn(N_3)_3]_n$ compound.

Summary

The azido ligand has been shown to be a good bridging agent generating a wide range of structural architectures of predetermined dimensionality. The increasing of coordination vacancies around the metal ions leads to an increase of the dimensionality of the resulting systems. Concerning the magnetic behavior observed in these species, it can be deduced that antiferromagnetic interactions are associated with the EE bridging as well as the EO bridges with bridge angles larger than 108°. Ferromagnetic interactions are found in EO bridging mode with angles lower than 108°. A "ferrimagnetic" behavior can be deduced for one Cu(II) compound with triple (two EE and one EO) bridging. Finally, the compound exhibiting alternating EO and EE bridging modes leads to alternating ferro and antiferromagnetic interactions.

References

1. Kahn, O. *Molecular Magnetism;* VCH Publishers: New York, 1993.
2. Miller, J. S.; Epstein, A. J. *Angew. Chem. Int. Ed. Engl.* **1994**, *33*, 385.
3. Entley, W. R.; Girolami, G. S. *Inorg. Chem.* **1994**, *33*, 5165 and references therein.
4. Wagner, F.; Mocella, M. T.; D'Aniello, M. J.; Wang, A. J. H.; Barefield, E. K. *J. Am. Chem. Soc.* **1974**, *96*, 2625.
5. Pierpont, C. G.; Hendrickson, D. N.; Duggan, D. M.; Wagner, F.; Barefield, E. K. *Inorg. Chem.* **1975**, *14*, 604.
6. Sikorav, R. E.; Bkouche-Waksman, I.; Kahn, O. *Inorg. Chem.* **1984**, *23*, 490.
7. Arriortua, M. I.; Cortés, R.; Lezama, L.; Rojo, T.; Solans, X. *Inorg. Chim. Acta* **1990**, *174*, 263.
8. Cortés, R.; Ruiz de Larramendi, J. I.; Lezama, L.; Rojo, T.; Urtiaga, K.; Arriortua, M. I. *J. Chem. Soc. Dalton Trans.* **1992**, 2723.
9. Cortés, R.; Pizarro, J. L.; Lezama, L.; Arriortua, M. I.; Rojo, T. *Inorg. Chem.* **1994**, *33*, 2697.
10. Kahn, O. *Comments Inorg. Chem.* **1984**, *3*, 105.
11. Thompson, L. K.; Tandon, S. S.; Manuel, M. E. *Inorg. Chem.* **1995**, *34*, 2356.
12. Aebersold, M.; Bergerat, P.; Gillon, B.; Kahn, O.; Pardi, L.; Tukzet, F.; Öhrström, L.; Grand, A. *NATO Advanced Research Workshop on Magnetism- A Supramolecular Function*, Carcans-Maubuisson (France), Sept 16-21, 1995.
13. Escuer, A.; Vicente, R.; El Fallah, M. S.; Ribas, J.; Solans, X.; Font-Bardia, M. *J. Chem. Soc. Dalton Trans.*, **1993**, 2975.
14. Escuer, A.; Vicente, R.; Ribas, J.; El Fallah, M. S.; Solans, X.; Font-Bardia, M. *Inorg. Chem.* **1993**, *32*, 3727.
15. Cortés, R.; Urtiaga, K.; Lezama, L.; Pizarro, J. L.; Goñi, A.; Arriortua, M. I.; Rojo, T. *Inorg. Chem.* **1994**, *33*, 4009.
16. Ribas, J.; Monfort, M.; Diaz, C.; Bastos, C.; Solans, X. *Inorg. Chem.* **1994**, *33*, 484.
17. Ribas, J.; Monfort, M.; Ghosh, B. K.; Solans, X. *Angew. Chem. Int. Ed. Engl.* **1994**, *33*, 2087.
18. Cortés, R.; Lezama, L.; Pizarro, J. L.; Solans, X.; Arriortua, M. I.; Rojo, T. *Angew. Chem. Int. Ed. Engl.* **1994**, *33*, 2488.
19. Mautner, F. A.; Cortés, R.; Lezama, L.; Rojo, T. *Angew. Chem. Int. Ed. Engl.* **1996**, *35*, 78.
20. Rushbrook, G. S.; Wood, P. J. *Mol. Phys.* **1958**, *1*, 257.

Chapter 13

Molecular Magnetic Materials and Small Clusters Containing *N*-Donor Chelated Metal Species Combined with Hexacyanometallate, Tris-oxalatometallate, and Related Bridging Groups

K. S. Murray[1], G. D. Fallon[1], D. C. R. Hockless[2], K. D. Lu[1], B. Moubaraki[1], and K. Van Langenberg[1]

[1]Department of Chemistry, Monash University, Clayton, Victoria 3168, Australia
[2]Research School of Chemistry, Australian National University, Canberra, Australian Capital Territory 0200, Australia

New molecular-based ferromagnetic materials based on the hexacyanoferrate(III) building block are described. They are formed by reaction of bispyrazolylmethane (bpm) chelates of copper(II) and nickel(II) with $K_3[Fe(CN)_6]$ and have formulae $[Ni^{II}(bpm)(H_2O)_2]_3[Fe^{III}(CN)_6]_2.2H_2O$ (**1**) and $K[Cu^{II}(bpm)][Fe^{III}(CN)_6]$ (**2**). Detailed studies of their magnetic susceptibilities and magnetizations showed that long-range ordering occurs at 19K for **1** and 10K for **2**. Attempts to make bimetallic oxalato compounds having 1D, 2D or 3D-bridged structures, and incorporating cationic trispyrazolylmethane (tpm) metal chelates, led unexpectedly to double salt formation. The crystal structures of three of these hydrated materials $[Cu(tpm)_2]_3[Cr(C_2O_4)_3]_2.20H_2O$ (**4**), $[Fe^{II}(tpm)_2]_3[Fe^{III}(C_2O_4)_3]_2.20H_2O$ (**5**), $[Cu^{II}(Me_2tpm)_2]_3[Fe^{III}(C_2O_4)_3].20H_2O$ (**7**) all show the individual cationic and anionic molecules with H-bonded water intermolecular contacts occurring between $[M(C_2O_4)_3]^{3-}$ groups. These contacts are deemed responsible for the weak exchange effects noted at low temperatures in the susceptibilities of these compounds. **5** is rather novel in that it shows spin-crossover behaviour of the $[Fe^{II}(tpm)_2]^{2+}$ centres above 250K whilst displaying field-dependent weak exchange effects below 20K. Finally, the importance of measuring multifield/multitemperature magnetization data on small clusters is stressed.

It is always interesting to look at the way in which a sub-area of chemistry develops. Those brave chemical souls who, in the late 60's and early 70's, began to explore solid-state molecular materials of the optical or magnetic types (*1,2,3*) were regarded with some suspicion by the chemical fraternity. "Too much like physics" was often

0097–6156/96/0644–0201$15.00/0

the catch cry, particularly when neutron experiments were involved or, "we know a lot about short-range order in exchange-coupled cluster complexes, so what is the interest in long-range and/or spontaneous 3-dimensional order?". One of the main reasons why the subject has eventually developed into an important, exciting and very active one, has been the new found ability of chemists, physicists and materials scientists to collaborate together, talk each others language and even understand each others' units. The ready availability of crystal structure determinations and of Squid or Faraday magnetometers capable of automatically varying temperature and magnetic field values over wide ranges has improved the experimental side of the subject enormously. A by no means exhaustive list of recent books and review articles can be found within the following references (*4-9*).

What is a Molecularly Based Magnetic Material?

Arguments rage over the answer to this question. In essence, it is a crystalline material containing discrete molecular species, often called building blocks, which are joined together to make up the total structure. The material displays spontaneous three-dimensional long-range magnetic ordering characteristic of the bulk material. The molecular species can be d- or f-block coordination complexes, organic (p-orbital) radicals, metal-complexes ligated to organic radicals, donor-acceptor complexes or metal-complex clusters. These molecular species can be joined by covalent bridges, intermolecular bridges, through-space effects or lattice forces. Such a definition perhaps excludes some of the most magnetic materials. Thus, if the classical transition metal oxides and metal fluorides are regarded as semi-ionic extended lattice structures, then the Prussian-blue $C^{I}A^{II}[M^{III}(CN_6)]$ family might be regarded likewise even though they contain $M^{III}(CN)_6$ octahedra bridged by CN groups to A^{II} ions, the latter occupying octahedral sites (see similar comments in ref. *10*). The definitions are of little consequence, however, when one consider that some new members of the Prussian-blue family have ferrimagnetic ordering temperatures near, or above, room temperature (see Table I and a very recent review (*10*) for selected examples (*11-15*)). In our work, described below, and that of other groups (*16-18*), the A^{II} ion in the above stoichiometry has been replaced by d-block metal-chelate fragments to yield new molecular ferromagnets with lower symmetry than the parent. Some polymeric species of stoichiometry $[M^{II}(chelate)]_3[M^{III}(CN)_6]_2$ are also described.

Table I. Some Recent Examples of Molecular Magnets Involving $[M(CN)_6]^{n-}$ Building Blocks

	Example	*Type of Magnet*	*$T_c(K)$*	*Reference*
1.	*Lattice Structures*			
	$Cs_2Mn^{II}[V^{II}(CN)_6]$	ferrimagnet	125	*11*
	$Mn^{II}[Mn^{IV}(CN)_6]$	ferrimagnet	49	*12*
	$Cr_3^{II}[Cr^{III}(CN)_6]_2.10H_2O$	ferrimagnet	240	*13*
	$Fe_4^{III}[Fe^{II}(CN)_6]_3.xH_2O$	ferrimagnet	5.6	*14*
	$CsNi^{II}[Cr^{III}(CN)_6].2H_2O$	ferromagnet	90	*15*
2.	*Chain or Sheet Structures*			
	$[Nien_2]_3[Fe(CN)_6]_2.2H_2O$	ferromagnet[a]	18.6	*16*
	$[Nipn_2]_2[Fe(CN)_6](ClO_4).2H_2O$	metamagnet	10	*17*
	$K[Mn(3\text{-}MeOsalen)]_2[Fe(CN)_6].2DMF$	metamagnet	9	*18*

[a] A ground-up sample of single crystals did not show any ordering -see script.
en = 1,2-diaminoethane; pn = 1,3-diaminopropane
3-MeOsalen = N,N'-ethylenebis(3-methoxysalicylaldehydeiminato) dianion

We, and others (*19*), have also "snipped-out" discrete heptanuclear clusters from the parent lattice structure to yield species of type [M((CN)M'(chelate))$_6$] (collaboration with L. Spiccia, Monash University). The synthesis and characterization of medium-to-large sized 'high-spin' coupled cluster species has grown apace in recent years, the interest in such compounds being driven by attempts to (i) make synthetic model for the iron storage protein ferritin, a natural magnetic material i.e. make μ-oxo, μ-hydroxo iron clusters of sizes up to Fe_{19} (*20-22*) (ii) make building blocks capable of being stitched together to yield long-range magnetic ordering effects - there are some philosophical difficulties with this interesting approach which are briefly alluded to later (iii) design new ligands and clusters and understand the intramolecular magnetic coupling theory for, say, M_6 to M_{20} sized clusters (*21*). In this short article we will restrict discussion to a few of our recently studied small clusters and point out some important experimental and theoretical aspects which are not always followed.

Clear expositions of the basic physics of spontaneously ordered ferro-, ferri- (and meta-) magnets have been given recently (*4,5,7,8*) and will not be repeated here. Experimentally, a different protocol is used compared to that employed for paramagnetic and small cluster complexes. In particular, variable-field/variable temperature studies are important, together with field-cooled (FCM), zero-field cooled (ZFCM) and remnant magnetization (RM) studies of the type shown below. In practice, observation of "spontaneous" ordering below T_c usually requires a small applied-field of a few gauss on account of domain randomization effects.

The design rules for successfully making 3-D ordered molecularly-based magnetic materials are still being formulated and serendipity often plays a part. The rules, which have evolved to date, have been given elsewhere (*4,7-9*). Suffice to say that many are based on correlations and experiences gained in exchange-coupled metal-cluster or chain magnetochemistry, e.g. the need for ferromagnetically coupled electron spins to occupy orthogonally oriented magnetic orbitals with a suitable covalent bridging group linking the metal centers together. In cases where covalent links are not present, such as the ion radical salt $[FeCp_2^*]^{\bullet+}[TCNE]^{\bullet-}$, then the challenging concepts of spin-polarization/configuration interaction and other "through-space" effects need to be mastered (*7,8,23,24*).

New Molecular Magnets Formed Between Bis-pyrazolylmethane Metal Chelates and K_3 [Fe(CN)$_6$]

The compounds described here were obtained in an ongoing project and serve to illustrate the principles and *modus operandi* referred to in the previous sections. Our work involves use of bis- and trispyrazolylmethane metal chelates. Polypyrazolymethanes are the neutral analogues of the much studied polypyrazolylhydridoborate anions (*25*); their structure and labels are shown in Figure 1. Guided by our early studies on ferromagnetic coupling in Fe^{III}-μ-CN-Cu^{II} binuclear porphyrin species (*26*) and by an early structural study of Morpurgo et al.(*27*) on the polymeric compound $[Cu^{II}(dien)]_3[Fe^{III}(CN)_6]_2$-.6$H_2O$, we anticipated the formation of magnetically ordered complexes with possible formulations $[M^{II}(bpm)_x]_3$ $[Fe^{III}(CN)_6]_2.yH_2O$ and $K[M^{II}(bpm)_x][Fe^{III}(CN)_6].yH_2O$. Compounds of these stoichiometries were indeed formed upon reaction of trans-$[M(bpm)_2(H_2O)_2](NO_3)_2$, where M^{II}=Ni, Cu, with aqueous solutions of $K_3[Fe(CN)_6]$.

The magnetic properties of the compounds are summarized in Table II. Samples can be reproducibly made and crystals of **1** have recently been obtained using solvent diffusion methods. Fe- and Ni-EXAFS studies on powdered samples of the starting materials and on $[Ni(bpm)(H_2O)_2]$ $[Fe(CN)_6]_2.2H_2O$ (**1**) clearly show that the $Fe(CN)_6$ group remains intact and a peak in the Fourier transform at *ca* 5Å is indicative of an Fe-Ni distance within Fe-CN-Ni chains (Kennedy, B.J., University of Sydney, unpublished data). The plot of magnetic moment versus temperature of **1**

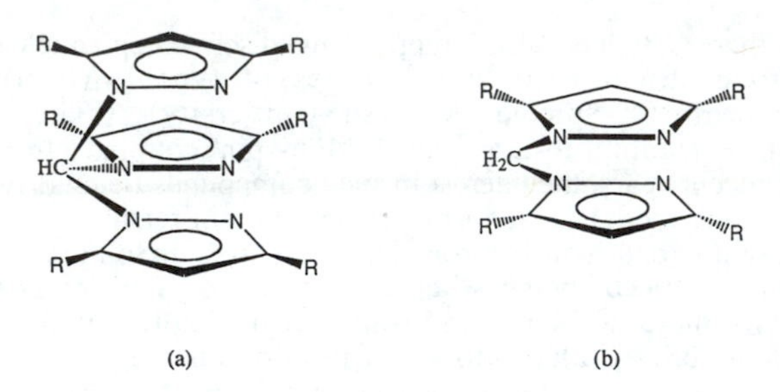

Figure 1. Structures of (a) trispyrazolylmethane ligands; R = H is tpm; R = CH_3 is Me_2tpm; (b) bispyrazolylmethane ligands; R = H is bpm; R = CH_3 is Me_2bpm.

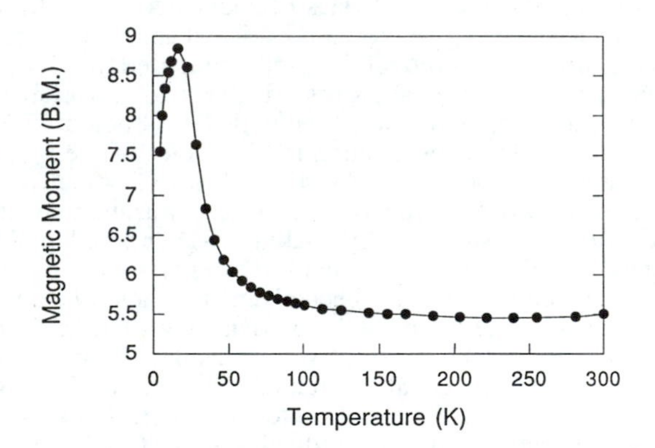

Figure 2. Plot of magnetic moment, per molecular formula, versus temperature for $[Ni(bpm)(H_2O)_2]_3[Fe(CN)_6]_2.2H_2O$ in a field of 1 Tesla.

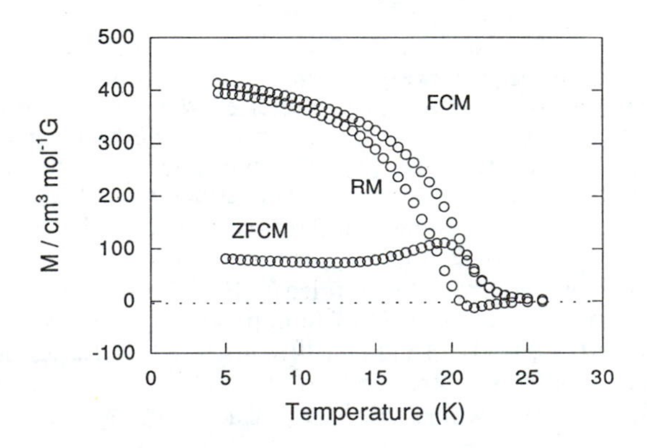

Figure 3. Magnetization versus temperature plots for a powdered sample of $[Ni(bpm)(H_2O)_2]_3[Fe(CN)_6]_2.2H_2O$ in a field of 10 gauss (RM = remnant magnetization, FCM = field-cooled magnetization, ZFCM = zero-field cooled magnetization).

Table II. Magnetic Behaviour of Bis-Pyrazolymethane and Related Metal Chelate Derivatives Bridged to $Fe^{III}(CN)_6^{3-}$

No.	*Complex*	*Type of Magnet*	$T_c(K)$
1	$[Ni^{II}(bpm)(H_2O)_2]_3[Fe^{III}(CN)_6]_2.2H_2O$	ferromagnet	19
2	$K[Cu^{II}(bpm)][Fe^{III}(CN)_6]$	ferromagnet	10
3	$[Cu^{II}(dien)]_3[Fe^{III}(CN)_6]_2.6H_2O$	a	

[a] The crystal structure shows a complicated polymeric plus dinuclear (dien)Cu-NC-$Fe(CN)_5$ fragments (*27*).
Preliminary magnetic data obtained here suggest either no ordering or weak ferrimagnetic ordering with $T_c < 4.2$ K.

obtained in a field of 1 Tesla, shows a rapid increase from the constant value of 5.5 μ_B noted in the range 300-50 K to reach 8.9 μ_B at *ca* 18K, then a rapid decrease towards 7.5 μ_B at 4.2K (Figure 2). This is symptomatic of ferromagnetic coupling occurring, although, the μ_{max} value is close to that expected for a Ni^{II}_3 Fe^{III}(low spin)$_2$ fragment, with S=4, and lower than anticipated for a 3-dimensionally ordered material. The material is, nevertheless, ordered as can be seen in Figure 3, where the FCM, ZFCM and RM versus temperature plots in a field of 10 gauss, clearly show a T_c value of 19K. The apparent negative values of the RM at ca. 22K are probably not real but more likely reflect uncertainties in the very small measured values. A plot of magnetization versus field (± 0.6 Tesla), at 4.2K, displays hysteresis and a coercive field of 275 gauss was evident. A detailed study of magnetization versus field (0-5 Tesla) in the temperature range 4.2 to 24K shows some unusual features which still need to be explained. The material begins to saturate at high fields but not completely and the highest magnetization value, at 4.2K, whilst close to the expected S=4 value, requires a zero-field splitting of this state, with $D \approx 5\ cm^{-1}$, to get good agreement of the M *vs* H curve. The magnetization curves at temperatures above 4.2K are closer to those expected for S=3 and thus an overlap of M_s sublevels, from S=4 and S=3 states seems likely.

The magnetic behaviour of **1** is similar, but not identical, to that recently reported (*16*) for a compound of related stoichiometry, $[Nien_2]_3[Fe(CN)_6]_2.2H_2O$. Notably different is the μ_{max} value of 26 μ_B observed in the en complex, but the T_c values are similar. A very interesting observation on the $Nien_2$-based material was reported during this symposium (Okawa, H., Ohba, M. Kyushu University, unpublished data), the results of which confirm that we should all proceed with caution in this area. Thus, use of a ground sample of individual single crystals of the complex, rather than the bulk preparation whose magnetism has been reported (*16*) and briefly described above, did not show any magnetic ordering but rather a gradual increase in μ, from its 300K value, reaching a μ_{max} value of 8.8μ_B at ca. 15K. Rigorous checking of different batches different crystalline sizes and the amount of grinding should, therefore, become general practice. Complexes **1** and **2** are two such examples being so checked.

It is possible that **1** has a similar 'rope-like' structure to that observed in the en complex. However, as noted in Table I, small changes in the chelate group bonded to Ni can lead to different structural motifs, the pn compound being one such example. The recently reported Mn(3-MeOsalen) derivative (*18*) possesses a somewhat similar stoichiometry to our complex **2**, and to the parent Prussian-blue formula. It displays a 2D-network structure and metamagnetic-like curvature in the low-field region of the hysteresis plot, i.e. a change from antiferromagnetic coupling (between layers) to ferromagnetic coupling (within the ... Mn-Fe-Mn .. layers). There is much scope for using other metal chelates, with different spin values, to obtain new molecular magnets incorporating different $M(CN)_6$ or $M(NCS)_6$ linking groups.

Some Surprises in Bimetallic Tris-Oxalatometallate Complexes; Intermolecular Coupling in Some Double-Salts

The magnetic behaviour of the 2-D layer structures and 3-D extended network structures of types $[M^{II}M^{III}(C_2O_4)_3]^-$ and $[M^{II}_2(C_2O_4)_3]^{2-}$ studied by Okawa et al. (*28*), Decurtins et al. (*29*) and Day et al. (*30*) has recently been summarized by Kahn (*10*). Subtle variations have been induced in the structural topography and in the resultant magnetism by use, for instance, of chiral tris-bipy metal cations such as $Fe^{II}(bpy)_3^{2+}$ which was able to induce a 3-D network in the anionic portion of $[Fe^{II}bipy_3][Fe^{II}(Fe^{II}(C_2O_4)_3)]$ because of its size and its similar chirality to $Fe^{II}(C_2O_4)_3^{4-}$ (*8,10,29*). The nature of the magnetic ordering in these materials depends on the combination of metal ions. Thus, in a bimetallic example such as $(Bu_4N)[Mn^{II}Cr^{III}(C_2O_4)_3]$, the orthogonally disposed metal orbitals, $t_{2g}{}^3e_g{}^2$ and $t_{2g}{}^3$, lead to ferromagnetic ordering below a T_c of 6K (*28*). Quite unexpected and rather rare observations of negative FCM occurred, at temperatures below a T_c of 43K, in the ferrimagnet $Bu_4N[Fe^{II}(Fe^{III}(C_2O_4)_3]$ and this has been ascribed to possible spin-glass behaviour (*10,30*). The antiferromagnetically coupled $Cr^{II}Cr^{III}$ analog does not show any ordering (*31*). Interestingly, the first structurally characterized $[M^{II}_2(C_2O_4)_3]^{2-}$ compound, in which M^{II}=Cu, was obtained by a somewhat bizarre route and does not appear to have had its magnetism studied (*32*).

Our own work in this area, which is at an early stage of development, was based on similar reasoning and approaches to those described above for the $Fe(CN)_6^{3-}$–$M(bpm)^{2+}$ systems. It was hoped that incorporation of cationic bis-trispyrazolylmethane complexes into anionic $[M^{II}M^{III}(C_2O_4)_3]^-$ networks would lead to novel structures and magnetic ordering in a related way to that recently discussed by Kahn (*10*). In retrospect, we probably chose cations of the wrong charge and, in the case of Decurtins $[M^{II}_2(C_2O_4)_3]^-$ and $[A^IM^{III}(C_2O_4)_3]^{2-}$ type networks (*29*), cations which lacked chirality. Reaction, for instance, of $[Cu(tpm)_2](NO_3)_2$, $MnCl_2.4H_2O$ and $K_3[Cr(C_2O_4)_3]3H_2O$ in 1:2:2 ratio in water and air yielded a mixture of large purple crystals (**4**) and thin blue prisms of $[Cu(tpm)(C_2O_4)]$. The same purple crystals were obtained under anaerobic conditions and, after the structure was known, when Mn^{II} was omitted from the reaction. In fact, a double salt of formula $[Cu(tpm)_2]_3[Cr(C_2O_4)_3]_2.20H_2O$ (**4**) had been obtained, rather than the desired bridged-network structure, presumably because of the principle of charge neutralization, combined with size effects (*33*) and chirality mismatches. Kahn has also recently made a 1:1 salt $[Ru(bipy)_3][Cu(oxamate)].9H_2O$ and used it as a precursor to a $M^{II}_2Cu^{II}_3$ ferrimagnet (*33*).

We obtained a group of such double salts by even more unexpected routes. For instance, when following the successful synthesis and structural elucidation of a zig-zag oxalato-bridged linear chain complex $[(bpm)Cu(C_2O_4)]_n$, which displayed weak intramolecular ferromagnetic coupling ($J=+1.0cm^{-1}$, $g=2.17$) (Van Langenberg, K.; Moubaraki, B.; Murray, K.S.; Hockless, D.C.R., Monash and The Australian National Universities, unpublished data), our related attempts to make oxalato-bridged Fe^{III}tpm compounds gave some surprises. Reaction of $(tpm)FeCl_3$ with oxalic acid yielded $[Fe^{II}(tpm)_2]_3[Fe^{III}(C_2O_4)_3]_2.20H_2O$ (**5**) whilst with $K_2[Cu^{II}(C_2O_4)]_2$ it yielded $[Cu^{II}(tpm)_2]_3[Fe^{III}(C_2O_4)_3]_2.20H_2O$ (**6**). The crystal structure was solved on complex **7** the dimethyl-tpm analog of **6**. Ligand-exchange and reduction processes have occurred in these reactions.

Table III. Homo- and Bimetallic Double Salts of Type $[M(tpm)_2]_3[M'(C_2O_4)_3]_2 \cdot xH_2O$

No.	*Complex*	μ_{obs}*(300K)* (μ_B)	μ_{calc} *(spin-only)*	*Features of Magnetism (4-300K)*
4	$[Cu^{II}(tpm)_2]_3 [Cr^{III}(C_2O_4)_3]_2 \cdot 20H_2O$	6.23	6.25	Curie-like between 300-20K; weak ferromagnetic coupling below 20K; no long range ordering in this temp. range.
5	$[Fe^{II}(tpm)_2]_3 [Fe^{III}(C_2O_4)_3]_2 \cdot 20H_2O$	10.3	11.92[a] 8.36[b]	Curie-like between 230-20K; field dependent μ values below 20K; increase in μ between 230-350K due to crossover Fe^{II} low-spin to high-spin
6	$[Cu^{II}(tpm)_2]_3 [Fe^{III}(C_2O_4)_3]_2 \cdot 20H_2O$	9.85	8.89	Curie-like between 300-4K weak antiferromagnetic coupling, and/or zero-field splitting noted by decrease in μ below 50K. Magnetization data obtained between 2-10K and 0.5 to 5 Tesla confirm this.

[a] $\{Fe^{III}, S=5/2; Fe^{II}, S=2\}$ [b] $\{Fe^{III}, S=5/2; Fe^{II}, S=0\}$

Crystal structures have been solved on **4**, **5** and **7** (Hockless, D.C.R., Australian National University; Fallon, G.D., Monash University, unpublished data). Some structural features are shown in Figures 4-6. Any magnetic interactions between nearest $[M^{III}(C_2O_4)_3]^{3-}$ anions, or between cation and anion would be expected to be very weak (see Kahn, ref. *10*, p.251, Carlin, Palacio et al. (*35*) and Figgis et al. (*36*)), and would likely be propagated by intermolecular effects such as those occurring via H-bonded water molecules. A brief qualitative survey of the observed magnetism is given in Table III.

The very weak ferromagnetic coupling in **4** probably occurs via $Cr^{III}...Cr^{III}$ interactions mediated by H-bonding of water to neighbouring oxalate groups. Closest intermolecular contact distances include; $O(1)...O_w$, 2.90(1); $O(2)...O_w$, 2.94(2); O(2)...O(2), 3.40(2), whilst $Cr(1)...O_w$, 3.999(8)Å. The nearest Cu...Cr distance is 6.814(2) and Cr...Cr ca. 10.16Å. Likewise, the very weak antiferromagnetic coupling in **6**, which leads to a small decrease in μ below 50K, is mediated by H-bonded water interactions occurring between $Fe^{III}(C_2O_4)_3{}^{3-}$ groups, as shown in the structure of the $Cu(Me_2tpm)^{2+}$ analog, **7** (Figure (6b)).

Complex **5** shows the most intriguing magnetic behaviour of these double-salts. As is evident in Figure 7, μ remains constant between 200-230K, irrespective of the size of the applied-field, then it increases linearly between 230 and 350K. This increase is due to a spin-crossover between $^1A_{1g}$ and $^5T_{2g}$ states in the $[Fe^{II}(tpm)_2]^{2+}$ moieties, a phenomenon already noted in the ClO_4^- and NO_3^- salts of this six-coordinate FeN_6 species (*37* and Moubaraki, B.; Murray, K.S.; White, A.H., Universities of Monash and Western Australia, unpublished data). The average Fe-N bond lengths (1.95Å) and N-Fe-N angles (87.2°) in the $[Fe(tpm)_2]^{2+}$ components of **5**

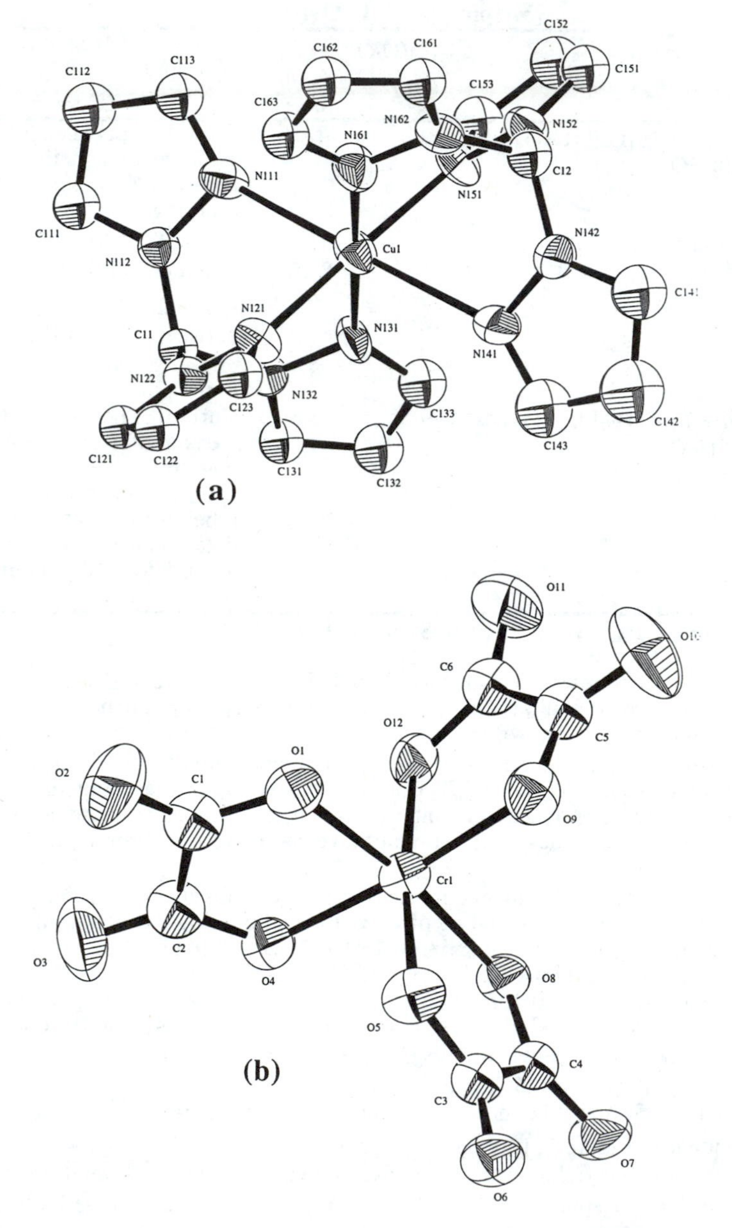

Figure 4. (a) Molecular structure of one of the two independent $[Cu(tpm)_2]^{2+}$ cations in $[Cu(tpm)_2]_3[Cr(C_2O_4)_3]_2.20H_2O$ (**4**) (b) Molecular structure of the $[Cr(C_2O_4)_3]^{3-}$ moiety in **4** (c) Packing diagram, including H_2O molecules, of **4**.

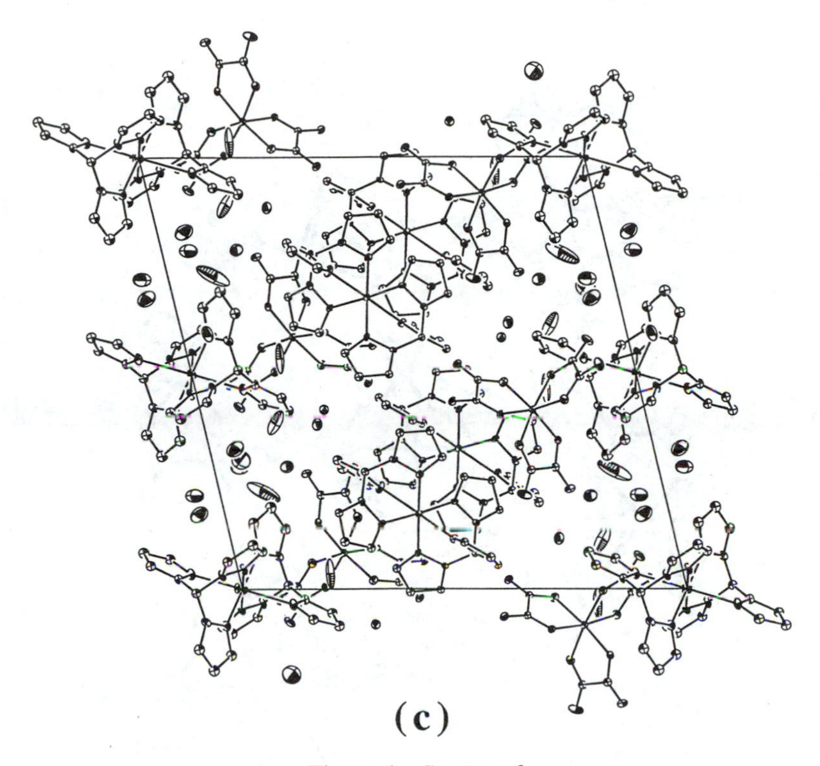

Figure 4. *Continued*

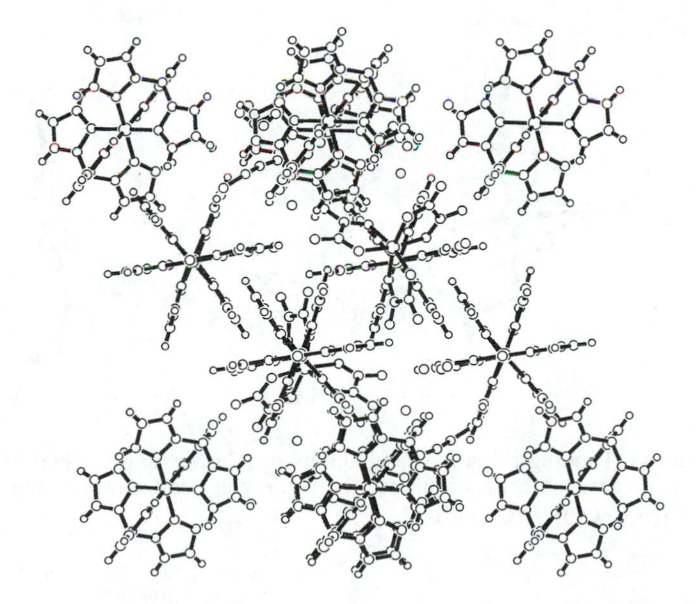

Figure 5. Crystal packing diagram of $[Fe(tpm)_2]_3[Fe(C_2O_4)_3]_2.20H_2O$ (**5**) with water molecules omitted for clarity.

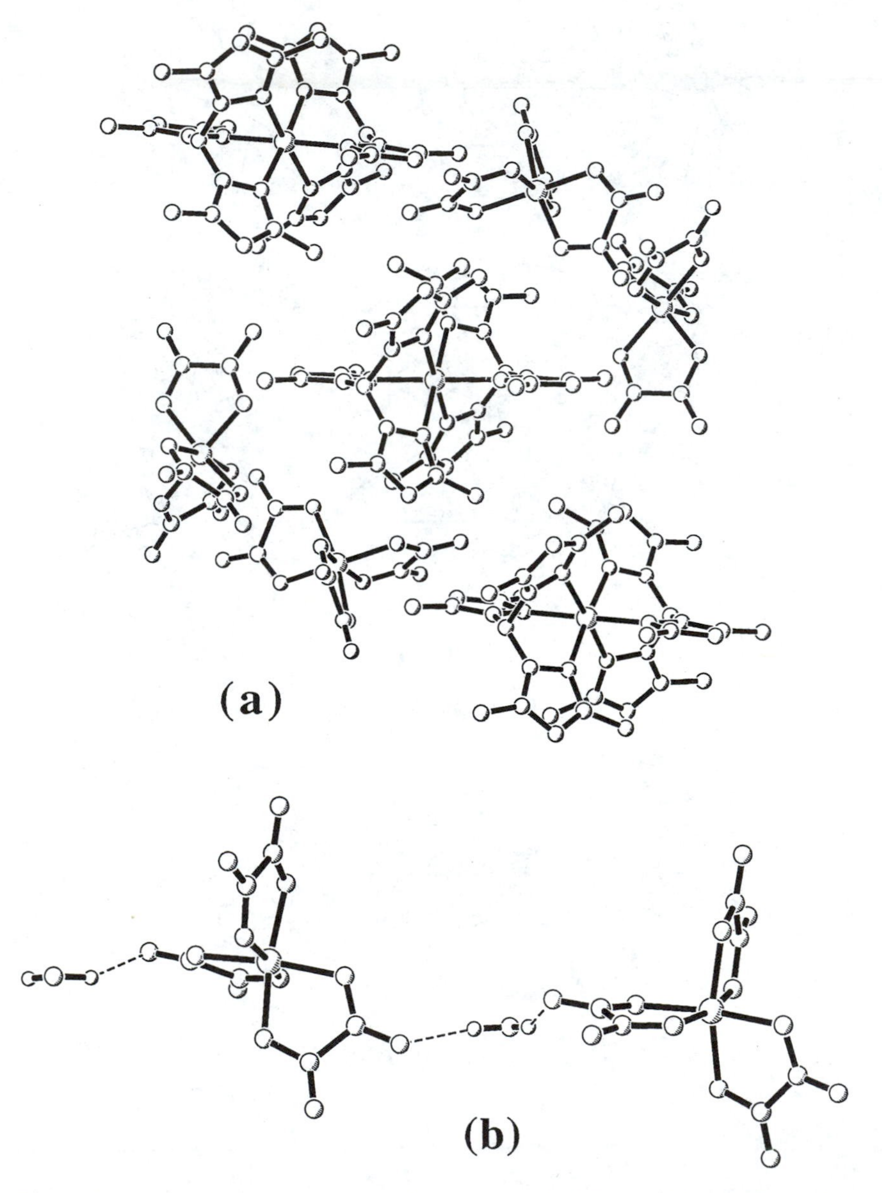

Figure 6. (a) Packing of cations and anions in $[Cu(Me_2tpm)_2]_3[Fe(C_2O_4)_3]_2$·-$20H_2O$ (**7**) water molecules omitted; (b) Hydrogen-bonded water interactions between $[Fe(C_2O_4)_3]^{3-}$ anions in **7**.

are similar to those in the NO_3^- salt. The Mössbauer spectrum of **5** at 77K shows two superimposed quadrupole doublets with parameters δ=0.47mm s^{-1}, ΔE_Q=0.27mm s^{-1} and δ=0.45mm s^{-1}, ΔE_Q= 0.23mm s^{-1}, again very similar to those observed for the low-spin forms of $[Fe^{II}(tpm)_2](PF_6)_2$ (*38*) and, coincidentally, very similar to those of high-spin $[Fe^{III}(C_2O_4)_3]^{3-}$ (*39*). The value of μ in the plateau portion of the μ/T plot of **5** is higher than that anticipated for {3 x Fe^{II} (low-spin) + 2 x Fe^{III} (high-spin)} viz 8.7μ_B, even allowing for a value of 1.25μ_B per Fe^{II} noted in $[Fe(tpm)_2](ClO_4)_2$ (*38*). This could be due to the $[Fe(tpm)_2]^{2+}$ moieties retaining more of the high-spin character in the present case or to partial dehydration of these highly hydrated crystals; there is no evidence for either effect at this point. Certainly, future magnetic studies on fully dehydrated samples of this series of hydrated salts are warranted. The low temperature field dependence of μ, shown in Figure 7, is intriguing. Behaviour of this kind, in which an increase in μ occurs with decreasing temperature, under conditions of low applied-field, changing to a decrease in μ under high-fields, is indicative of Zeemans mixing and thermal depopulation of a closely spaced series of low-lying M_S levels usually brought about by weak ferromagnetic exchange-coupling effects. $Fe^{III}(C_2O_4)...(C_2O_4)Fe^{III}$ intermolecular contacts, mediated by H-bonded water interaction are, again, evident from the crystal packing diagram of **5** and are deemed responsible. Zero-field splitting of the $^6A_{1g}$ states in individual $[Fe^{III}(C_2O_4)_3]^{3-}$ groups could not give this field dependent behaviour.

Some Features of Intramolecularly Coupled Cluster Complexes

Great advances in an understanding of the magnetism of spin-coupled clusters of sizes up to about M_{20} have been made in recent years and some aspects are described elsewhere in this symposium and in various reviews (*5,21*). Theoretical methods needed to calculate the ground energy levels and susceptibilities of clusters of high-spin metal centres such as Fe_8^{III}, etc., are being successfully developed by Gatteschi and his group (*21*). Heisenberg-type vector-coupling models, which have been so successful in small clusters calculation, can not cope in such cases.

Experimental approaches have also advanced in concert with theoretical developments. As well as making detailed magnetization and susceptibility measurements over wide ranges of field (DC) and temperature, it has recently been shown that frequency dependent out-of-phase AC susceptibility (*41,42*) and very high frequency (250-550 GHz) E.S.R. techniques (*43*) are required to gain a full understanding of cluster magnetism (*21,43*). This is particularly so, for instance, in manganese (III/IV) clusters containing μ-oxo and μ-carboxylato bridged fragments which are actually magnetized at low temperatures and behave virtually like single domains or superparamagnets. They display hysteresis in DC fields and frequency shifts in very small AC fields (*41*). The key reason behind such behaviour is the way that domains move in an applied-field which, in turn, relates to the degree of magnetic anisotropy and zero-field splitting in the clusters (*21,42*).

It is interesting to note that the ability of individual cluster molecules to become magnetized, in a superparamagnetic manner, has led to a modification of the expectations for such molecules in the context of molecular-based magnetic materials. As indicated at the beginning of this article, the original aims of a number of groups was to try and stitch together, via covalent bridges, clusters having high-spin ground-states thus yielding long-range ferromagnetic ordering between cluster spins. By its very nature this is, in retrospect, probably not feasible since the desirability of producing equal interactions in a 3-D manner is not likely to occur, unless in a ferrimagnetic-like fashion.

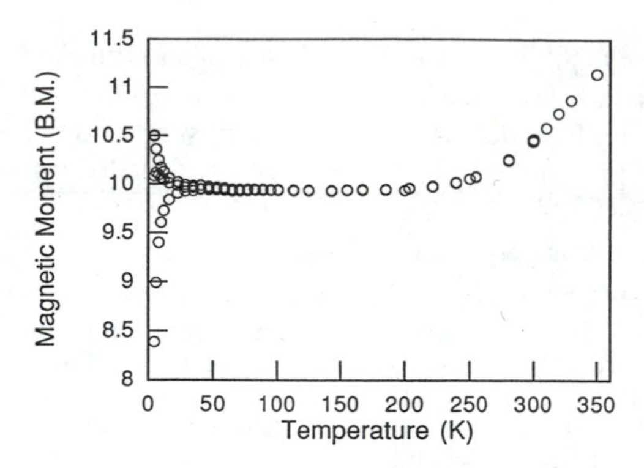

Figure 7. Plots of magnetic moment versus temperature, per mol of $[Fe(tpm)_2]_3[Fe(C_2O_4)_3]_2 \cdot 20H_2O$, at various applied-field values. The field-dependent portion below 20K are for fields of 0.1 Tesla (top) 1 Tesla, and 3 Tesla (bottom)

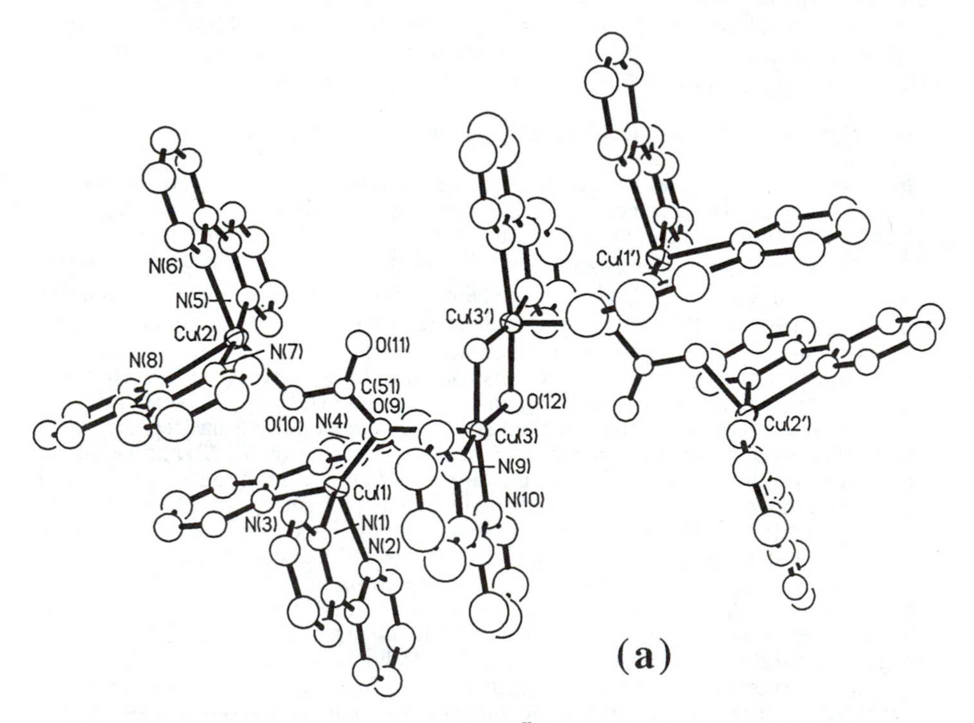

Figure 8. (a) Molecular structure of $[Cu^{II}_6(bipy)_{10}(\mu\text{-}CO_3)_2(\mu\text{-}OH)_2](ClO_4)_6 \cdot 4H_2O$. Hydrogen atoms and the ClO_4^- and H_2O groups are not shown; (b) Plots of effective magnetic moment (per Cu_6) of $Cu^{II}_6(bipy)_{10}(\mu\text{-}CO_3)_2(\mu\text{-}OH)_2](ClO_4)_6 \cdot 4H_2O$ versus temperature (2-10K) as a function of applied field (field values in Tesla): ❑, 0.1; ▲, 0.5, ●, 1.0; ◆, 2.0; ■, 3.0). The solid lines are the calculated values using the model and best-fit parameters given in ref (*40*). (Reproduced with permission from ref. 40)

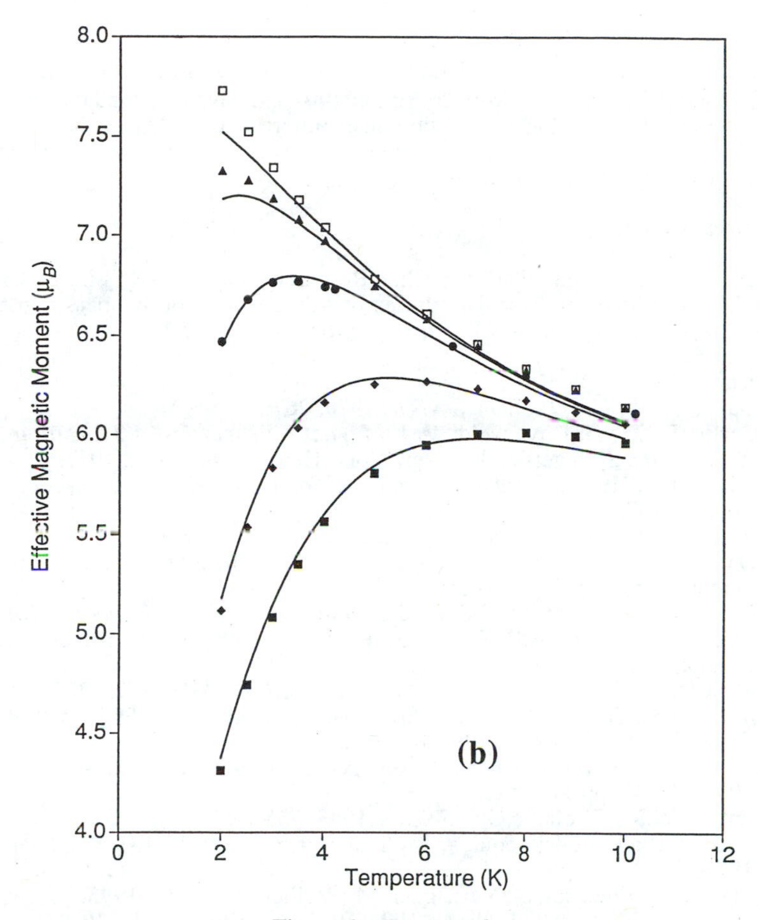

Figure 8. *Continued*

In addition to the abovementioned advances we present, briefly, some pertinent features of a novel ferromagnetically coupled hexanuclear Cu^{II} bipy cluster which, as shown in Figure 8(a), contains hydroxo and carbonato bridging groups (*40*). A detailed study of the cluster ground-state, S=3, and of low-lying energy levels was made by means of variable-field/variable-temperature measurements in combination with calculations employing the thermodynamic, field-dependent form of susceptibility, rather than the commonly used Van Vleck equation. At low temperatures and high-fields , the use of Van Vleck's equation is fraught with large errors on account of Zeeman mixing between, and thermal depopulation of, closely spaced M_S levels. As is clearly demonstrated in Figure 8(b), this kind of treatment, when applied to a series of isofield plots of μ vs. temperature, some of which show maxima in μ whilst other do not, gives a unique set of J and g values. There is no need to invoke effects such as superimposed antiferromagnetic ordering, a rather common procedure which might or might not be operative.

Finally, we note the importance of calculating the energies of coupled spin-states in clusters which contain competing exchange pathways and/or different metal ions, in terms of a *ratio* of exchange parameters. This can lead to the occurrence of different ground-states for different ratios, and to the phenomenon of spin-frustration. We have recently made such calculations for a novel series of linear trinuclear $M^{III}(\mu$-O)$Ru^{IV}(\mu$-O)M^{III} complexes, M = Cr, Mn, Fe (*44*) and have reviewed the situation in various tetranuclear topologies e.g. rhombus, 'butterfly', etc. (*45*). Others have also realized the importance of using correlation energy diagram of E vs J/J' and specific applications are given in reference *45*.

Acknowledgements

This work was supported by an Australian Research Council (ARC) Large Grant, to KSM. The authors thank Associate Professor J.D. Cashion for the measurement and fitting of Mössbauer spectra and Dr Qi Zhang for synthesis of **2**.

Literature Cited

1. Day, P.; Hall, I.D. *J. Chem. Soc. (A)* **1970**, 2679.
2. Carlin, R.L.; van Duyneveldt, A.J. *Magnetic Properties of Transition Metal Compounds*; Springer-Verlag: New York, Heidelberg, Berlin, 1977.
3. Stiddard, M.H.B. *Elementary Language of Solid State Physics* ; Academic Press: London, 1975.
4. Kahn, O *Molecular Magnetism*; V.C.H.: New York, 1993.
5. O'Connor, C.J., Ed. *Research Frontiers in Magnetochemistry* ; World Scientific: Singapore, 1993.
6. Gatteschi, D.; Kahn, O.; Miller, J.S.; Palacio, F., Eds. *Magnetic Molecular Materials*, NATO ASI Series E; Kluwer Academic: Dordrecht, Boston, London, 1991.
7. Miller, J.S.; Epstein, A.J. *Angew Chem. Intl. Ed. Engl.* **1994**, *33*, 385.
8. Miller, J.S.; Epstein, A.J. *Chem. Eng. News* , October 2nd, **1995**, pp.30-41.
9. Gatteschi, D. *Adv. Mater.* **1994**, *6*, 635.
10. Kahn, O. *Adv. Inorg. Chem.*, Sykes, A.G., Ed.; Academic Press: San Diego, 1995, Vol 43, pp 179-259.
11. Entley, W.R.; Girolami, G.S. *Science* **1995**, *268*, 397.
12. Klenze, R.; Kanellakopoulos, B.; Trageser, G.; Eysel, H.H. *J. Chem. Phys.* **1980**, *72*, 5819.
13. Mallah, T.; Thiebaut, S.; Verdaguer, M.; Veillet, P. *Science* **1993**, *262*, 1554.
14. Herren, F.; Fischer, P.; Ludi, A.; Hälg, W. *Inorg. Chem.* **1980**, *19*, 956.
15. Gadet, V.; Mallah, T.; Castro, I.; Vergaguer, M. *J. Am. Chem. Soc.* **1992**, *114*, 9213.
16. Ohba, M.; Maruono, M.; Okawa, H.; Enoki, T.; Latour, J.M. *J. Am. Chem. Soc.* **1994**, *116*, 11566.
17. Ohba, M.; Okawa, H.; Ito, T.; Ohto, A. *J. Chem. Soc. Chem. Commun.* **1995**, 1545.
18. Miyasaka, H.; Matsumoto, N.; Okawa, H.; Re, N.; Gallo, E.; Floriani, C. *J. Am. Chem. Soc.* **1996**, *118*, 981.
19. Mallah, T.; Auberger, C.; Vergaguer, M.; Veillet, P. *J. Am. Chem. Soc. Chem. Commun.* **1995**, 61.
20. Heath, S.L.; Powell, A. *Angew. Chem. Int. Ed. Engl.* **1992**, *31*, 191.
21. Gatteschi, D.; Caneschi, A.; Pardi, L.; Sessoli, R. *Science* **1994**, *265*, 1054.
22. Taft, K.L.; Papaefthymiou, G.C.; Lippard, S.J. *Science* **1993**, *259*, 1302.
23. Kollmar, G.; Kahn, O. *Acc. Chem. Res.* **1993**, *26*, 259.
24. Miller, J.S.; Epstein, A.J.; Reiff, W.M. *Chem. Rev.* **1988**, *88*, 201.
25. Trofimenko, S. *Progr. Inorg. Chem.* **1986**, *34*, 115.
26. Gunter, M.J.; Berry, K.J., Murray, K.S. *J. Am. Chem. Soc.* **1984**, *106*, 4227.
27 Morpurgo, G.O.; Mosini, V.; Porta, P.; Dessy, G.; Fares, V. *J. Chem. Soc. Dalton Trans.* **1981**, 111.

28. Tamaki, H.; Zhong, Z.J.; Matsumoto, N.; Kida, S.; Koikawa, M.; Achiwa, N.; Hashimoto, Y.; Okawa, H. *J. Am. Chem. Soc.* **1992**, *114*, 6974.
29. Decurtins, S.; Schmalle, H.W.; Schneuwly, P.; Ensling, J.; Gütlich, P. *J. Am. Chem. Soc.* **1994**, *116*, 9521.
30. Mathoniere, C.; Carling, S.G.; Yusheng, D.; Day, P. *J. Chem. Soc. Chem. Commun.* **1994**, 1551.
31. Nuttall, C.J.; Bellitto, C.; Day, P. *J. Chem. Soc. Chem. Commun.* **1995**, 1513.
32. Sundberg, M.R.; Kivekäs, R.; Koskimies, J.K. *J. Chem. Soc. Chem. Commun.* **1991**, 526.
33. Mingos, D.M.P.; Rohl, A.L. *Inorg. Chem.* **1991**, *30*, 3769.
34. Turner, S.S.; Michaut, C.; Kahn, O.; Ouahab, L.; Amouyal, E. *New J. Chem.* **1995**, *19*, 773.
35. Moron, M.C.; Palacio, F.; Pons, J.; Casabo, J.; Solans, X.; Merabet, K.E.; Huang, D.; Shi, X.; Teo, B.K.; Carlin, R.L. *Inorg. Chem.* **1994**, *33*, 746.
36. Reynold, P.A.; Delfs, C.D.; Figgis, B.N.; Moubaraki, B.; Murray, K.S. *Austral. J. Chem.* **1992**, *45*, 1899.
37. McGarvey, J.J.; Toftlund, H.; Al-Obaidi, A.H.R.; Taylor, K.P.; Bell, S.E.J. *Inorg. Chem.* **1993**, *32*, 2469.
38. Winkler, H.; Trautwein, A.X.; Toftlund, H. *Hyperfine Inter.* **1992**, *70*, 1083.
39. Bancroft, G.M.; Dharmawardhena, K.G.; Maddock, A.G. *J. Chem. Soc. (A)* **1969**, 2914.
40. Kruger, P.E.; Fallon, G.D.; Moubaraki, B.; Berry, K.J.; Murray, K.S. *Inorg. Chem.* **1995**, *34*, 4808.
41. Wemple, M.W.; Adams, D.M.; Hagen, K.S.; Folting, K.; Hendrickson, D.N.; Christou, G. *J.. Chem. Soc. Chem. Commun.* **1995**, 1591.
42. Squire, R.C.; Aubin, S.M.J.; Folting, K.; Streib, W.E.; Christou, G.; Hendrickson, D.N. *Inorg. Chem.* **1995**, *34*, 6493.
43. Barra, A.L.; Caneschi, A.; Gatteschi, D.; Sessoli, R. *J. Am. Chem. Soc.* **1995**, *117*, 8855.
44. Berry, K.J.; Moubaraki, B.; Murray, K.S.; Nichols, P.J.; Schulz, L.D.; West, B.O. *Inorg. Chem.* **1995**, *34*, 4123.
45. Murray, K.S. *Adv. Inorg. Chem.,* Sykes, A.G. Ed. Academic Press: San Diego, 1995, vol 43, pp 261-358.

EXTENDED ORGANIC SYSTEMS

Chapter 14

Design of Organic-Based Materials with Controlled Magnetic Properties

Paul M. Lahti

Department of Chemistry, University of Massachusetts, Box 34510, Amherst, MA 01003–4510

Abstract. Organic chemistry has a strong role to play in delineating theoretical, computational, and experimental strategies for designing new three-dimensional materials with interesting and potentially useful magnetic properties.

The past decade has witnessed a tremendous growth in research aimed at applying chemical principles to the design of electronic materials. One of the most ambitious of these research areas is the design of molecular magnetic materials. Modern civilization has come to depend upon magnetic technologies. Credit cards identify us, cassette tapes amuse us, and on-line computer access services inform us. All of these capabilities -- and many, many more -- would not exist in their present form without magnetic technology. All of this technology is based upon materials of the ferromagnetic triad of inorganic elements in the periodic table, which have been known since the days that early explorers used "lodestone" to detect Earth's magnetic north pole. The use of these materials has matured in the past century to become a modern industry generating billions of US dollars in sales.

For many years, scientific research on magnetism consisted of measuring the magnetic properties of various elements and alloys as functions of temperature and composition. That work led to the present technologies, which are quite impressive. In 1987, it was possible to purchase a 250 Megabyte hard drive computer disk occupying about 6 cubic feet for about US$22,000. In 1995, it is possible to purchase a 9000 Megabyte disk occupying about 40-50 cubic *inches* for US$2,500. Although it would seem that such improvements continue apace every year, some believe that the end is in sight for further optimization of current technology.[1] Hence, scientific research on magnetism is evolving to the next level, from the elucidation and optimization of properties in existing materials, to the use of molecular principles to design new materials with novel properties. This is the ambitious goal of researchers in the area of molecular magnetic materials.

The design of new magnetic materials is necessarily multidisciplinary. It is necessary to combine strategies from many "traditional" areas such as inorganic-organic-polymer chemistry, physics, and engineering. Space considerations preclude detailed description of the full scope of background ideas and design strategies being employed in this research area. This chapter will specifically focus on general theoretical concepts and synthetic design strategies that are being employed in efforts to design organic molecules with controlled magnetic properties.

0097–6156/96/0644–0218$15.00/0

Theoretical Design Strategies

Models of Intramolecular Exchange. Various approaches have been used to predict and understand the ground state spin multiplicities and exchange interactions of organic molecules. Most of the following discussion will be concerned with neutral conjugated open-shell molecules, often termed non-Kekulé molecules.

Early Work on Intramolecular Exchange. One of the earliest models for open-shell organic molecules was the direct application of Hund's rule[2] for atomic ground state spin multiplicities. By this model, a triplet ground state is expected for any diradical. More generally, any number of unpaired electrons would be expected to be spin parallel to give the highest spin state. This model was applied to the few stable diradicals known in the 1930's, when Hund's rule was a fairly new concept. The Schlenk and Schlenk-Brauns hydrocarbons, for example were found to be paramagnetic, in accord with the expectation of a triplet ground state.[3] Experimental limitations prevented extensive testing of the Hund's rule approach, but its apparent success in these early tests led to the application of atomic spin multiplicity rules to large molecular systems. Although this was known to be an approximation, the data known during the 1930's to mid-1950's seemed to support such treatment.

During this early phase of theoretical developments for non-Kekulé molecules a remarkably prescient paper by E. Hückel appeared. Hückel suggested[4] that Hund's rule was not necessarily an over-riding principle governing ground state multiplicity. Using the equivalent of modern fragment molecular orbital (MO) analysis, Hückel showed that the Schlenk-Brauns hydrocarbon had a connectivity that might tend to isolate its two unpaired spins in noninteracting portions of the molecule (Scheme 1), such that a singlet ground state or near degeneracy of triplet and singlet states would be possible. Hückel's paper appears to have been forgotten for nearly a half-century after its publication, but we shall see below that the notions he presented were independently promulgated in the 1970's, and are commonly accepted today.

Scheme 1

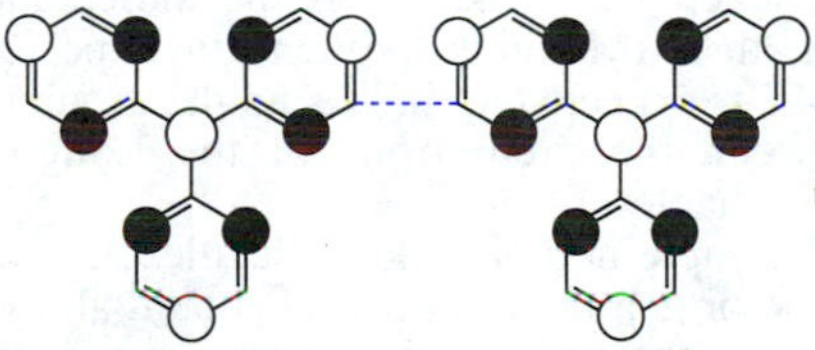

Schlenk-Brauns hydrocarbon modeled by weak interaction of triphenylmethyl radical singly-occupied MO's

Parity-based Models. In the 1950's Longuet-Higgins[5-6] described a connectivity-based approach for predicting the ground state multiplicity of alternant conjugated organic molecules (Scheme 2, equation 1). This model was based on application of Hund's spin criteria, but was generalized by use of parity (electron pairing into pi-bonds) to determine the number of electrons that should occur in a non-Kekulé molecule. Equation (1) gives the number of nonbonding molecular orbitals (NBMOs, assumed = number of unpaired electrons), when N = the number of p_π-orbitals in the conjugated system, and T = the maximum number of double bonds that may be drawn for the system. As shown in Scheme 2, trimethylenemethane (TMM), tetramethyleneethane (TME) and *m*-xylylene are all expected to have triplet high-spin ground states by Longuet-Higgins model. Secondary effects are not taken into account for this qualitative approach. We shall see below that indiscriminate application of Hund's rule to organic molecules is not a justifiable strategy.

In 1968 a paper was published by Mataga[7] describing a general approach toward high-spin molecules, based upon the use of *m*-phenylene to link unpaired spin in either one or two-dimensional arrays. By using *m*-phenylene units, the spins connect

Scheme 2

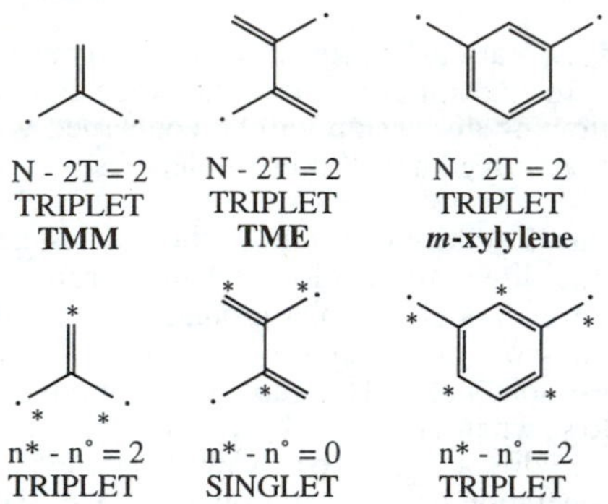

N - 2T = 2
TRIPLET
TMM

N - 2T = 2
TRIPLET
TME

N - 2T = 2
TRIPLET
***m*-xylylene**

$$\#NBMOs = N - 2T \quad (1)$$

n* - n° = 2
TRIPLET

n* - n° = 0
SINGLET

n* - n° = 2
TRIPLET

$$S = 0.5\,(n^* - n^\circ) \quad (2)$$

together in a way that prevents their spin-pairing. As a result, the *m*-phenylene unit acts as a universal ferromagnetic (FM) exchange unit. This model implicitly assumes the operation of Hund's rule, based on the observation that phenyl carbenes have a strong preference for an FM coupled, high-spin triplet ground state having both delocalized and localized electrons. The success of this approach has been confirmed by the work of the Itoh and Iwamura groups on *m*-phenylene linked polycarbenes, with spin quantum numbers $S \geq 6$.[8-9] Although polycarbenes are not sufficiently stable to allow their practical use as magnetic materials, their observed FM coupled nature demonstrates the power of theoretical design in choosing systems for synthesis.

In 1977-1978 Ovchinnikov[10-11] described a model for alternant conjugated non-Kekulé molecules that is based upon a reorganization of the Pariser-Parr-Pople Hamiltonian describing electron exchange. Unlike the previous models, the Ovchinnikov model makes no external application of the Hund criteria. Instead, the pi-sites in the molecule are divided into alternating α- and β- sites in a typical valence bond manner, by designating one spin type as starred sites, and the other not starred. One then applies equation 2 (Scheme 2), where S is the molecular spin quantum number. Using this approach, both TMM and *m*-xylylene are expected to have triplet high-spin ground states, but TME is expected to have a singlet ground state. We shall return to this difference between the predictions of the Longuet-Higgins and Ovchinnikov models later in this chapter.

Ovchinnikov went beyond simple non-Kekulé molecules as putative targets of his model. He suggested that the principle of the model should allow for heteroatom substitution without changing the expected ground state spin, a suggestion later borne out by experiment. In addition, he pointed out that one should be able to use this model to design organic very high-spin polymers and related ferromagnetic materials, *e.g.* polymer **1**. This paper was very important in stimulating experimental chemists to pursue polyradicals as potential organic ferromagnets.

In 1982 Klein[12-14] described a model for alternant conjugated non-Kekulé molecules that is similar in principle to that of Ovchinnikov. In this and following work, Klein gave a more rigorous theoretical justification to the derivation of equation 2 and demonstrated its application to a large variety of non-Kekulé types of connectivity.[15] Because of its ease of use and its ability to demonstrate the pairing of pi-electrons in a manner that makes intuitive sense, equation 2 has become one of the

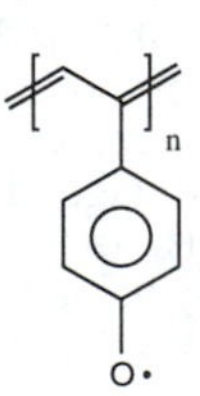

n

O•

1

more popular models for predicting ground state spin multiplicity. However, equation 2 assumes geometric planarity of any system treated, such that all connected sites have similar magnitudes of exchange. When this criterion is not met, violations of equation 2 can occur, as we shall see later.

In 1983 Tyutyulkov and coworkers[16-17] extended the parity approach to nonalternant systems, which they termed as "non-classical polymers". Related theoretical work has also been carried out by Hughbanks.[18-19] Such nonalternant polymeric systems have potentially interesting properties, but experimental tests of this class of polymers remain to be carried out in a systematic fashion.

The Disjoint Model. In 1977 Borden and Davidson[20] first described the disjointness classification of non-Kekulé molecules. A non-Kekulé molecule is disjoint if its singly occupied molecular orbitals (SOMOs) may be linearly recombined to formulate a non-symmetry adapted set that has no overlap. In such a case, it may be shown that the exchange interactions which favor a triplet ground state in non-Kekulé molecules will be much reduced, and may even allow the singlet state to fall below the triplet energetically.[21] Thus, a molecule is disjoint if $n^*=n^\circ$ in equation 2, since this automatically assures that the SOMOs of the molecule may be confined to non-overlapping sites, the n^* set and the n° set. If a molecule is non-disjoint ($n^* > n^\circ$), then a non Kekulé molecule is expected to have considerable exchange interaction and a high spin ground state.

Because equation 2 may be used to discern disjoint systems, its predictions hold true for the disjointness model. However, a few disjoint systems are predicted by equation 2 to have high spin ground states, a point of contradiction between these two models. The pentamethylenepropane (PMP) system and its analogs are examples of such a contradiction. PMP can be formulated by fragment analysis as having two allyl radicals connected to a 1,1-ethenediyl linker unit, with both allyl radicals connected at nodal (inactive) sites. This treatment (Scheme 3) is very similar to that used by Hückel[4] decades before, but the Borden-Davidson model was independently derived and given the rigor of modern terminology and methodology. As Borden and Davidson point out, linkage across inactive sites will result in very small exchange interactions, a near-degeneracy of spin states, and the possibility of a low spin ground state. TME may similarly be treated by fragment analysis, but is predicted by equation 2 to have a low spin ground state, while PMP is predicted to have a high

Scheme 3

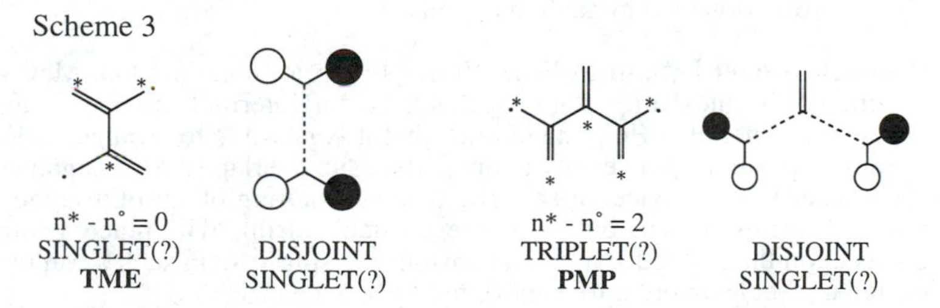

spin ground state. Iwamura has termed systems with the PMP type of connectivity to be doubly disjoint,[22] to distinguish them from the disjoint class represented by TME. Both the Iwamura[22] and Lahti[23-25] groups have found PMP connectivity to give low spin or apparently near-degenerate behavior, consistent with the disjointness criterion. The disjointness criterion has proven to be a powerful predictor of systems in which experimental exchange coupling is very small, and is a very successful model if not over-interpreted to *require* that disjoint systems will have low spin ground states.

The Spin Polarization Approach. The use of spin polarization (SP) arguments to treat non-Kekulé molecules is powerful, general, and can be done almost by inspection. The method may be related to equation 2 designating alternant α- and β-spin sites in a pi-conjugated system following a valence bond (VB) approach, analogous to the designation of the n* and n° sets. Thus, predictions of the SP model are similar in most cases to those of equation 2. However, the SP model allows generalization beyond radicals and alternant systems. One may incorporate atoms with two electrons in a single pi-orbital center (2e/1c) such as oxygen and nitrogen, to give superexchange coupling between the connected sites. One may treat nonalternant systems, if one considers that any two connected sites with the same sign of electron spin will have near-zero spin density. One may also treat systems with more than one type of exchange, such as carbenes and nitrenes, in which a large localized sigma spin density site polarizes the pi-spin density on the same atom to have the same sign.

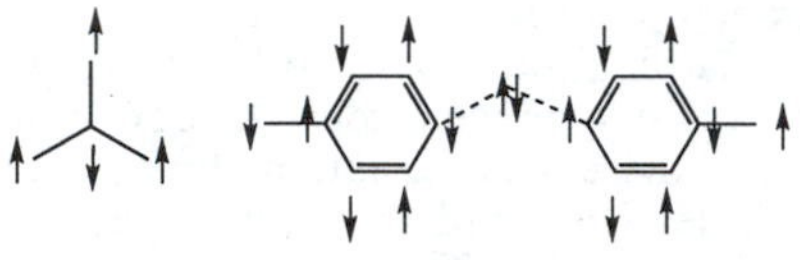

linkage by 2e/1c site

Itoh has shown that a valence bond SP approach is quantitatively successful in describing ground state spin multiplicities and spin density distributions for polycarbenes.[26] Lahti[24,27-29] and Iwamura[30-31] have independently demonstrated the qualitative success of the SP model for a variety of dinitrenes. Overall, the SP model is simple and intuitive to apply, and successfully describes the qualitative ground state of the large majority of conjugated non-Kekulé molecules examined to date.[28]

Models of Intermolecular Exchange. The previous section has shown that various approaches may be used to treat exchange within an organic molecule. Experimental results based on all the above models have produced useful results regarding isolated molecule ground state multiplicity. However, in order to design organic-based materials with bulk magnetic properties, control of intermolecular exchange coupling is critically important, since it will be the weakest exchange term. This is the hardest part of the material design problem, since it depends critically upon crystal lattice structure and polymer morphology. Although experimental control of intermolecular exchange is still problematic, some design models have proven quite successful, especially for hybrid organic/inorganic materials.

McConnell Model I (Spin Polarization). In a brief note in 1963, McConnell described one of the most important mechanisms for intermolecular exchange in organic magnetic materials. He pointed out[32] that it is possible to arrange molecules such that sites of positive spin density at one lattice site overlap sites of negative spin density at another site, and vice versa. The resultant pairing of spin throughout the bulk of the lattice will then be favorable. Such pairing will place geometric requirements upon the lattice for a typical anisotropic spin system such as an organic pi-conjugated molecule, in order to achieve bulk FM exchange.

Scheme 4 demonstrates this model for the simple case of two interacting benzyl radicals stacked with rings directly above one another. In a pseudo-para and pseudo-ortho arrangement, a high spin ground state is expected when the McConnell model is followed. However, in a pseudo-meta arrangement, a low spin ground state is expected. This prediction was confirmed by the classic work of Iwamura and coworkers, in which the paracyclophanes **2-3** were found to have ground state spin multiplicities in accord[33] with the simple model.

Other experimental work on crystalline organic radicals has confirmed the applicability of the so-called McConnell I mechanism of Scheme 4, when bulk magnetic behavior is compared to available crystal structures. Theoretical computations have also been carried out to investigate the effects of various other

Scheme 4

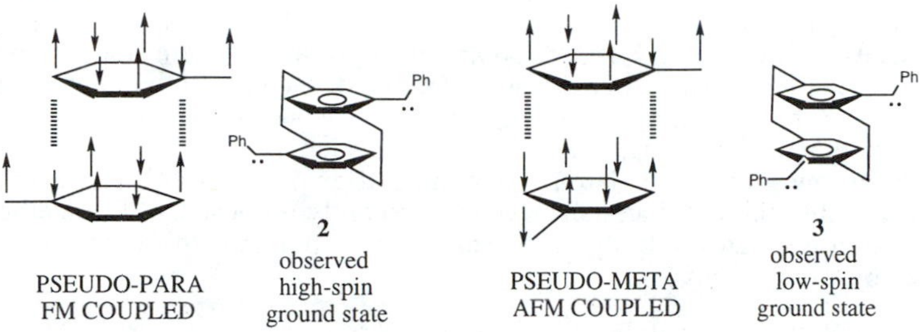

PSEUDO-PARA FM COUPLED

2 observed high-spin ground state

PSEUDO-META AFM COUPLED

3 observed low-spin ground state

geometric orientations upon through-spin pi-spin interactions besides the simple examples shown. The model is simple in concept, but difficult to execute due to lack of control in solid state crystal packing. For instance, the triplet state dications **4-5** of Breslow[34-35], apparently excellent candidates as a building blocks for bulk magnetic behavior, are AFM coupled in the solid state due to structural distortion from crystal packing forces. Likewise, the majority of solid state sample of organic radicals show AFM coupling, when cooperative behavior is observed in the bulk at all.

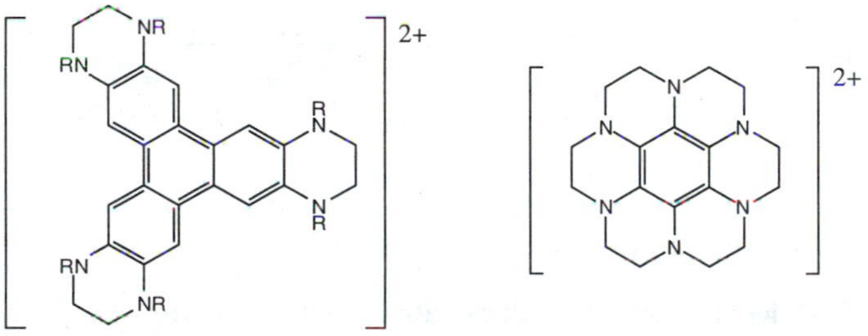

4

5 (HOC)

McConnell Model II (Charge Transfe). Various closely related models have shown how charge transfer from a donor to an acceptor molecule can lead to radical anion / radical cation pairs that are FM exchange coupled. One of the earliest, the so-called McConnell II mechanism, was described in 1967.[36] Scheme 5 shows an example of this general approach, by which charge transfer between diamagnetic species can lead

Scheme 5

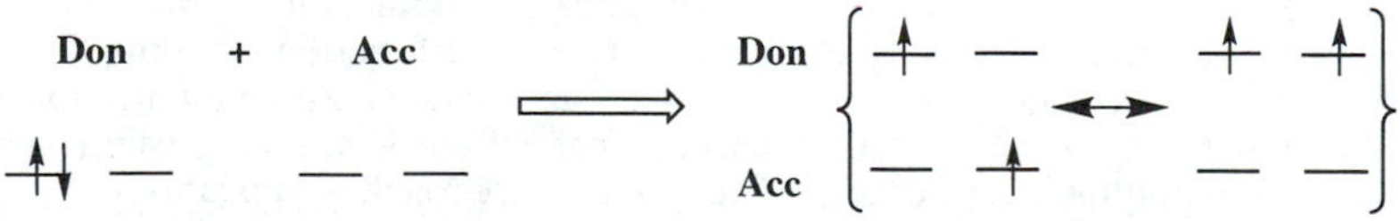

Don + Acc ⟹ Don / Acc

to an overall high spin mixed system. Modifications of this approach have been suggested by Breslow,[37-38] Torrance,[39] and Wudl,[40-41] where the central idea has been retained of finding systems in which the charge-transfer configuration with high spin occupancy is favored by orbital degeneracy. Miller and coworkers have found this approach to intermolecular FM exchange to yield some outstanding materials, such as the bulk ferromagnet[42] $Fe(III)(C_5Me_5)_2^{+\bullet}$ $[TCNE]^{-\bullet}$, with $T_c = 4.8$ K. Because this strategy has been extensively described elsewhere, we refer the reader to the literature for further information on charge-transfer FM materials.[43-45]

A variant of the charge-transfer magnetic materials are the polaronic ferromagnets described by Fukutome and coworkers in 1987,[46] in which dopable conjugated systems are connected by linker groups that do not permit the formation of spinless bipolarons. Scheme 6 demonstrates this for a *meta* and a *para* connectivity, which should give a polaronic high spin system and a bipolaronic system, respectively. This strategy has not been extensively explored, although some model systems have been made. Dougherty and coworkers have shown that one can get FM coupling in polaronic polymeric systems such as **6**, although these materials suffer from low yields of polaronic spin sites after p-doping.[47-49] Further putative examples of this class of materials have been recently proposed.[50] The method could be promising if greater stability of the spin-bearing polarons in the materials can be achieved by some means.

Scheme 6

FM coupled polarons

6

bipolaron

Experimental Design Strategies.

Open Shell Molecular Building Blocks and Model Systems. The previous section has shown that various approaches may be used to model exchange within an organic molecule. Strategies for designing high-spin state organic molecules in isolation have been fairly successful. As a result, a fairly large portion of the rule book for single molecule building blocks has now been discerned, using both transient and stable open-shell systems. We shall briefly summarize below some of the approaches being used to examine test model systems.

Variation of Exchange Linker Groups and Connectivity. In a fairly general experimental approach described by several groups, one may use a variety of exchange linker groups -X- to couple together open shell spin bearing sites. The spin bearing sites may be neutral radicals or polyradicals, or radical ions. By investigating the ground state spin multiplicity of many systems as a function of both -X- and the spin bearing sites, a "rule-book" of intermolecular exchange may be built up. While it is not entirely clear that lessons learned from dimeric model systems may be extrapolated to polymeric systems, the logic of this approach is appealing.

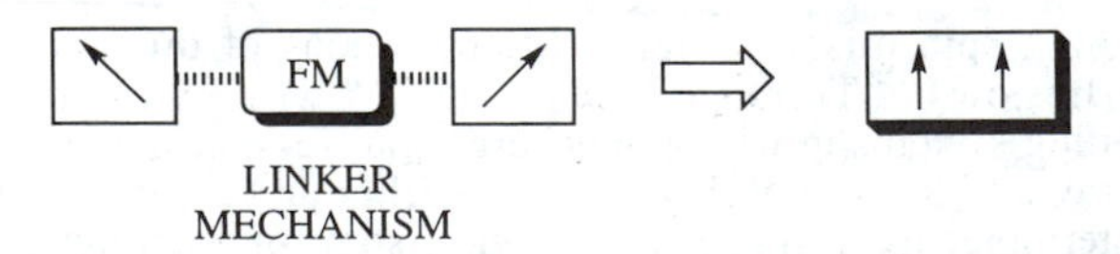

A large number of diradical systems has been studied, starting from the early days when stable diradicals such as the Schlenk-Brauns molecule had to be studied by magnetic susceptibility, with a considerable acceleration after the 1960's, when cryogenic electron spin resonance (ESR) techniques became available. Even at the present time, model diradicals remain of interest. For example, only in 1986 was

TME identified as ground state triplet diradical,[51] despite its being disjoint; only in 1993 was TME modeled by high level ab initio computations to have a nonplanar triplet[52] ground state, after a number of earlier ab initio computations had suggested a singlet[53-54] ground state. This is an example in which molecular geometry confounded basic parity expectations. Clearly, fundamental questions remain to be answered about even simple open-shell organic systems Still, many fundamentally FM exchange coupling units -X- have been identified during diradical work, such as *m*-phenylene and 1,4-cyclobutanediyl. Work on conjugated diradicals has been summarized in various books and symposia, hence we refer the reader to these for further information.[21,55-56]

Dicarbenes and polycarbenes have been the subject of considerable study, since the pioneering, independent studies by Wasserman[57] and Itoh,[58] and have culminated in the fascinating oligomeric linear and branched polycarbenes studied by Itoh, Iwamura and their colleagues.[8,59] Using *m*-phenylene linkages and the strategy described in Mataga's paper,[60] many electrons can be FM coupled as polycarbenes such as **7**. Two-dimensional polycarbene analogs of **7** have also recently been made and shown to have FM coupling.[9,61-62] On the other hand, when disjoint connectivity

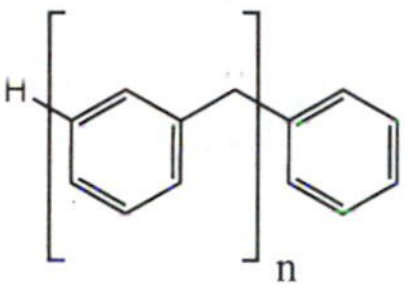

7 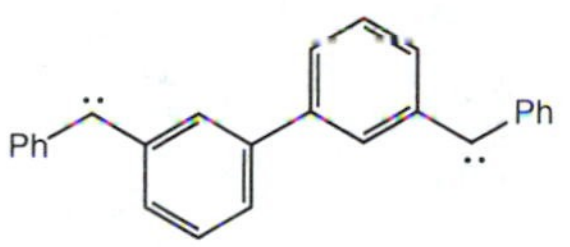 **8**

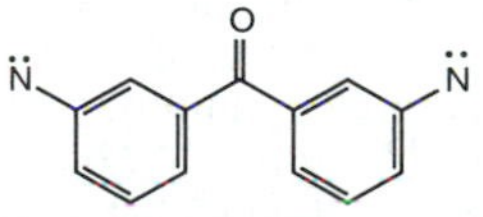

9 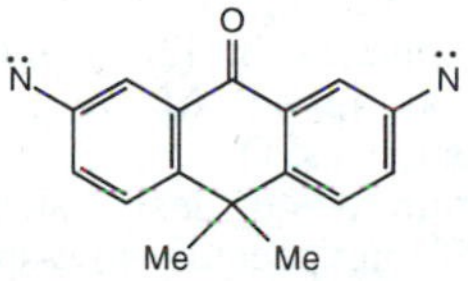 **10**

patterns are used, as in **8**, near-degeneracy of spin states is observed, with a singlet ground state being close in energy to thermally populated triplet and quintet states.[29,59,63] Fragment analysis of **8** shows it to have poor exchange as expected, since it can be formulated as the fusion of a pair of weakly interacting radicals.. Itoh has shown that modeling such disjoint dicarbenes by weak interaction of fragments allows prediction of their spin density distributions and inter-state energy gaps.[26] Such successes are powerful support for the study of model compounds as guides for larger, oligomeric high spin systems, the first step in design of extended, three-dimensional materials.

Various groups have studied dinitrenes as model systems. Early work in this area was done by Wasserman[57,64] and by Singh,[65] but recent work with varying linker groups and connectivities has been done by our group,[24-25,27-28,66] by Iwamura's group,[30-31,67-68] and by Yabe's[69-71] group. Dinitrenes have an advantage over dicarbenes, in that they have monovalent spin bearing sites of known geometry, but like dicarbenes they are not stable in solution or under high concentration conditions. In general, dinitrene model systems have yielded results comparable to those for the more closely-studied dicarbenes. Disjoint systems have low spin ground states with low-lying high-spin states, while nondisjoint systems -- such as those incorporating *m*-phenylene -- have high-spin ground states. Like dicarbenes, dinitrenes may show considerable changes in behavior when conformational torsion decreases conjugation and confounds parity considerations. For example **9** and **10** have quite different quintet ESR spectra and singlet-quintet energy gaps, although both are qualitatively disjoint ground state singlet species with low lying excited quintet states. This is apparently due to the fact that **9** has multiple possible conformations with different ESR spectra, while **10** has only one conformation available.[72] A summary of the ground state behavior of dinitrenes as a function of connectivity and of linker group -X- has been given.[72] Some work has also been done on polynitrene systems.[73-74]

Dougherty and coworkers have studied the effect of altering linker groups -X- and connectivity upon the exchange coupling between TMM spin bearing units.[75-76] For example, they have shown conclusively that torsional deconjugation can decrease FM coupling of a *m*-phenylene unit, by studying the series of compounds **11-13**. They have also shown that the nominally non-conjugating linker 1,4-cyclobutanediyl is a robust FM coupling unit in **14**.[77-78] However, other nonconjugated linkers such as cyclopentanediyl and adamantanediyl (**15** and **16**) are found to be AFM in nature.[75] TMM spin bearing units have problems related to their conformational flexibility, but Dougherty's group has shown them to be quite useful as part of studies aimed at elucidating FM coupling linker groups and connectivities.

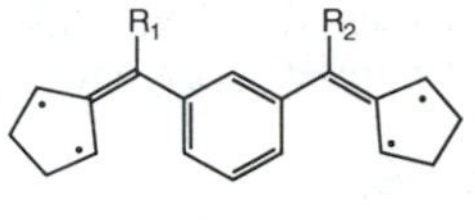

11 $R_1 = R_2 = Me$
12 $R_1 = Me$, $R_2 = t\text{-}Bu$
13 $R_1 = R_2 = t\text{-}Bu$

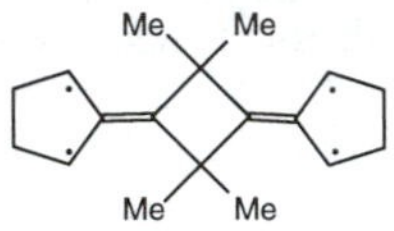

14 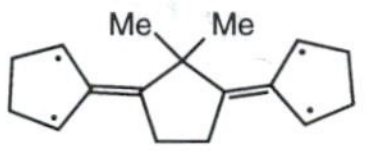 **15** 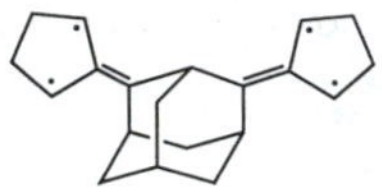 **16**

Overall, the systematic variation of linker groups and connectivity has proven a powerful approach toward filling out various "pages" of the desired "rule-book" of organic-based exchange behavior. While it is specifically suited only toward the design of the smallest building blocks of bulk magnetic materials, the approach has the intellectual appeal of rigor, and has elucidated a number of interesting possibilities such as Dougherty's 1,4-cyclobutanediyl linker. This approach has elucidated limitations of design strategies based purely upon connectivity by turning up a variety of unexpectedly low-spin ground state diradicals such as **17**[79]-**18**[80]. Such "anomalous" systems are explained[81] by a combination of various effects combining to overcome a "natural" connectivity-based preference for high-spin FM coupling, and illustrate that increasing the complexity of molecular systems typically increases the difficulty of achieving overall control of magnetic and exchange behavior. Even apparently subtle effects such as protonating a pyridine ring are now known to affect exchange coupling.[82] Therefore, important explorations apparently remain even for apparently simple model compounds as part of the organic exchange rule book.

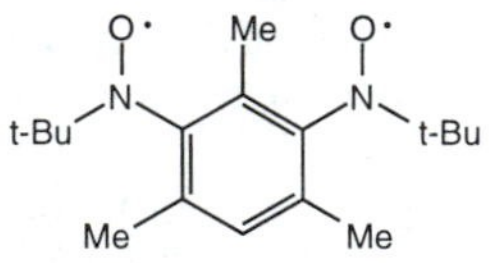

17 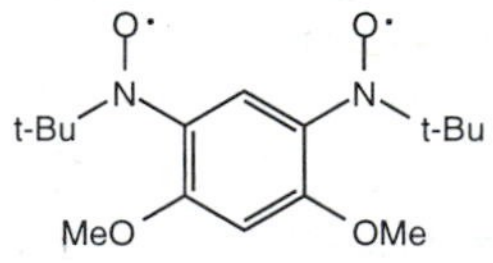 **18**

Backbone Coupled Polyradicals. Once stable spin-bearing units have been identified by model studies, one may proceed to devise large magnetic domains having multiple spins, using the same design precepts described in the previous section. As pointed out above, systems based upon the triphenylmethyl radical have been of interest since the earliest studies of paramagnetic organic species. In recent years considerable success has been realized in making large polyradicals of the Mataga type, using triphenylmethyl radical as a spin-bearing unit.

In a recent review,[56] Rajca summarized the syntheses of polyethers which may be converted to polyanions, and then oxidized to poly(triarylmethyl) polyradicals.[56,83-85] Rajca's group has by this methodology made systems of controlled architecture with the highest spin quantum numbers yet achieved for organic polyradicals. The ultimate goal of this strategy is to make dendritic polyradicals (*e.g.*, **19**) large enough to constitute mesoscopic domains in a bulk, purely organic material. Despite the

19 **20**

torsion expected in these systems, strong FM exchange through direct, backbone conjugated frameworks maintains high spin ground states in the smaller versions of Rajca's molecules. In recently synthesized larger variants, the ground state spin multiplicities, while high, have been less than expected.[56] The reasons for the failure of this strategy to be extended are apparently related both to the extreme torsional deconjugation in the larger dendrimers, and to the extraordinary requirements for defectless spin-site production in order for the spin quantum number to continue to increase in backbone conjugated systems (termed Type I by Rajca in his review,[56] see also Scheme 7). In these polyradicals, exchange interactions are direct and very strong, but defects in spin-site production interrupt conjugation, leading to the division of the desired mesoscopic multispin domain into two or more domains with a smaller number of spins. In such a case where perfection is required, a breakdown of the strategy is a near-certainty for the larger dendrimers.

Scheme 7

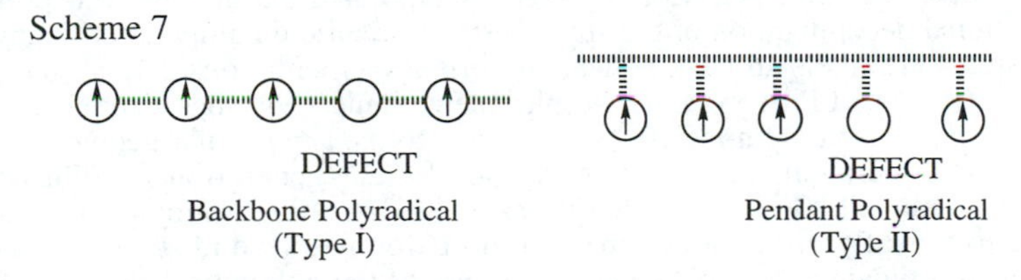

Rajca has devised a strategy to circumvent this problem by making polyradicals with multiple conjugation paths, such as **20**.[56] By incorporating redundant conjugation pathways into a single molecule, exchange throughout the mesoscopic domain may be maintained even when defects are present. While synthetically challenging, this approach offers a possibility for extensions of dendritic polyradicals.

Veciana and coworkers have investigated dendritic poly(triphenylmethyl) systems which are stabilized by perchlorination.[86-87] The polyradicals are generated by treatment of appropriate polychloride precursors with strong Lewis acids such as SbF_5. A number of such systems show impressive stability at ambient temperatures, and so are quite interesting despite the rigor of the conditions under which they are generated. These backbone conjugated systems have the same advantages and drawbacks described above for Rajca's molecules. Other work has extended the use of perchlorinated poly(triphenylmethyl) polyradical systems using different connectivity schemes,[88] showing the general utility of this approach.

Pendant Coupled Polyradicals. An alternative to the backbone exchange coupling connectivities described are pendant polyradicals (Rajca's Type II, Scheme 7). This approach has attracted the efforts of a number of groups, due to the plethora of synthetic possibilities for achieving it. Pendant systems have a natural disadvantage in that exchange from one pendant spin-site to another in a conjugated polyradical is naturally weaker than in a backbone conjugated system. Exchange between the spins-sites is truncated or even eliminated by torsion of either the backbone or the pendant site. However, pendant systems have an advantage over backbone systems, in that spin-site defects do not eliminate exchange along the overall conjugated system, but merely extend the length of the conjugated unit across which the spins must be coupled. Naturally, such defects would "dilute" the number of spins per polyradical domain, but FM coupling could in principle still be maintained within the unit. The critical design parameters to establish in such a case are: (1) what spin-bearing units and connectivities are most effective in allowing pendant spins-site exchange, regardless of torsional deconjugation; (2) what efficiency of spin-site generation can be achieved to minimize "spin-dilution".

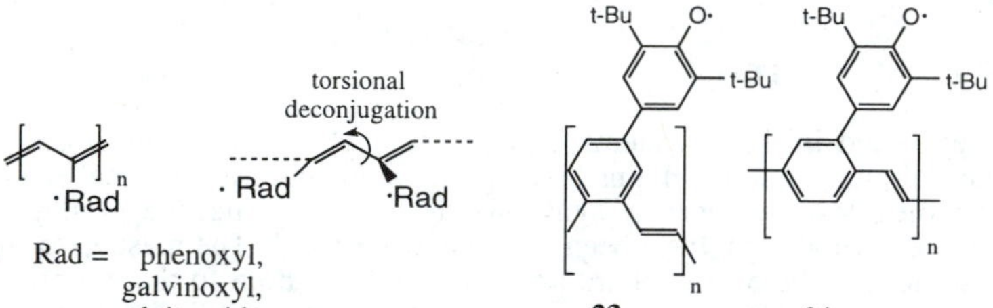

torsional deconjugation

·Rad · Rad ·Rad

t-Bu O· t-Bu t-Bu O· t-Bu

22 Rad = phenoxyl, galvinoxyl, arylnitroxide

23 **24**

Much attention has been given to polyphenylacetylene (PPhA) based systems, due to their synthetic availability. Nishide,[89-94] Itoh,[97-98] and their respective coworkers, have investigated variants of PPhA with the general structure **22**, using spin sites <u>Rad</u> such as galvinoxyl, phenoxyl, arylnitroxides, and aryl nitronylnitroxides. Polymers with spin-yields of about 90% were achieved in some cases by solution oxidation of appropriate precursors, but in none of these cases was any cooperative exchange behavior observed. We pointed out that this was probably due to torsional deconjugation of the spin sites.[99] Force field computations show that PPhA systems are likely to have helical conformations with considerable backbone torsion.[100] Thus, all PPhA systems, despite their connectivity based preference for exchange, tend to act as spin-isolated systems due to their molecular geometries. We attempted to circumvent this problem by solid state generation of PPhA-based polyphenoxyl radicals (Rad was a diaryl oxalate[101] derivative), hoping to be able to achieve a sufficiently different polymer conformation in the solid state to allow FM coupling. Unfortunately, the solid state generation of the polyradical (Rad = 2,6-di-*t*-butylphenoxyl) also gave paramagnetic behavior.[102] Overall, prospects for obtaining FM exchange coupled PPhA-based polyradicals appear to be poor.

A variety of other conjugated polyradicals have been made with varying spin-yields,[103-106] but until recently all showed no evidence of cooperative exchange coupling, despite some early results[39,107-108] of poor reproducibility which were reported to show ferromagnetic behavior. In order to achieve longer range FM exchange in pendant architectures, efficient conjugation must be achieved. In 1989 we predicted that polyphenylene vinylene (PPV) based polyradicals should be good targets, based on their expected planarity.[99,109] During 1993-1995 Nishide and coworkers confirmed this by making poly(*p*-phenylene vinylenes) and poly(*o*-phenylene vinylenes) with pendant phenoxyl radicals, **23-24**.[110-112] These systems have S = 4-5, showing FM coupling of up to 8-10 pendant spins. Thus, the viability of the pendant polyradical approach for making high-spin was confirmed for the first time. There are prospects for further improving this strategy by use of more stable spin-sites, or by solid state photochemical generation of spin-sites by methods such as those used by us[101,113-114] for phenoxyl radical generation. It remains for future work to establish further possibilities for the organic polyradical approach.

Organic Molecular Magnetic Compounds -- Pure Solid Compounds. The previous section describes the use of multispin organic magnetic building blocks. However, the simplest organic building blocks -- stable monoradicals -- have proven to be effective building blocks when crystal packing geometries are appropriate.[115] Pure organic radicals have even turned out to be bulk ferromagnets in the solid state.

The bulk magnetic susceptibilities of organic radicals have been studied since the 1930's, as described in the introduction. These studies intensified during the 1960's. Pure radicals were typically found to exhibit no cooperative behavior or AFM bulk exchange at low temperatures, with occasional exceptions. Mukai carried out numerous susceptibility investigations of phenoxyl-based radicals such as galvinoxyl.[116-118] Kinoshita showed that mixed crystals of galvinoxyl with hydrogalvinol (the non-oxidized precursor form) exhibit FM behavior.[119-120] This is one of the early examples of the notion that one can control the bulk exchange behavior of solid state radicals by altering the crystal structure, in accord with the McConnell I model. Chiang and coworkers showed[121] that further modification of the galvinoxyl crystal structure through use of different crystallization procedures led to magnetic behavior that was different yet. While phenoxyl based radicals have been somewhat studied since Mukai and Kinoshita's work (*e.g.*, by our group[122]), their limited stability in the bulk has limited their interest to other workers.

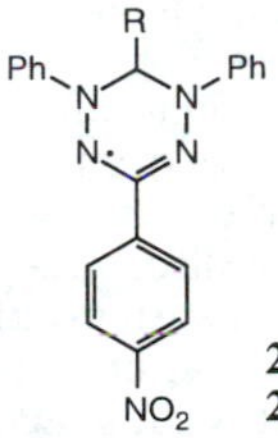

25 R = H
26 R = Ph

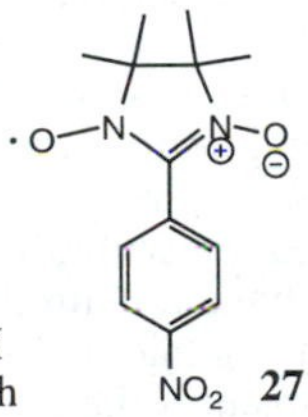

27

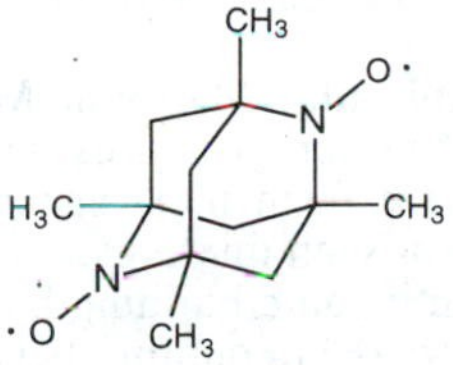

28

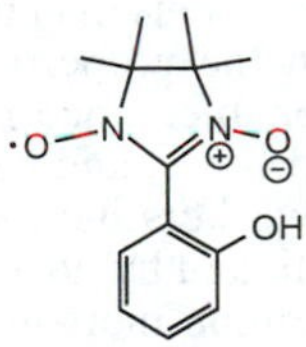

29

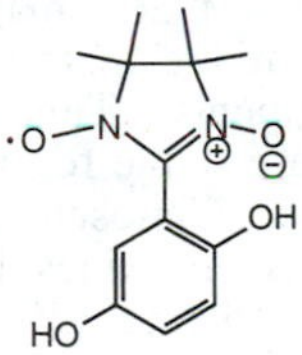

30

Wudl has shown that the stable verdazyl radical system can be subjected to crystal engineering techniques to control its bulk magnetic behavior.[123] Radical **25** is found to have AFM behavior at low temperature, but the closely related **26** is a bulk ferromagnet with $\theta = 1.3$ K. The alteration in crystal packing caused an alignment of spin densities for **26** that favors FM coupling by the McConnell I model. Although it is still not possible to *predict* the effect that substitution will have on crystal packing within a series of related molecules, Wudl's study shows that systematic variation of substitution is a viable (if tedious) strategy to find organic ferromagnetic systems.

In terms of magnetic behavior, nitroxide-based radicals have proven to be most successful to date. These radicals are very stable, readily synthesized, tolerate a wide variety of substitution patterns, and are sufficient polar to have strong crystal packing structures. This family of radicals includes the nitroxides, α-nitronylnitroxides, and iminoylnitroxides. Kinoshita first showed that a pure organic compound could be a bulk ferromagnet by demonstrating that nitronylnitroxide **27** has an ordering transition at T_C = 0.6 K.[124-126] Since this discovery, a number of pure organic ferromagnets have been discovered with related structures; we refer the reader to the proceedings of this and previous related meetings[127-128] for details of a number of molecular crystals that have promising magnetic properties. Rassat's dinitroxide **28** has the highest ordering temperature for a pure organic substance, with T_C = 4.6 K.[129] Sugawara's group has shown[130] that combination of a bis-nitroxide and a nitronyl-nitroxide can give favorable magnetic interactions in the mixed material. Given the large number of possible nitroxides, more such systems will presumably be discovered. The promising behavior of these systems has inspired computational[131] and neutron diffraction studies[132-133] of spin density distributions and interactions in these solid materials, with an eye to correlating various models of through-space spin density interaction and the experimental facts. Despite the low ordering temperatures found to date, the finding of ferromagnetic phases in these molecules shows them to be worthy of the intensive study they are presently undergoing.

Of particular interest is the possibility for using molecular recognition and crystal engineering strategies to force appropriate geometries between organic radicals. The use of strongly polar bonds to direct crystal packing appears particularly promising. For example, the groups of both Veciana and Sugawara have shown that attachment of varied hydrogen bonding substituents to nitronylnitroxides leads to variation in bulk magnetic behavior. For example compound **29** has a weak network of hydrogen bonding in the crystal that appears be important in making this material a bulk ferrimagnet (T_C = 0.45 K).[134-135] Compound **30** has strong hydrogen bonding, and is a bulk ferromagnet (T_C = 0.43 K).[136] Presently there is an ad hoc nature to such crystal engineering strategies, but advances in this area will be crucial to allow the control and prediction of bulk solid state magnetic behavior in organic open-shell molecules.

Organic Molecular Magnetic Materials -- Mixed Materials. In terms of achieving high ordering temperatures and magnetic moments, mixed organic/inorganic materials have been most successful. A very wide variety of strategies may be envisioned for such mixtures. This chapter will arbitrarily limit discussion to systems in which the organic building blocks have unpaired spin. Such systems may exhibit either FM or AFM coupling between organic and inorganic units, but either case is likely to yield an overall magnetic moment if the inorganic unit is paramagnetic, since organic systems typically have smaller magnetic moments than inorganic ions.

The large majority of mixed materials with the properties just described use nitroxide type organic fragments. This is reasonable, since nitroxides are readily synthesized in most cases, are stable for extended times, and have sufficient polarity to chelate to inorganic ions. Gatteschi and coworkers have shown the wealth of possibilities for use of nitronylnitroxides that are linked by nonconjugated spacers, as demonstrated pictorially in Scheme 8.[137-139] Common inorganic units are transition metal ions such as V^{2+}, Mn^{2+}, Cu^{2+}.[140] This approach is somewhat at the mercy of whatever crystal structure is chosen by a particular material, but successive variation of inorganic ions with one type of organic unit can give fascinating changes.

Scheme 8

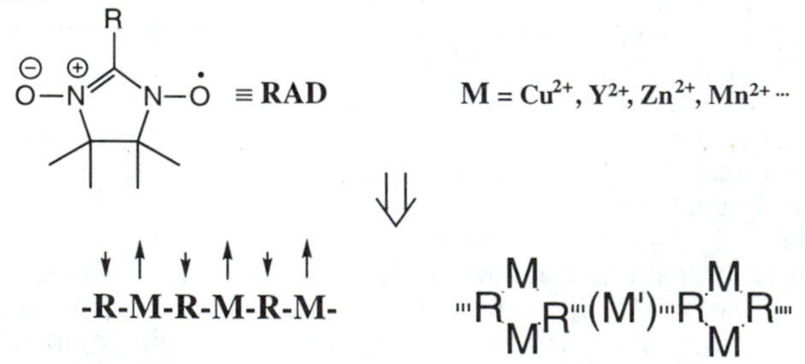

≡ **RAD** **M = Cu^{2+}, Y^{2+}, Zn^{2+}, Mn^{2+} ...**

⇓

-R-M-R-M-R-M-

$\cdots R \cdots (M') \cdots R \cdots$

Iwamura and coworkers have used conjugated polyradicals as building blocks for magnetic materials.[141-142] The strategy has the advantage of using organic fragments of higher magnetic susceptibility than Gatteschi's, since the organic fragments are conjugated di- and tri-nitroxides with appropriate connectivities to give high spin ground states. These organic building blocks in some cases combine with Mn^{2+} and

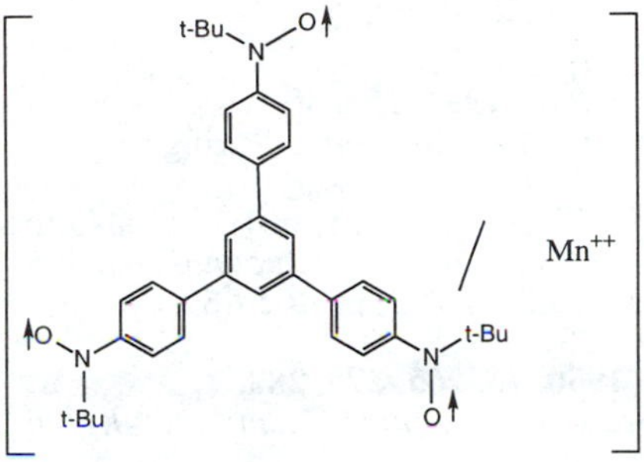

31 (2:3 complex)

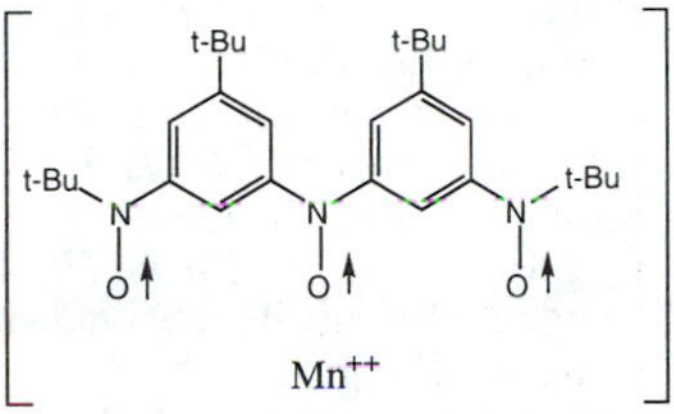

32 (2:3 complex)

other paramagnetic ions to give highly crystalline linear and network materials. The 2:3 complex of **31** with Mn^{2+} was one of the earliest reported by this group using this strategy, and has T_c = 3.4 K at a field of 1 Oe.[141] The 2:3 complex of trinitroxide **32** with Mn^{2+} has T_c = 46 K.[142] Not all such materials are successful -- *e.g.*, 1,3,5-benzene-tris(t-butylnitroxide) is too hindered[142] to give chelates with manganese.

Given the intensive efforts going on to investigate the magnetic properties of pure radicals, it is clear that further efforts will be made to explore the properties of the radicals mixed with various inorganic ions.[143] While such endeavors are still descriptive chemistry with outcomes that cannot be confidently predicted, the successes achieved to date make the exploration of organic/inorganic magnetic materials one of the most promising, in this author's opinion.

Organic Chemistry in the Search for Magnetic Materials -- Present and Future. The organization and relative lengths of the sections above represent the present status and future directions for organic magnetic materials. Considerable progress has been made toward understanding *and* predicting ground state multiplicities of simple organic polyradicals. Work continues in this area, but much of the "rule-book" for isolated organic open-shell systems is known. Work is still ongoing to establish whether the rule-book may be simplistically extended to polymeric organic systems,[144] since torsional deconjugation and inefficient long-range exchange of multiple spins along a large conjugated system may limit the effective size of a purely organic multispin domain.

Although crystal engineering is still a descriptive rather than a predictive science, there is considerable interest in pure radicals as FM materials, since these serve as models for intermolecular exchange interaction. The results to date suggest that pure radicals will have low ordering temperatures, in part due to the small magnetic moments of organic molecules. Hybrid organic/inorganic materials offer multifold possibilities for making materials with useful magnetic properties. Predictive certainty is not possible, but recent results indicate that efforts to investigate magnetic properties as a function of systematically varying the choice of organic and inorganic building blocks in mixed materials have the promise to be well rewarded.

Acknowledgments. I acknowledge support by the National Science Foundation (CHE-9204695 and 9521594) and the Donors of the Petroleum Research Fund, administered by the American Chemical Society (PRF 21150-AC4 and 29379-AC4). I appreciate the support of my research group, many of whom are named in various citations within this chapter. I also acknowledge the inspiration of many people within the field of molecular magnetic materials. The design of magnetic materials resembles the community of people working on the design -- many interactions must fit together in proper proportions to achieve the goal, requiring much work and much exchange (of information, that is) between scientists. Such favorable interactions between people have seemed the rule, rather than the exception.

References

1. See the discussion in the *Mater. Res. Soc. Bull.* **1990**, *15*, 28ff.
2. Hund, F. *Linienspektren und periodisches del Elemente*; Springer-Verlag OHG: Berlin, 1927.
3. Schlenk, W.; Brauns, M. *Ber. Deutsch. Chem.* **1915**, *48*, 661. Cf. also the summary by Berson, J. A. In *The Chemistry of Quinoid Compounds*; Patai, S.; Rappaport, Z.; Ed.; John Wiley and Sons: 1988; Vol. 2; pp 462-469.
4. Hückel, E. *Z. phys. Chem. (B)* **1936**, *34*, 339.
5. Longuet-Higgins, H. C. *J. Chem. Phys.* **1950**, *18*, 265, 275, 283.
6. Longuet-Higgins, H. C. In *Theoretical Organic Chemistry, Kekulé Symposium*; Butterworth: London, 1958; pp 9-19.
7. Mataga, N. *Theor. Chim. Acta* **1968**, *10*, 273.
8. Fujita, I.; Teki, Y.; Takui, T.; Kinoshita, T.; Itoh, K.; Iwamura, H.; Izuoka, A.; Sugawara, T. *J. Am. Chem. Soc.* **1990**, *112*, 4074.
9. Nakamura, N.; Inoue, K.; Iwamura, H.; Fujioka, T.; Sawaki, Y. *J. Am. Chem. Soc.* **1992**, *114*, 1484.
10. Misurkin, I. A.; Ovchinnikov, A. A. *Russ. Chem. Res. (Engl).* **1977**, *46*, 967.
11. Ovchinnikov, A. A. *Theor. Chim. Acta* **1978**, *47*, 297.
12. Klein, D. J.; Nelin, C. J.; Alexander, S.; Matsen, F. E. *J. Chem. Phys.* **1982**, *77*, 3101.
13. Klein, D. J. *Pure Appl. Chem.* **1983**, *55*, 299.
14. Klein, D. J.; Alexander, S. A. In *Graph Theory and Topology in Chemistry*; King, R. B.; Rouvray, D. H., Ed.; Elsevier: Amsterdam, 1987; Vol. 51; p 404.
15. Klein, D. J.; Alexander, S. A. *Mol. Cryst. Liq. Cryst.* **1993**, *232*, 219.
16. Tyutyulkov, N. *Theor. Chim. Acta* **1983**, *63*, 291.
17. Tyutyulkov, N.; Karabunarliev, S.; Ivanov, C. *Mol. Cryst. Liq. Cryst.* **1989**, *176*, 139.
18. Hughbanks, T.; Kertesz, M. *Mol. Cryst. Liq. Cryst.* **1989**, *176*, 115.
19. Hughbanks, T.; Yee, K. A. In *Magnetic Molecular Materials*; Gatteschi, D.; Kahn, O.; Miller, J. S.; Palacio, F., Ed.; Kluwer: Dordrecht: 1991, p 133.
20. Borden, W. T.; Davidson, E. R. *J. Am. Chem. Soc.* **1977**, *99*, 4587.
21. Cf. *Diradicals*; Borden, W. T., Ed.; Wiley: New York, 1982.
22. Matsumoto, T.; Ishida, T.; Koga, N.; Iwamura, H. *J. Am. Chem. Soc.* **1992**, *114*, 9952.
23. Kearley, M. L.; Lahti, P. M. *Tet. Lett.* **1991**, *32*, 5869.
24. Ling, C.; Minato, M.; Lahti, P. M.; van Willigen, H. *J. Am. Chem. Soc.* **1992**, *114*, 9959.
25. Kearley, M. L.; Ichimura, A. S.; Lahti, P. M. *J. Am. Chem. Soc.* **1995**, *117*, 5235.
26. Itoh, K. *Pure Appl. Chem.* **1978**, *50*, 1251.
27. Ling, C.; Lahti, P. M. *J. Am. Chem. Soc.* **1994**, *116*, 8784.
28. Minato, M.; Lahti, P. M. *J. Phys. Org. Chem.* **1994**, *5*, 495.
29. Lahti, P. M.; Minato, M.; Ling, C. *Mol. Cryst. Liq. Cryst.* **1995**, *271*, 147.
30. Iwamura, H.; Murata, S. *Mol. Cryst. Liq. Cryst.* **1989**, *176*, 33.
31. Matsumoto, T.; Ishida, T.; Koga, N.; Iwamura, H. *J. Am. Chem. Soc.* **1992**, *114*, 9952.
32. McConnell, H. M. *J. Chem. Phys.* **1963**, *39*, 1910.
33. Izuoka, A.; Murata, S.; Sugawara, T.; Iwamura, H. *J. Am. Chem. Soc.* **1985**, *107*, 1786.
34. Breslow, R. *Mol. Cryst. Liq. Cryst.* **1985**, *125*, 261.
35. LePage, T. J.; Breslow, R. *J. Am. Chem. Soc.* **1987**, *109*, 6413.

36. McConnell, H. M. *Proc. R. A. Welch Found. Conf.* **1967**, *11*, 1144.
37. Breslow, R. *Pure & Appl. Chem.* **1982**, *54*, 927.
38. Breslow, R.; Jaun, B.; Klutz, R.; Xia, C.-Z. *Tetrahedron* **1982**, *38*, 863.
39. Torrance, J. B.; Oostra, S.; Nazzal, A. *Synth. Met.* **1987**, *19*, 709.
40. Dormann, E.; Nowak, M. J.; Williams, K. A.; Angus, R. O., Jr.; Wudl, F. *J. Am. Chem. Soc.* **1987**, *109*, 2594.
41. Wudl, F.; Closs, F.; Allemand, P. M.; Cox, S.; Hinkelmann, K.; Srdanov, G.; Fite, C. *Mol. Cryst. Liq. Cryst.* **1989**, *176*, 249.
42. Miller, J. S.; Epstein, A. J.; Reiff, W. M. *Mol. Cryst. Liq. Cryst.* **1986**, *120*, 27.
43. Miller, J. S.; Epstein, A. J. *Mol. Cryst. Liq. Cryst.* **1993**, *233*, 133.
44. Miller, J. S.; Epstein, A. J.; Reiff, W. M. *Accts. Chem. Res.* **1988**, *21*, 114.
45. Miller, J. S.; Epstein, A. J.; Reiff, W. M. *Chem. Rev.* **1988**, *88*, 201.
46. Fukutome, H.; Takahashi, A.; Ozaki, M. *Mol. Cryst. Liq. Cryst.* **1987**, *133*, 34.
47. Dougherty, D. A.; Kaisaki, D. A. *Mol. Cryst. Liq. Cryst.* **1990**, *183*, 71.
48. Kaisaki, D. A.; Chang, W.; Dougherty, D. A. *J. Am. Chem. Soc.* **1991**, *113*, 2764.
49. Murray, M. M.; Kaszynski, P.; Kaisaki, D. A.; Chang, W.; Dougherty, D. A. *J. Am. Chem. Soc.* **1994**, *116*, 8152.
50. Mizouchi, H.; Ikawa, A.; Fukutome, H. *J. Am. Chem. Soc.* **1995**, *117*, 3260.
51. Dowd, P.; Chang, W.; Paik, Y. H. *J. Am. Chem. Soc.* **1986**, *108*, 7416.
52. Nachtigall, P.; Jordan, K. D. *J. Am. Chem. Soc.* **1993**, *115*, 270.
53. Du, P.; Borden, W. T. *J. Am. Chem. Soc.* **1987**, *109*, 930.
54. Nachtigall, P.; Jordan, K. D. *J. Am. Chem. Soc.* **1992**, *114*, 4743.
55. Michl, J.; Bonacic-Koutecky, V. *Tetrahedron* **1988**, *24*, 7559.
56. Rajca, A. *Chem. Rev.* **1994**, *94*, 871.
57. Wasserman, E.; Murray, R. W.; Yager, W. A.; Trozzolo, A. M.; Smolinsky, G. *J. Am. Chem. Soc.* **1967**, *89*, 5076.
58. Itoh, K. *Chem. Phys. Lett.* **1967**, *1*, 235.
59. Itoh, K. In *Magnetic Molecular Materials*; Gatteschi, D.; Kahn, O.; Miller, J. S.; Palacio, F., Ed.; Kluwer: Dordrecht: 1991; p 67.
60. See Takui, T.; Teki, Y.; Sato, K.; Itoh, K. *Mol. Cryst. Liq. Cryst.*, **272**, *21*, reference 15 concerning the Itoh-Mataga approach for spin polarization.
61. Furukawa, K.; Matsumura, T.; Teki, Y.; Kinoshita, T.; Takui, T.; Itoh, K. *Mol. Cryst. Liq. Cryst.* **1993**, *232*, 251.
62. Nakazawa, S.; Sato, K.; Kinoshita, T.; Takui, T.; Itoh, K.; Fukuyo, M.; Higuchi, T.-I.; Hirotsu, K. *Mol. Cryst. Liq. Cryst.* **1995**, *271*, 163.
63. Matsushita, M.; Nakamura, T.; Momose, T.; Shida, T.; Teki, Y.; Takui, T.; Kinoshita, T.; Itoh, K. *J. Am. Chem. Soc.* **1992**, *114*, 7470.
64. Trozzolo, A. M.; Murray, R. W.; Smolinsky, G.; Yager, W. A.; Wasserman, E. *J. Am. Chem. Soc.* **1963**, *85*, 2526.
65. Singh, B.; Brinen, J. S. *J. Am. Chem. Soc.* **1971**, *93*, 540.
66. Minato, M.; Lahti, P. M. *J. Phys. Org. Chem.* **1993**, *6*, 483.
67. Matsumoto, T.; Koga, N.; Iwamura, H. *J. Am. Chem. Soc.* **1992**, *114*, 5448.
68. Doi, T.; Ichimura, A. S.; Koga, N.; Iwamura, H. *J. Am. Chem. Soc.* **1993**, *115*, 8928.
69. Ohana, T.; Kaise, M.; Nimura, S.; Kikuchi, O.; Yabe, A. *Chem. Lett.* **1993**, 765.
70. Nimura, S.; Kikuchi, O.; Ohana, T.; Yabe, A.; Maise, M. *Chem. Lett.* **1993**, 837.
71. Ohana, T.; Kaise, M.; Yabe, A. *Chem. Lett.* **1992**, 1397.
72. Ling, C.; Lahti, P. M. *Chem. Lett.* **1994**, 1357,2447.
73. Sasaki, S.; Iwamura, H. *Chem. Lett.* **1992**, 1759.
74. Iwamura, H. *Pure Appl. Chem.* **1993**, *65*, 57.
75. Jacobs, S. J.; Shultz, D. A.; Jain, R.; Novak, J.; Dougherty, D. A. *J. Am. Chem. Soc.* **1993**, *115*, 1744.
76. Dougherty, D. A.; Jacobs, S. J.; Silveman, S. K.; Murray, M. M.; Shultz, D. A.; West, A. P., Jr.; Clites, J. A. *Mol. Cryst. Liq. Cryst.* **1993**, *232*, 289.
77. Jain, R.; Sonsler, M.; Coms, F. D.; Dougherty, D. A. *J. Am. Chem. Soc.* **1988**, *110*, 1356.
78. Novak, J. A.; Jain, R.; Dougherty, D. A. *J. Am. Chem. Soc.* **1989**, *111*, 7618.

79. Dvolaitzky, M.; Chiarelli, R.; Rassat, A. *Angew. Chem. Int. Ed. Engl.* **1992**, *31*, 180.
80. Kanno, F.; Inoue, K.; Koga, N.; Iwamura, H. *J. Am. Chem. Soc.* **1993**, *115*, 847.
81. Borden, W. T.; Iwamura, H.; Berson, J. A. *Acc. Chem. Res.* **1994**, *27*, 109.
82. West, A. P., Jr.; Silverman, S. K.; Dougherty, D. A. *J. Am. Chem. Soc.* **1996**, *118*, 1452.
83. Rajca, A. *J. Am. Chem. Soc.* **1990**, *112*, 5891. Rajca, A. *J. Org. Chem.* **1991**, *56*, 2557.
84. Rajca, A.; Utamapanya, S.; Xu, J. *J. Am. Chem. Soc.* **1991**, *113*, 9235.
85. Rajca, A.; Utamapanya, S. *J. Am. Chem. Soc.* **1993**, *115*, 2396.
86. Veciana, J.; Rovira, C.; Crespo, M. I.; Armet, O.; Domingo, V. M.; Palacio, F. *J. Am. Chem. Soc.* **1991**, *113*, 2552.
87. Veciana, J.; Rovira, C. In *Magnetic Molecular Materials*; Gatteschi, D.; Kahn, O.; Miller, J. S.; Palacio, F., Ed.; Kluwer: Dordrecht: 1991; p 121.
88. Carilla, J.; Julia, L.; Riera, J.; Brillas, E.; Gariddo, J. A.; Labarta, A.; Alcala, R. *J. Am. Chem. Soc.* **1991**, *113*, 8281.
89. Nishide, H.; Yoshioka, N.; Inagaki, K.; Tsuchida, E. *Macromolecules* **1988**, *21*, 3120.
90. Nishide, H.; Yoshioka, N.; Kaneko, T.; Tsuchida, E. *Macromolecules* **1990**, *23*, 4487.
91. Yoshioka, N.; Nishide, H.; Tsuchida, E. *Mol. Cryst. Liq. Cryst.* **1990**, *190*, 45.
92. Nishide, H.; Kaneko, T.; Gotoh, R.; Tsuchida, E. *Mol. Cryst. Liq. Cryst.* **1993**, *233*, 89.
93. Nishide, H.; Kaneko, T.; Igarashi, M.; Tsuchida, E.; Yoshioka, N.; Lahti, P. M. *Macromolecules* **1994**, *27*, 3082.
94. Nishide, H.; Kaneko, T.; Nii, T.; Katoh, K.; Tsuchida, E.; Lahti, P. M. manuscript to be published.
95. Fuji, A.; Ishida, T.; Koga, N.; Iwamura, H. *Macromol.* **1991**, *24*, 1077.
96. Inoue, K.; Koga, N.; Iwamura, H. *J. Am. Chem. Soc.* **1991**, *113*, 9803.
97. Miura, Y.; Inui, K.; Yamaguchi, F.; Inoue, M.; Teki, Y.; Takui, T.; Itoh, K. *J. Polym. Sci., Polym. Chem.* **1992**, *30*, 959.
98. Miura, Y.; Ushitani, Y.; Matsumoto, M.; Inui, K.; Teki, Y.; Takui, T.; Itoh, K. *Mol. Cryst. Liq. Cryst.* **1993**, *232*, 135.
99. Lahti, P. M.; Ichimura, A. S. *Mol. Cryst. Liq. Cryst.* **1989**, *176*, 125.
100. Lahti, P. M.; Ling, C.; Yoshioka, N.; Rossitto, F. C.; van Willigen, H. *Mol. Cryst. Liq. Cryst.* **1993**, *233*, 17.
101. Modarelli, D. A.; Lahti, P. M. *Chem. Comm.* **1990**, 1167.
102. Rossitto, F.; Lahti, P. M. *Macromolecules* **1993**, *26*, 6308.
103. Alexander, C.; Feast, W. J. *Polym. Bull.* **1991**, *26*, 245.
104. Nishide, H.; Kaneko, T.; Kuzumaki, Y.; Yoshioka, N.; Tsuchida, E. *Mol. Cryst. Liq. Cryst.* **1993**, *232*, 143.
105. Iwamura, H.; Koga, N. *Acc. Chem. Res.* **1993**, *26*, 346.
106. Crayston, J. A.; Iraqi, A.; Walton, J. C. *Chem. Soc. Rev.* **1994**, 147.
107. Korshak, Y. V.; Medveda, T. V.; Ovchinnikov, A. A.; Spector, V. N. *Nature* **1987**, *126*, 370.
108. Cao, Y.; Yang, P.; Hu, Z.; Li, S.; Zhang, L. *Synth. Met.* **1988**, B625.
109. Lahti, P. M.; Ichimura, A. S. *J. Org. Chem.* **1991**, *56*, 3030.
110. Nishide, H.; Kaneko, T.; Nii, T.; Katoh, K.; Tsuchida, E.; Yamaguchi, K. *J. Am. Chem. Soc.* **1995**, *117*, 548.
111. Takui, T.; Sato, K.; Shiomi, D.; Itoh, K.; Kaneko, T.; Tsuchida, E.; Nishide, H. *Mol. Cryst. Liq. Cryst.* **1995**, *271*, 191.
112. Kaneko, T.; Toriu, S.; Nii, T.; Tsuchida, E.; Nishide, H. *Mol. Cryst. Liq. Cryst.* **1995**, *272*, 153.
113. Modarelli, D. A.; Rossitto, F. C.; Lahti, P. M. *Tetrahdron Lett.* **1989**, *30*, 4477.
114. Kalgutkar, R.; Ionkin, A.; Quin, L. D.; Lahti, P. M. *Tetrahedron Lett.* **1994**, *35*, 3889.

115. Palacio, F. In *Magnetic Molecular Materials*; Gatteschi, D.; Kahn, O.; Miller, J. S.; Palacio, F., Ed.; Kluwer: Dordrecht: 1991; p 1.
116. Mukai, K.; Nishiguchi, H.; Deguchi, Y. *J. Phys. Soc. Japan* **1967**, *23*, 125.
117. Mukai, K. *Bull. Chem. Soc. Japan* **1969**, *42*, 40.
118. Mukai, K.; Sakamoto, J. *J. Chem. Phys.* **1978**, *68*, 1432.
119. Awaga, K.; Sugano, T.; Kinoshita, M. *Solid State Comm.* **1986**, *57*, 453.
120. Awaga, K.; Sugano, T.; Kinoshita, M. *J. Chem. Phys.* **1986**, *85*, 2211.
121. Chiang, L. Y.; Upasani, R. B.; Sheu, H. S.; Goshorn, D. P.; Lee, C. H. *J. Chem. Soc.Chem. Comm.* **1992**, 959.
122. Jung, K.; Lahti, P. M.; Sheridan, P.; Britt, S.; Zhang, W.-r.; Landee, C. P. *J. Mater. Chem.* **1994**, *4*, 161.
123. Allemand, P. M.; Srdanov, G.; Wudl, F. *J. Am. Chem. Soc.* **1990**, *112*, 9391.
124. (a) Takahashi, M.; Turek, P.; Nakazawa, Y.; Tamura, M.; Nozawa, K.; Shiomi, D.; Ishikawa, M.; Kinoshita, M. *Phys. Rev. Lett.* **1991**, *67*, 746. (b) Kinoshita, M. *Mol. Cryst. Liq. Cryst.* **1993**, *232*, 1.
125. Kinoshita, M. In *Magnetic Molecular Materials*; Gatteschi, D.; Kahn, O.; Miller, J. S.; Palacio, F., Ed.; Kluwer: Dordrecht: 1991, p 87.
126. Nakazawa, Y.; Tamura, M.; Shirakawa, N.; Shiomi, D.; Takahashi, M.; Kinoshita, M.; Ishikawa, M. *Phys. Rev.* **1992**, *B46*, 1.
127. Iwamura, H., Miller, J. S. (Eds.) *Mol. Cryst. Liq. Cryst.* **1993**, *232-233*, 1.
128. Miller, J. S.; Epstein, A. J. (Eds.) *Mol. Cryst. Liq. Cryst.* **1995**, *271-274*, 1.
129. Chiarelli, R.; Rassat, A.; Rey, P. *J. Chem. Soc., Chem. Comm.* **1992**, 1081.
130. Izuoka, A.; Fukada, M.; Kumai, R.; Itakura, M.; Hikami, S.; Sugawara, T. *J. Am. Chem. Soc.* **1994**, *114*, 2609.
131. (a) Okumura, M.; Mori, W.; Yamaguchi, K. *Mol. Cryst. Liq. Cryst.* **1993**, *232*, 35. (b) Yamanaka, S.; Kawakami, T.; Nagao, H.; Yamaguchi, K. *Mol. Cryst. Liq. Cryst.*, **1995**, *271*, 19. (c) Novoa, J. J.; Mota, F.; Veciana, J.; Cirujeda, J. *Mol. Cryst. Liq. Cryst.*, **1995**, *271*, 79.
132. Gillon, B.; Schweizer, J. In *Molecules in Physics, Chemistry, and Biology*; Maruani, J.; Ed.; Kluwer: Dordrecht: 1989; Vol. 2; p 111.
133. Bonnet, M.; Luneau, D.; Ressouche, E.; Rey, P.; Schweizer, J.; Wan, M.; Wang, H.; Zheludev, A. *Mol. Cryst. Liq. Cryst.* **1995**, *271*, 35.
134. Cirujeda, J.; Ochando, L. E.; Amigó, J. M.; Rovira, C.; Rius, J.; Veciana, J. *Angew. Chem. Int. Ed. Engl.* **1995**, *34*, 55.
135. Cirujeda, J.; Hernàndez-Gasió, E.; Panthou, F. L.-F.; Laugier, J.; Mas, M.; Molins, E.; Rovira, C.; Novoa, J. J.; Rey, P.; Veciana, J. *Mol. Cryst. Liq. Cryst.* **1995**, *271*, 1.
136. Sugawara, T.; Matsushita, M. M.; Izuoka, A.; Wada, N.; Takeda, N.; Ishikawa, M. *J. Chem. Soc., Chem. Comm.* **1994**, 1723.
137. Caneschi, A.; Gatteschi, D.; Sessoli, R.; Rey, P. *Accts. Chem. Res.* **1989**, *22*, 392.
138. Caneschi, A.; Gatteschi, D.; Sessoli, R In *Magnetic Molecular Materials*; Gatteschi, D.; Kahn, O.; Miller, J. S.; Palacio, F., Ed.; Kluwer: Dordrecht: 1991; p 215.
139. Benelli, C.; Caneschi, A.; Gatteschi, G.; Pardi, L In *Magnetic Molecular Materials*; Gatteschi, D.; Kahn, O.; Miller, J. S.; Palacio, F., Ed.; Kluwer: Dordrecht: 1991; p 233.
140. Rey, P.; Luneau, D.; Cogne, A In *Magnetic Molecular Materials*; Gatteschi, D.; Kahn, O.; Miller, J. S.; Palacio, F., Ed.; Kluwer: Dordrecht: 1991; p 203.
141. Inoue, K.; Iwamura, H. *J. Am. Chem. Soc.* **1994**, *116*, 3173.
142. Inoue, K.; Hayamizu, T.; Iwamura, H. *Mol. Cryst. Liq. Cryst.* **1995**, *273*, 67.
143. Cf. for example Jang, S.-H.; Bertsch, R. A.; Jackson, J. E.; Kahr, B. *Mol. Cryst. Liq. Cryst.* **1992**, *211*, 289 for work on phenoxyl based materials.
144. Yoshizawa, K.; Hoffmann, R. *J. Am. Chem. Soc.* **1995**, *117*, 6921.

Chapter 15

Progress Toward 'Conditional' Magnetic Materials

Magnetic Properties of (*m*-Pyridyl Nitronyl Nitroxide)$_2$HBr and $Cu_2(OH)_3$ (alkylcarboxylate)

Kunio Awaga

Department of Pure and Applied Sciences, University of Tokyo, Komaba, Meguro, Tokyo 153, Japan

We report our efforts toward a 'conditional' magnet, *e.g.* a magnetic material which drastically changes properties upon perturbation, such as irradiation, pressure, electric field, etc. First, we describe coexistence of an intermolecular ferromagnetic interaction and $[NHN]^+$ hydrogen bond in (*m*-pyridyl nitronyl nitroxide)$_2$HBr. To our knowledge, this is the first example of "naked proton" to be found in organic radical crystals. Secondly, we show an drastic change in the magnetic properties of the layered material, $Cu_2(OH)_3(n\text{-}C_mH_{2m+1}COO)$, depending on the alkyl-chain length. The m=0 and 1 materials are metamagnetic, while the m=7-9 materials are weak ferromagnets below 22 K.

The search for molecule-based magnetic materials has intensified in recent years (*1*). One of the self-evident characters of the material is that it consists of molecules, and there are many molecules which change the molecular and electronic structures upon perturbation, such as irradiation, pressure, temperature, electric field etc. If the magnetic properties, namely the intramolecular spin state and/or the intermolecular interaction are modified according to the change, it would mean realization of controllable, changeable and functional magnetic materials. Let us call them 'conditional' magnetic materials. In this report we describe two different efforts toward the conditional magnetic materials.

Coexistence of Intermolecular Ferromagnetic Interaction and $[NHN]^+$ Hydrogen Bond in *N*-Protonated *m*-Pyridyl Nitronyl Nitroxide. (*2*)

Controllable change, motion, switch, etc. in characteristic sizes of molecules has been proposed for molecular electronic devices, based on the proton transfer in a double minimum potential well (*3*). Polaron-type electric conduction has been also proposed, through a modulation caused by the proton motion (*4*). To find cooperative phenomena between magnetism and proton motion is interesting in the field of molecular magnetism as well.

0097–6156/96/0644–0236$15.00/0

A stable organic radical family, nitronyl nitroxide, is attracting much interest, because of potential ferromagnetic properties. Various nitronyl nitroxide derivatives have been found to exhibit ferromagnetic intermolecular interactions in their bulk crystals (*5-15*) since the discovery of the first pure organic ferromagnet, *p*-nitrophenyl nitronyl nitroxide (*16*). Recently we have embarked upon the study of pyridyl nitronyl nitroxide radical family, making various chemical modifications. After *N*-alkylation, the iodide salts of *p*-*N*-alkylpyridinium nitronyl nitroxides were found to exhibit magnetic properties which systematically depend on the length of the *N*-alkyl chain (*17*). We also found an antiferromagnetic *Kagome* lattice in the crystals of *m*-*N*-methylpyridinium nitronyl nitroxide salts (*18*). In this section, we describe *N*-protonation of *m*-pyridyl nitronyl nitroxide (*m*-PYNN).

Materials. The *N*-protonation of *m*-PYNN was carried out by the following procedure: flowing HBr gas into the dry benzene solution of *m*-PYNN resulted in precipitation of (*m*-PYNN)$_2$HBr. The reaction was completed within 2 hrs. The obtained microcrystalline powder was washed for several times with the solvent. Recrystallization from the CH_2Cl_2 solution resulted in polycrystalline needles, but growth of the single crystal, suitable for X-ray crystal analyses, was unsuccessful. The HBr complex was rather sensitive to moisture, but was stable for a few weeks in a dry refrigerator. The elemental analyses indicated that obtained was the 2:1 complex between *m*-PYNN and HBr.

There are some precedents for such 2:1 ratio between nitrogen-containing base and proton, where it is known that the two base molecules are connected by an intermolecular $[NHN]^+$ hydrogen bond (*19*). This bond has been studied, with respect to homoconjugation, proton dynamics, and so on. In fact, the infrared spectrum suggests that (*m*-PYNN)$_2$HBr also involves an intermolecular $[NHN]^+$ hydrogen bond, such as is shown below: the spectrum shows an intense NH stretching band at *ca*. 2500 cm^{-1}, implying the protonation of the pyridyl rings, and has no band which is ascribed to that of *m*-PYNN. The presence of the protonation and the absence of the parent *m*-PYNN can be understood in terms of proton sharing by the two pyridyl rings in the $[(m\text{-PYNN})_2H]^+$ unit. To our knowledge, this is the first example of an intermolecular $[NHN]^+$ hydrogen bond found in organic radical solids.

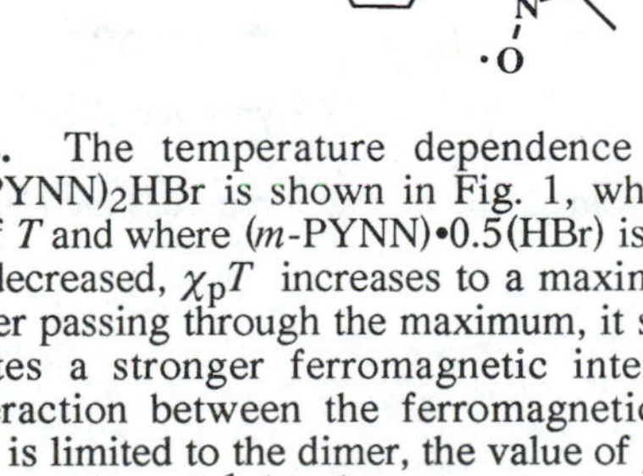

Magnetic Properties. The temperature dependence of the paramagnetic susceptibilities of (*m*-PYNN)$_2$HBr is shown in Fig. 1, where $\chi_p T$ is plotted as a function of logarithm of T and where (*m*-PYNN)•0.5(HBr) is taken as the molar unit. As the temperature is decreased, $\chi_p T$ increases to a maximum of *ca*. 0.65 emu K mol^{-1} at 3.5 K, but, after passing through the maximum, it shows a slight decrease. This behavior indicates a stronger ferromagnetic interaction and a weaker antiferromagnetic interaction between the ferromagnetic units. Since, if the ferromagnetic coupling is limited to the dimer, the value of $\chi_p T$ should approach an upper limit of 0.501 emu K mol^{-1} (g=2) at the absolute zero temperature, the

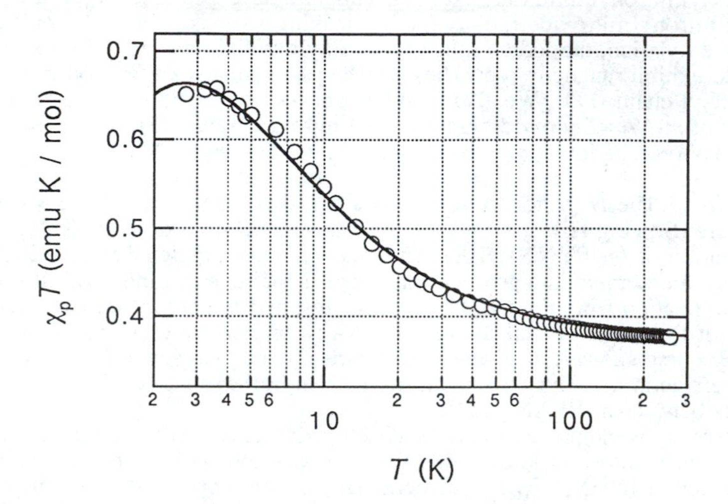

Figure 1. Temperature dependence of the paramagnetic susceptibilities of $(m\text{-PYNN})_2HBr$. The solid curves are theoretical ones. See the text.

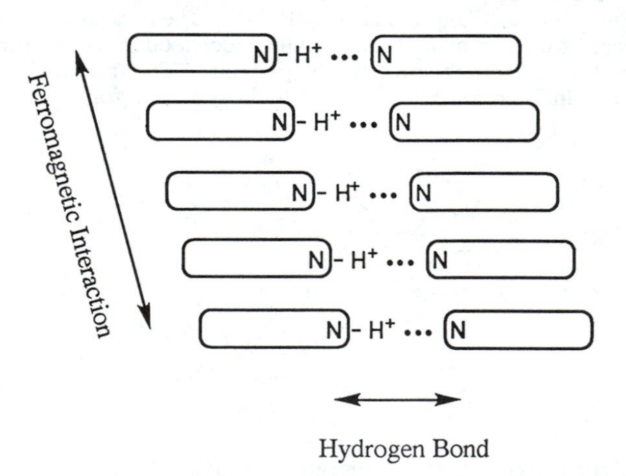

Figure 2. Ferromagnetic interaction across the $[NHN]^+$ dimers.

ferromagnetic interaction in (*m*-PYNN)$_2$HBr is concluded to cover more than three molecules. In fact, the observed temperature dependence can be quantitatively interpreted in terms of a 1-D ferromagnetic column with a weak antiferromagnetic interchain coupling (*11*). The solid curve going through the plots for (*m*-PYNN)$_2$HBr in Fig. 1, is the best fit obtained with g=2 (fixed), an intrachain ferromagnetic interaction of J/k_B=5.7 K and an interchain antiferromagnetic interaction of J'/k_B=-0.24 K.

The intermolecular [NHN]$^+$ hydrogen bond suggested by the infrared spectrum measurements, cannot be ascribed to be a pathway of the ferromagnetic coupling, because it would be limited in the dimer. Generally speaking, the magnetic interaction of the nitronyl nitroxide, either ferromagnetic or antiferromagnetic, is not governed by the intermolecular contact between the substituents at the α-position (namely the pyridyl groups in this case), on which there is little density of the unpaired electron, but originates in the intermolecular contacts of the NO groups, on which the unpaired electron is localized (*20*). Therefore, the ferromagnetic interaction is expected to extend across the [NHN]$^+$ dimers, as is schematically shown in Fig. 2. This situation may open the possibility of a cooperative phenomenon. If *m*-PYNN and *m*-PYNN•H$^+$ prefer different intermolecular magnetic couplings, the position of the proton would critically affect the intermolecular magnetic interaction, and in its turn the bulk magnetism. We are looking for cooperative phenomena between the proton dynamics and the magnetism.

(*m*-PYNN)$_2$(Fumaric Acid). We have learned that it is possible to make molecular complexes of PYNN, based on the acid-base interaction. To extend this idea, we have obtained single crystals of (*m*-PYNN)$_2$(Fumaric Acid), which crystallize into the triclinic $P\bar{1}$ space group [a=10.115(1)Å, b=10.334(1)Å, c=7.525(1)Å, α=105.63(1)°, β=97.80(1)°, and γ=80.31(1)°]. Figure 3 shows the unit cell of (*m*-PYNN)$_2$(Fumaric Acid). The fumaric acid plays a role of bridge between the two pyridyl rings of *m*-PYNN. Temperature dependence of the magnetic susceptibility is found to make a maximum at 6K, suggesting a weak antiferromagnetic intermolecular interaction and magnetic low-dimensionality in it.

Summary. We have carried out the *N*-protonation of *m*-pyridyl nitronyl nitroxide. The obtained material, (*m*-PYNN)$_2$HBr, is found to be ferromagnetic, and also to include an intermolecular [NHN]$^+$ hydrogen bond, which will attract interest with respect to proton dynamics, homoconjugation and so on. Coexistence of the two unusual features, may lead to cooperative phenomena between them in future. Further, we have succeeded in the crystal growth of (*m*-PYNN)$_2$(Fumaric Acid), in which the bridge of the fumaric acid between two pyridyl rings of *m*-PYNN. The complex formation of PYNN with other acid molecules seems to take place.

Magnetic Properties of Organic/Inorganic Hybrid Layered Material, Cu$_2$(OH)$_3$(alkylcarboxylate). (*21*)

We are also interested in organic/inorganic hybrid layered materials. If a photochromic organic molecule, such as azobenzene etc., is inserted into the interlayer of a magnetic inorganic material, as is shown in Fig. 4, and if the magnetic properties of the inorganic layer is seriously affected by the structural change of the organic molecule, the obtained materials can be regarded as a conditional magnetic material controlled by irradiation. In this case, the organic molecule plays a role of sensor and the inorganic layer plays a role of center of magnetic change. At this stage it is important to look for a candidate for the inorganic layer.

Layered metal hydroxides are of great interest to both science and technology, because of their utility as an ion-exchange materials (*22*), catalysts(*23*), antacids

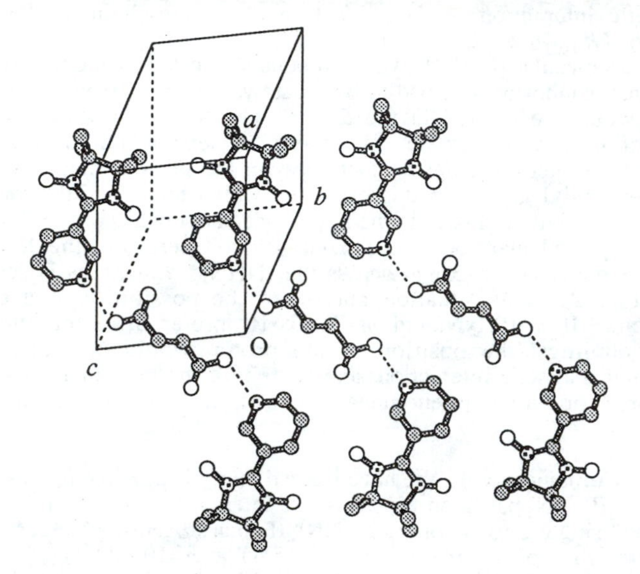

Figure 3. Crystal structure of (m-PYNN)$_2$(Fumaric Acid).

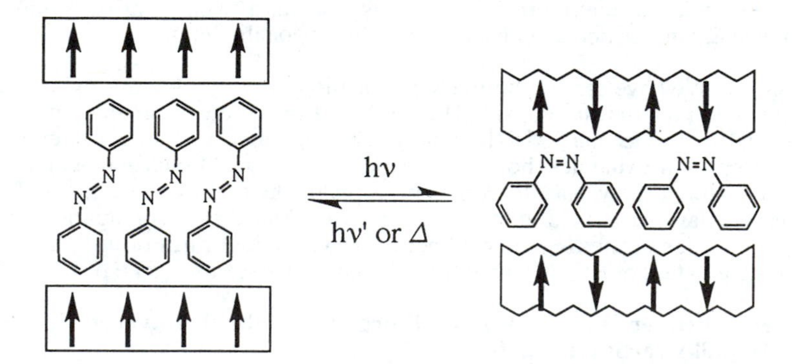

Figure 4. Conditional magnetic material based on organic/inorganic hybrid layered system.

(*24*), and modified electrodes (*25*). They also attract interest as two-dimensional magnetic materials (*26*). The copper hydroxy salts, $Cu_2(OH)_3X$ (X=exchangeable anion OAc, NO_3 etc.), exhibit a botallackite-type structure, in which two-crystallographically distinct copper atoms lie in 4+2 (oxygen+X) and 4+1+1 (oxygen+oxygen+X) environments (*27*). The anion is located in the interlayer, while the molecular end of it coordinates the copper ion. To see the effect of the molecular size of the intercalated anion on the magnetic behavior of the inorganic layer, we have embarked upon magnetic measurements on the $Cu_2(OH)_3(n\text{-}C_mH_{2m+1}COO)$ series.

Intercalation. The intercalation compounds were obtained, according to the procedure in ref. (*28*). The parent material, $Cu_2(OH)_3(CH_3COO){\bullet}H_2O$ (m=1), was prepared: slow titration of an aqueous solution of $Cu(OAc)_2$ (1 l, 0.1 M) with NaOH (0.1 M) to OH^-:Cu^{2+} ratio of 1 gave a precipitate of bluish-green crystals. The m=0, 7-9 materials were obtained by the anion exchanges: 200 mg of the parent material and 1 mol of the corresponding sodium alkylcarboxylate were dispersed in 0.1 l of water. The mixture was stirred for a day at room temperature, and the obtained material was filtrated, washed with water, and dried under vacuum. The elemental analyses indicated completeness of the ion-exchanges. However the exchanges for the m=2-6 materials were not successful because of decomposition (m=2-4) and of too slow exchange rate (m=5-6), in contrast to the stableness and immediate exchange for the m=0, 7-9 materials.

The X-ray powder diffraction patterns of the parent and the exchanged materials showed intense (00l) reflections, indicating the layered structure in them (*29*). The interlayer distance was estimated by subtracting the thickness of the inorganic host layer from the distance between two neighboring planes of class (00l). Figure 5 shows the dependence of the interlayer distance on m, namely the number of the carbons in the alkyl chain. The distance increases in the order of m. The bars in this figure show calculated molecular heights of the carboxylates. The distances of the m=0 and 1 materials are shorter than their molecular heights, respectively, presumably because of their coordinations to the intralayer copper ion. The distances of the m=7-9 materials are nearly double of their molecular heights, indicating a bilayer structure in them. The structural difference of the organic layer is also suggested by the fact that the plots of the m=0 and 1 materials appear to deviate from the fitted line to the plots for the m=7-9 materials.

Magnetic Properties. We have investigated the magnetic properties of the $Cu_2(OH)_3(n\text{-}C_mH_{2m+1}COO)$ series, focusing on the results on the m=0 and 7 materials. The open circles in Fig. 6 show the temperature dependence of χ_pT of the m=0 material, $Cu_2(OH)_3(HCOO)$. We adopt half of $Cu_2(OH)_3X$ as the molar unit. The value of χ_pT increases with decreasing temperature down to *ca.* 10 K, indicating dominance of a ferromagnetic interaction. The Weiss constant evaluated using the data T>100 K is 4 K. The intralayer interaction between the copper ions is ferromagnetic. However χ_pT decreases suddenly below 4 K, suggesting an interlayer antiferromagnetic coupling. The open circles Fig. 7 show the real component of the ac susceptibilities χ_{ac} of the m=0 material. χ_{ac} increases with decreasing temperature down to 5.8 K, and after making a rather sharp maximum, it decreases slightly. The open circles in Fig. 8 show the magnetization curve at 3.0 K, the lowest temperature attainable in our apparatus. One can see S-shape dependence on the field. This is consistent with the coexistence of the ferromagnetic and antiferromagnetic interactions. The observed behavior is quite similar to those of the transition metal hydroxides, $M(OH)_2$ with M=Fe, Co and Ni, whose structure can be understood as that of $M_2(OH)_3X$ with X=OH, and which are well characterized as metamagnets (*30-32*). The m=0 material is expected to be a metamagnet at a lower temperature which consists of an intralayer ferromagnetic interaction and an

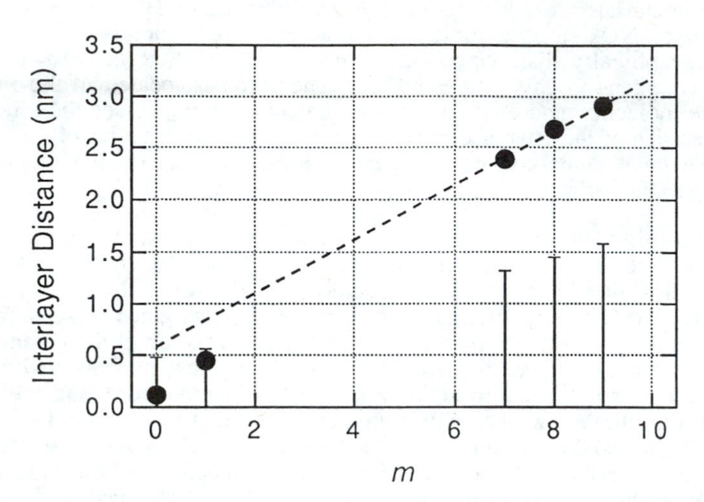

Figure 5. Interlayer distances of the $Cu_2(OH)_3(n\text{-}C_mH_{2m+1}COO)$ series.

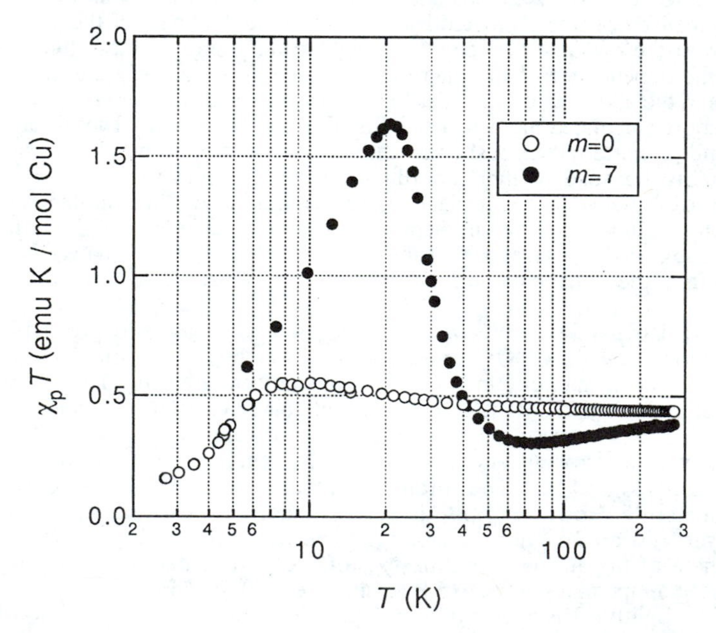

Figure 6. Temperature dependence of the paramagnetic susceptibilities χ_p of the m=0 and 7 materials. $\chi_p T$ is plotted as a function of temperature.

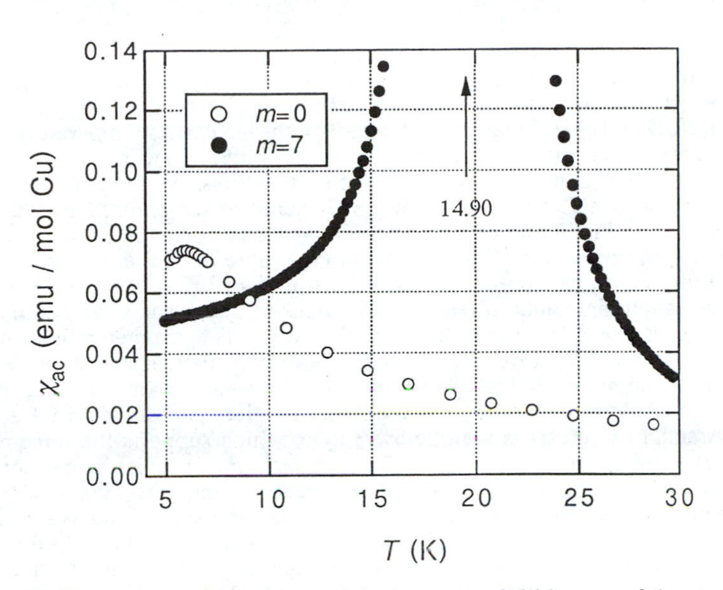

Figure 7. Temperature dependence of the ac susceptibilities χ_{ac} of the m=0 and 7 materials.

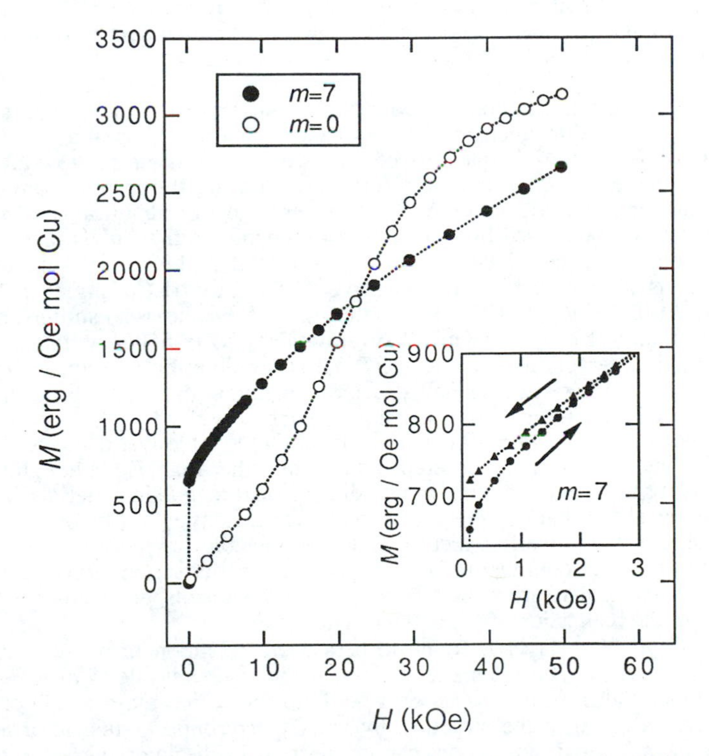

Figure 8. Field dependence of the magnetizations of the m=0 (4.2 K) and m=7 (3.0 K) materials. The inset shows the hesteresis of the magnetization of the m=7 material at 3.0 K.

interlayer antiferromagnetic interaction. The $m=1$ material, $Cu_2(OH)_3(CH_3COO)\cdot H_2O$, also shows metamagnetic behavior (not shown).

The closed circles in Fig. 6 show χ_pT of the $m=7$ material, $Cu_2(OH)_3(CH_3(CH_2)_6COO)$. χ_pT decreases with decreasing temperature down to 60 K, indicating an intralayer antiferromagnetic interaction, in opposition to the ferromagnetic ones in the $m=0$ and 1 materials. The Weiss constant is obtained to be -30 K with the data $T>100$ K. However χ_pT increases suddenly below 60 K. The closed circles in Fig. 7 show χ_{ac} of the $m=7$ material. The plots of χ_{ac} make an anomalous peak around 22 K whose maximum value is 14.9 emu mol^{-1}. The ac susceptibilities indicate a ferromagnetic order with $T_c=22$ K, in spite of the intralayer antiferromagnetic interaction. The closed circles in Fig. 8 show the magnetization curve measured with increasing field at $T=4.2$ K. The magnetization shows an abrupt increase at the lower fields, but it gradually increases without showing saturation at the higher fields. This is typical of a weak ferromagnet. The closed triangles in the inset of Fig. 8 show the magnetizations measured with decreasing field. A small hysteresis is seen below 3 kOe, which supports the ferromagnetic order. The residual magnetization is 710 erg Oe^{-1} mol^{-1} which corresponds to a canting angle of 7.0° between two antiferromagnetic moments. We speculate that the spin canting takes place between the two copper ion sites in the 4+2 and 4+1+1 environments. It is worth mentioning the significant decrease in χ' in the ordered state below 22 K. Similar temperature dependence of χ' is reported for the layered complexes, A_2MX_4 (A=alkylammonium, etc., M=Fe, Mn, and X=Cl, Br), which are characterized as weak ferromagnets (*33-35*). The phenomenon is explained in terms of the domain size effect on the ac magnetic susceptibility (*36*).

The $m=7$, 8 and 9 materials also show divergence of χ_{ac} at the same temperature, and are also indicated to be weak ferromagnets with a transition temperature of $T_c=22$ K.

Discussion. The intercalation compounds, $Cu_2(OH)_3(n\text{-}C_mH_{2m+1}COO)$, show the interesting variety of magnetism. The $m=0$ and 1 materials show the metamagnetic behavior, while the $m=7$-9 materials become weak ferromagnets below 22 K. The intralayer magnetic interaction is ferromagnetic in the former ones, while antiferromagnetic in the latter ones. The change of the intralayer magnetic interaction could be caused by some structural change in the CuOH layer with the intercalation. This could be related to the fact that the Cu-OH-Cu exchange interaction is known to be quite sensitive to the Cu-OH-Cu angle: a di-nuclei complex, $[Cu(EAEP)OH]_2(ClO_4)_2$, with the angle of 99° shows an antiferromagnetic coupling of $2J=-130$ cm^{-1}, while $[Cu(bipy)OH]_2SO_4\cdot 5H_2O$ with 97° does a ferromagnetic interaction of 48 cm^{-1} (*37*). It is possible that the drastic magnetic change can be explained by a small structural change in the CuOH layer caused by the difference in the intercalated organic molecules.

It is notable that the $m=7$-9 materials in which the interlayer distances are more than 20 Å have the magnetic ordered states with rather high T_c, although there are precedents for the magnetic order in two-dimensional magnetic materials (*38*). The interlayer weak exchange interaction and the intralayer anisotropic exchange interaction are theoretically expected to be responsible for the three dimensional order (*39*). It is no doubt that the $m=7$-9 materials involve an intralayer anisotropic exchange, because thay are weak ferromagnets. The anisotropic exchange could be essential for the realization of the ordered states in them.

The $[Cu_2(OH)_3]^-$ layer is found to be a useful component of the conditional magnetic material. The magnetic variety in the layer may lead to a functional magnetic material in future: if an organic molecule which plays a role of sensor, namely which changes the structure by small perturbation, such as irradiation, pressure, electric field, and so on, can be inserted in the interlayer, the magnetic properties may show drastic changes by the perturbation to the organic molecule.

Acknowledgments

The author would like to thank to Messrs. Tsunehisa Okuno, Takeo Otsuka, Wataru Fujita, and Profs. Allan E. Underhill of Wales university and Tamotsu Inabe of Hokkaido university for their experimental contributions and for helpful discussions. This work was supported by the Grant-in-aid for Scientific Research from the Ministry of Education, Science, and Culture, Japanese government. Support from New Energy and Industrial Technology Development Organization (NEDO) is also acknowledged.

References

1. Proceedings of international symposium of chemistry and physics of molecular based magnetic materials, Salt Lake City, 1994, in *Mol. Cryst. Liq. Cryst.,* vol. 273 and 274 (1995); Miller, J. S.; Epstein, A. J., Eds.
2. Okuno, T.; Otsuka, T.; Awaga, K. *J. Chem. Soc. Chem. Commun.*, **1995**, 827.
3. Molecular Electronic Devices,; Carter, F. L. Ed.; Dekker: New York, 1982.
4. Inabe, T. *New J. Chem.*, **1991**, *15*, 129.
5. Wang, H.; Zhang, D.; Wan, M.; Zhu, D. *Solid State Commun.,* **1993**, *85*, 685.
6. Turek, P.; Nozawa, K.; Shiomi, D.; Awaga, K.; Inabe, T.; Maruyama, Y.; Kinoshita, M. *Chem. Phys. Lett.,* **1991**, *180*, 327.
7. Sugano, T.; Tamura, M.; Kinoshita, M.; Sakai, Y.; Ohashi, Y. *Chem. Phys. Lett.,* **1992**, *200*, 235.
8. Awaga, K.; Inabe, T.; Maruyama, Y.; Nakamura, T.; Matsumoto, M. *Chem. Phys. Lett.,* **1992**, *195*, 21.
9. Hernàndez, E.; Mas, M.; Molins, E.; Rovira, C.; Veciana, J. *Angew. Chem., Int. Ed. Engl.,* **1993,** *32*, 882.
10. Tamura, M.; Shiomi, D.; Hosokoshi, Y.; Iwasawa, N.; Nozawa, K.; Kinoshita, M.; Sawa, H.; Kato, R. *Mol. Cryst. Liq. Cryst.,* **1993,** *232*, 45.
11. de Panthou, F. L.; Luneau, D.; Laugier, J.; Rey, P. *J. Am. Chem. Soc.,* **1993**, *115*, 9095.
12. Inoue, K.; Iwamura, H.; *Chem. Phys. Lett.,* **1993**, *207*, 551.
13. Sugawara, T.; Matsushita, M. M.; Izuoka, A.; Wada, N.; Takeda, N.; Ishikawa, M.; *J. Chem. Soc., Chem. Commun.,* **1994**, 1723.
14. Akita, T.; Mazaki, Y.; Kobayashi, K.; Koga, N.; Iwamura, H. *J. Org. Chem.,* **1995,** *60*, 2092.
15. Cirujeda, J.; Ochando, L. E.; Amigó, J. M.; Rovira, C.; Rius, J.; Veciana,. *Angew. Chem., Int. Ed. Engl,* **1995**, *34*, 55.
16. Kinoshita, M.; Turek, P.; Tamura, M.; Nozawa, K.; Shiomi, D.; Nakazawa, Y.; Ishikawa, M. Takahashi, M.; Awaga, K.; Inabe, T.; Maruyama, Y. *Chem. Lett.,* **1991**, 1225.
17. Awaga, K.; Okuno, T.; Yamaguchi, A.; Hasegawa, H.; Inabe, T.; Maruyama, Y.; Wada, N. *Phys. Rev. B,* **1994,** *49*, 3975.
18. Awaga, K.; Yamaguchi, A.; Okuno, T.; Inabe, T.; Nakamura, T.; Matsumoto, M.; Maruyama, Y. *J. Mater. Chem.,* **1994**, *4*, 1377.
19. Jones, D. J.; Brach, I.; Rozière, J. *J. Chem. Soc., Dalton Trans.*, **1984**, 1795. References are therein.
20. Awaga, K.; Inabe, T.; Nagashima, U.; Nakamura, T.; Matsumoto, M.; Kawabata, Y.; Maruyama, Y.; *Chem. Lett.,* **1991**, 1777.
21. Fujita, W.; Awaga, K. submitted.
22. Dutta, P. K.; Puri, M. *J. Phys. Chem*. **1989**, *93*, 376.
23. Reichle, W. T.; Kang, S. Y.; Everhardt, D. S. *J. Catal*. **1986**, *101*, 352.
24. Schmidt, P. C.; Benke, K. *Pharm. Acta. Helv*. **1988**, *63*, 188.
25. Itaya, K. Chang, H.-C.; Uchida, I. *Inorg. Chem*. **1987**, *26*, 624.

26. Rabu, P.; Angelov, S.; Legoll, P.; Belaiiche, M.; Drillon, M. *Inorg. Chem.* **1993**, *32*, 2463.
27. Jiménez-López, A.; Rodríguez-Castellón, E.; Olivera-Pastor, P.; Maireles-Torres, P.; Tomlinson, A. A. G.; Jones, D. J.; Roziére, J. *J. Mater. Chem.* **1993**, *3*, 303.
28. Yamanaka, S.; Sako T.; Hattori, M. *Chem. Lett.* **1989**, 1869. Yamanaka, S.; Sako T.; Seki, K.; Hattori, M. *Solid State Ionics.* **1992**, *53-56*, 527.
29. Prof. W. Mori (Osaka univ.), private communication: recently, the single crystals of the m=0 material were prepared with a different method. The results of the structural analysis are consistent with our data.
30. Miyamoto, H.; Shinjo, T.; Bando, Y.; Takada, T. *J. Phys. Soc. Jpn.* **1967**, *23*, 1421.
31. Takada, T.; Bando, Y.; Kiyama, M.; Miyamoto, H. *J. Phys. Soc. Jpn.* **1966**, *21*, 2726.
32. Takada, T.; Bando, Y.; Kiyama, M.; Miyamoto, H. *J. Phys. Soc. Jpn.* **1966**, *21*, 2745.
33. Epstein, A.; Gurewitz, E.; Makovsky, J.; Shaked, H. *Phys. Rev. B* **1970**, *2*, 3703.
34. Mostafa, M. F.; Willet, R. D. *Phys. Rev. B* **1971**, *4*, 2213.
35. Blake, A. B.; Hatfield, W. E. *J. Chem. Soc. Dalton Trans.* **1979**, 1725.
36. Groenendijk, H. A.; van Duyneveldt, A. J.; Willett, R. D. *Physica* **1980**, *101B*, 320.
37. Kahn, O. "Molecular magnetism"; VCH Publishers (UK) Ltd.: Cambridge, 1993; p. 159.
38. Haseda, T.; Yamakawa, H.; Ishizuka, M.; Okuda, Y.; Kubota, T.; Hata, M.; Amaya, K . *Solid State Commun.* **1977**, *24*, 599.
39. Oguchi, T. *Phys. Rev.* **1964**, *133*, A1098.

Chapter 16

Intrachain Ferromagnetic Spin Alignment in π-Conjugated Polyradicals with a Poly(phenylenevinylene) Chain

Hiroyuki Nishide

Department of Polymer Chemistry, Waseda University, Ohkubo 3, Shinjuku, Tokyo 169, Japan

Poly(phenylenevinylene) was selected as a linear polymer-type magnetic coupler in polyradicals, because of its developed π-conjugation and solvent-solubility even after substitution on the phenylene ring. An intrachain and through-bond ferromagnetic interaction between the side-chain built-in radical groups was described for the chemically stable poly[4-(3',5'-di-*tert*-butylphenoxy)-1,2-phenylenevinylene]. Effects of radical (spin) defects and molecular weight of the polyradical were discussed.

For π-conjugated and alternant, but non-Kekulé-type organic molecules bearing multiple radical centers, the spin quantum number (S) at the ground state has been theoretically predicted and experimentally substantiated, which is clearly related to the molecular connectivity or substitution positions of the radical centers on the conjugated coupler (*1-4*). Such results for di- and oligo-radical molecules with a high-spin ground state have been extended to organic π-conjugated polyradicals. The major synthetic approaches have been based on a cross-conjugated polyradical where radicals are formed in a main chain and are connected to each other with a ferromagnetic coupler (FC) to satisfy the ferromagnetic connectivity of the radicals (Scheme 1(a)). The typical example is the radical derivative of poly(1,3-phenylenemethine) reported by Rajca *et al.* (*5*). Although the cross-conjugated polyradicals displayed a strong intrachain and through-bond ferromagnetic coupling between two 1,3-phenylene connected radical electrons at low temperature, a small number of radical defects in the polyradicals significantly prevented an increase in the resulting S because the radicals were formed through cross-conjugated structures (*3,5*). In addition to this drawback, most of these molecules lacked chemical stability at room temperature.

Another synthetic approach is focused on π-conjugated linear polymers bearing side-chain or pendant radical groups, which are also π-conjugated with the polymer main chain and have substantial chemical stability (Scheme 1(b)). The typical example is the poly(phenylenevinylene) bearing 2,6-di-*tert*-butylphenoxy and *tert*-butyl nitroxide radicals recently reported by us (*6,7*). The intrachain and through-bond interaction between the side-chain spins has often been theoretically studied for

0097–6156/96/0644–0247$15.00/0

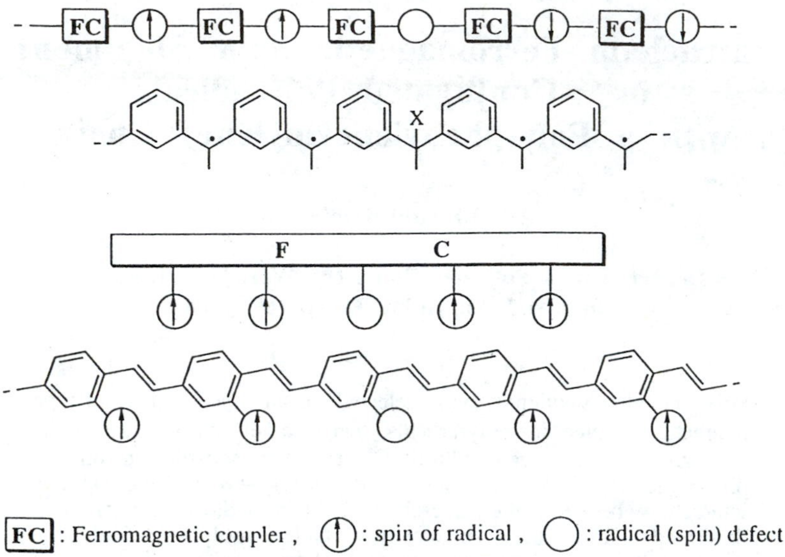

FC : Ferromagnetic coupler , ↑ : spin of radical , ○ : radical (spin) defect

Scheme 1

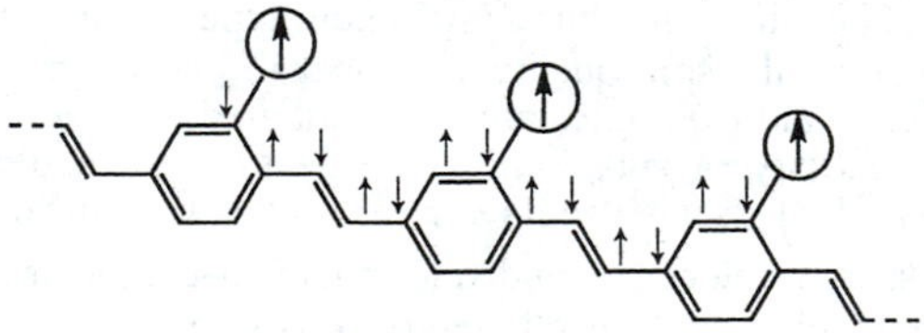

Scheme 2

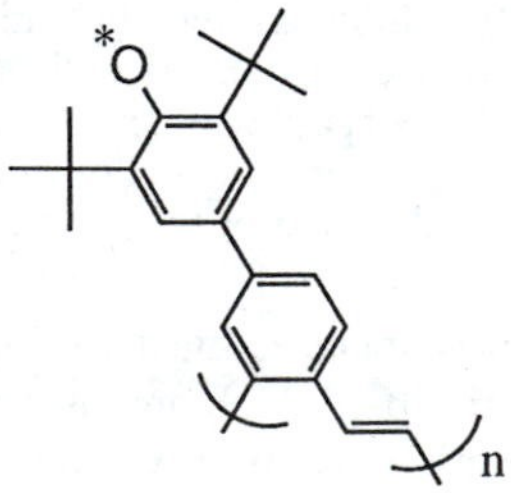

Scheme 3

this type of polyradical, *e.g.*, a high-spin alignment is expected for poly(1,4-phenylenevinylene) bearing a 3-substituted built-in radical by simple enumeration of the connectivity (represented by up-down small arrows in Scheme 2, big arrows = spin at the radical center). Lahti semiempirically calculated the significant stability of the high-spin ground state for the polyradicals with the poly(phenylenevinylene)-chain (*8*).

This paper describes the requisites needed to synthesize such a polyradical as represented by Schemes 1(b) and 2 using the example of poly[4-(3',5'-di-*tert*-butylphenoxy)-1,2-phenylenevinylene] (Scheme 3), and discussed the effects of spin defects and molecular weight of the polyradical on the intrachain ferromagnetic spin alignment.

Precise Synthesis of a Poly(phenylenevinylene) Radical

A restricted primary structure is essential for a ferromagnetic spin alignment in the Scheme 1(b)-type polyradical as illustrated in Scheme 4. A complete head-to-tail linkage of the monomer unit is the first requisite, since a synthetic error in the polymerization cancels out the high-spin alignment (on the left in Scheme 4). Radicals are generated *via* oxidation of the corresponding precursor polymer, which may cause defects in the radical or spin. The radical defect increases the distance between the side-chain spins and certainly decreases the spin interaction; however, the defect is not fatal for the polyradical because the interaction is transmitted through the π-conjugated backbone in contrast to the cross-conjugated polyradicals of Scheme 1(a), where a defect inhibits the π-conjugation itself. A conformational disorder, *e.g.* represented on the right of Scheme 4, presumably influences the stability of the high-spin state. The discussion in this paper focuses on the former two issues as well as the effect of polyradical's molecular weight.

A mono-substituted poly(phenylenevinylene) involves three linkage structures (head-to-tail, head-to-head, and tail-to-tail) to form the poly(phenylenevinylene)-chain. Although poly(phenylenevinylene)s with high molecular weights have usually been synthesized *via* sulfonium precursor polymers (*9*), these lack the restricted primary structure of head-to-tail, because the polymerization proceeds through a repeated elimination reaction of the dialkyl sulfide from α,α'-bis(dialkylsulfonio)xylenes (Scheme 5). We attempted and compared the synthesis of the head-to-tail linked poly(phenylenevinylene)s *via* the Wittig reaction (*10*) of the (formylphenyl)methyl phosphonium salts, and the Heck reaction (*11*) of bromostyrene derivatives using a palladium catalyst (Scheme 5). The Heck reaction was more favorable than the Wittig reaction to yield poly(phenylenevinylene)s with a relatively high molecular weight and an all *trans*-vinylene structure.

2-Bromo-4-(3',5'-*tert*-butyl-4'-acetoxyphenyl)styrene was polymerized using a catalyst of palladium acetate and tris-*o*-tolylphosphine in the presence of a small excess of triethylamine. The catalytic cycle involves four steps (Scheme 6); oxidative addition of the aryl bromide to the Pd(0) complex in which the bulky phosphine ligand provides an unsaturated coordination site for the addition, coordination of the styryl double bond to the Pd(II) complex, elimination of the β-hydrogen of the styryl group, and elimination of HBr from the Pd hydride with the amine base (regeneration of catalyst). The catalytic coupling reaction or the polymerization of the bromostyrene proceeded under moderate conditions.

The polymer was obtained as a yellow powder and found very soluble in common solvents such as $CHCl_3$, benzene, THF and acetone. The head-to-tail and *trans*-vinylene linkage structure of the polymer was confirmed by the extremely simple ^{13}C-NMR spectrum (Figure 1). The 12 aromatic carbon absorption peaks were ascribed to the phenylene rings and vinylene, and the six peaks at 130-155 ppm

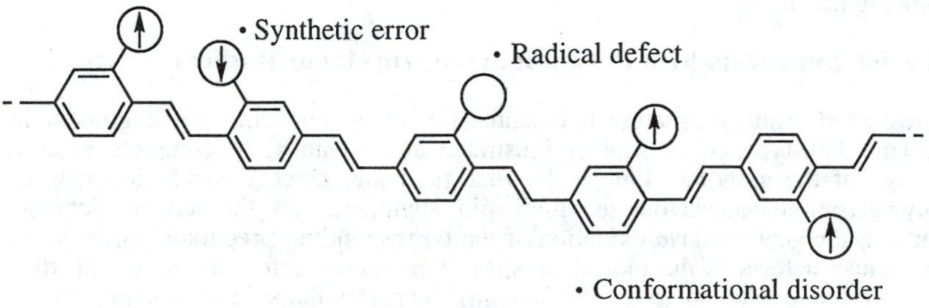
• Synthetic error
• Radical defect
• Conformational disorder

Scheme 4

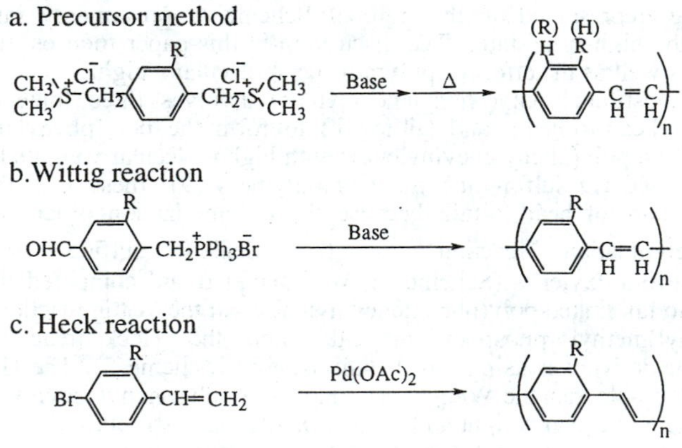
a. Precursor method
Base
Δ
b.Wittig reaction
OHC
$CH_2\overset{+}{P}Ph_3\overset{-}{Br}$
Base
c. Heck reaction
Br
CH=CH₂
$Pd(OAc)_2$

Scheme 5

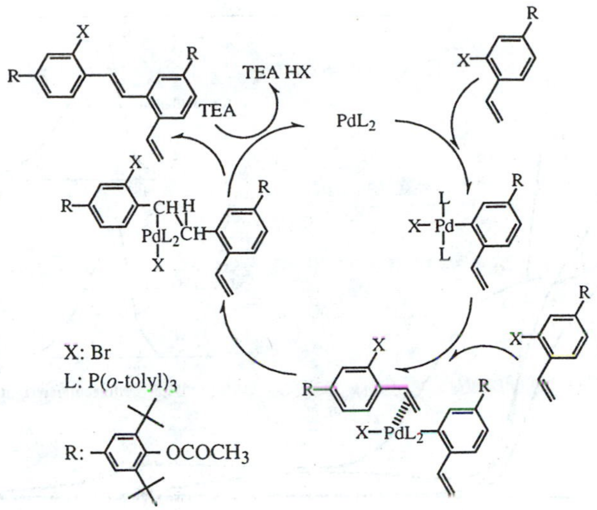

X: Br

L: P(*o*-tolyl)$_3$

R: OCOCH$_3$

Scheme 6

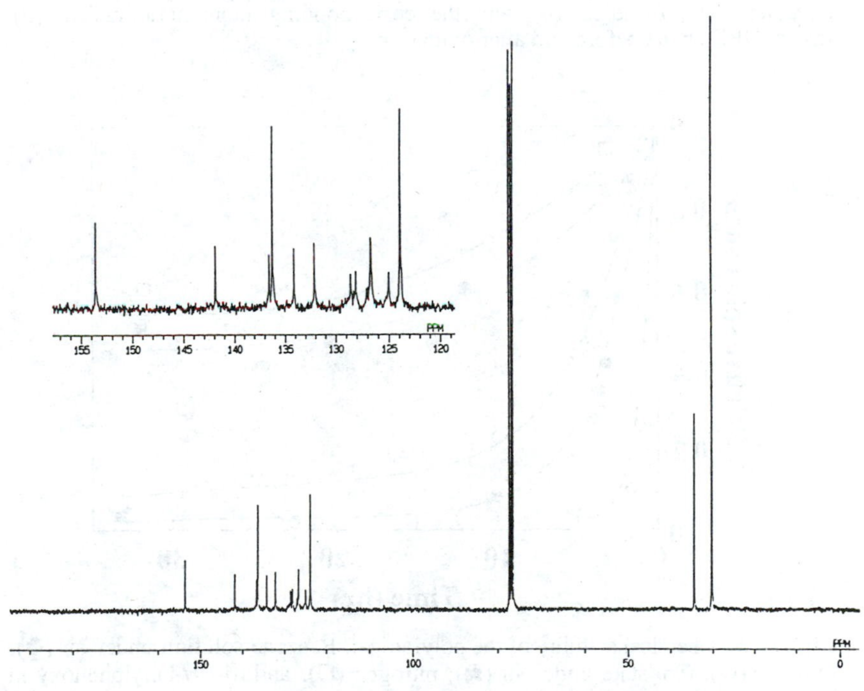

Figure 1. ^{13}C-NMR Spectrum of the hydroxy precursor polymer.

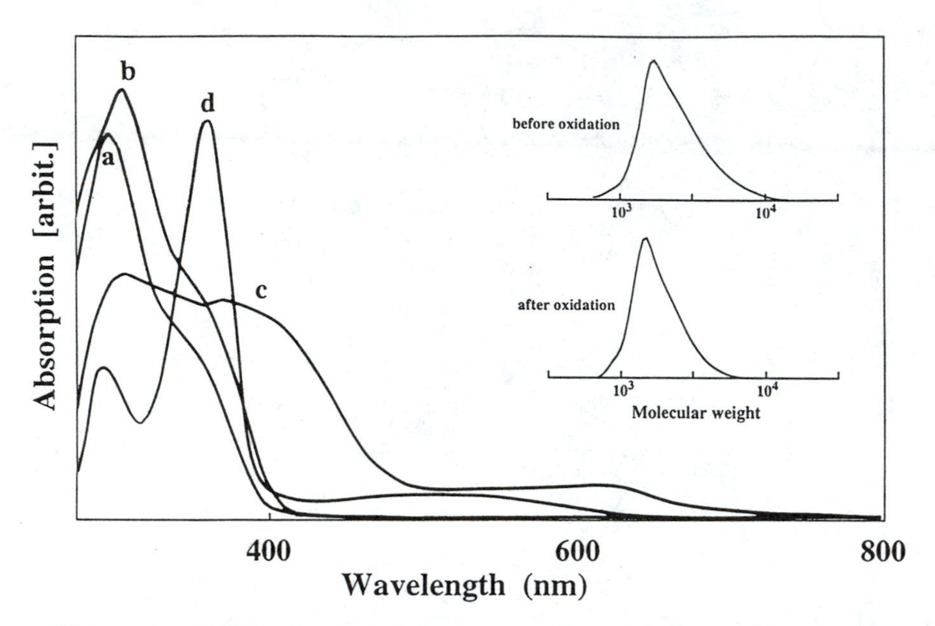

Figure 2. UV/Vis Spectra of the acetoxy (a) and hydroxy (b) precursor polymer, the polyradical (c), and the corresponding monomeric radical (d). Insert: GPC charts before and after oxidation.

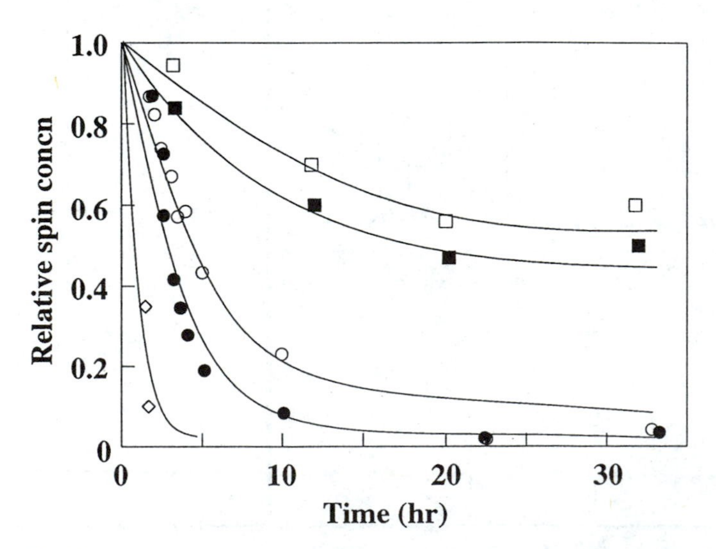

Figure 3. Chemical stability of the polyradical. Benzene solution under air (●), nitrogen (○), film state under air (■), nitrogen (□), and tri-*tert*-butylphenoxy in benzene solution (◇).

were assingned to the quarternary carbons using APT spectroscopy. The IR absorption for δ(*trans*-vinylene) = 961 cm^{-1} and fluorescence at 450 nm were also attributed to the *trans*-stilbene moiety. The polymer was converted to the corresponding hydroxy form and chemically oxidized to yield the polyradical, which was isolated as a brownish green powder.

The polyradical was also soluble in the common solvents. UV/ Vis absorption spectra (Figure 2) of the precursor polymers showed an absorption maximum of λ_{max} = *ca.*300 nm and λ(shoulder) = *ca.*370 nm, suggesting a developed π-conjugation in the poly(phenylenevinylene) backbone. The polyradical gave a broad absorption with maxima at *ca.*440 and 660 nm, which was assigned to a conjugated radical in comparison with that of the monomeric phenoxy radical (Figure 2). The ionization thresholds estimated using photoelectron spectroscopy (*12*) were 5.65 and 5.81 eV for the hydroxy precursor polymer and the polyradical, which also supports their π-conjugated structure. GPC elution curves of the polymer before and after radical generation (inserted charts in Figure 2) almost coincided with each other. This is consistent with the assumption that the radical generation does not bring about oxidative cleavage and crosslinking of the poly(phenylenevinylene)-chain.

The corresponding low molecular weight 2,6-di-*tert*-butyl-4-isopropylphenoxy radical rapidly disproportionates to guinone methide and the parent phenol (*13*), and the 2,6-di-*tert*-butyl-4-(*p*-styryl)phenoxy radical dimerizes to yield bisquinone methide(*14*). However the polyradical was ESR active for a relatively long time both in solution and in the solid state. Under atmospheric conditions at room temperture (Figure 3), the spin concentration of the polyradical in benzene solution gradually decreased with a half-life of several hours, although the spin concentration of 2,4,6-*tert*-butylphenoxy steeply decreased due to the formation of the peroxide (*15*). The polyradical in the solid or solvent-free film state was active even after standing for a few days at room temperature. The sterically crowded structure in the polyradical probably suppresses the disproportionation and coupling reaction between the radical groups.

Ferromagnetic Interaction and Effect of Spin Defect

Magnetization and static magnetic susceptibility of the polyradical were obtained in frozen 2-methyltetrahydrofuran solution to minimize intermolecular and through-space magnetic interactions, and were measured with a SQUID magnetometer as previously described (*16*). The frozen solution was in a typical amorphous state. Figure 4 shows the magnetization (*M*) plots normalized with saturated magnetization (M_S) *vs* the effective temperature (T-θ) for the polyradicals at 2K with spin concentrations of 0.15, 0.42, and 0.67 spin/monomer unit. All spin concentrations were determined both by careful integration of the ESR signals in comparison with those of the standard TEMPO solution and by saturated magnetization at 2K using a SQUID magnetometer. The normalized magnetization plots in Figure 4 are presented above the theoretical Brillouin curves for S=1/2, indicating a high-spin ground state. The plots increase to higher S positions with an increase in the spin concentration of polyradical and are close to the curve of S=4/2 for the polyradical with a spin concentration of 0.67 spin/unit.

Distribution of the generated radical or the radical defect along the poly(phenylenevinylene)-chain is presumed to be random and statistical, because the radical was chemically generated in a homogeneous solution of the precursor polymer (although a spin clustering phenomenon cannot be ruled out). The polyradical involves the contribution of various arrangements of the radical site along the polymer chain (inserted scheme in Figure 5). The distribution of the ground spin states was simulated for the random radical generation for the polyradical with a

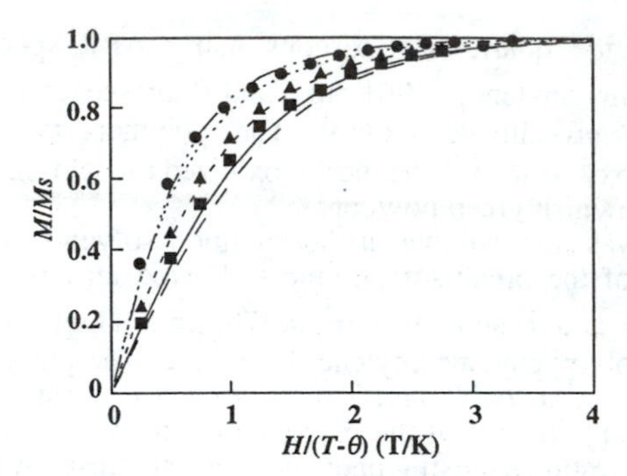

Figure 4. Normalized plots of magnetization (M/M_S) *vs* the ratio of magnetic field and temperature ($H/(T\text{-}\theta)$) for the polyradical with the spin concentration of 0.15(■), 0.42(▲), and 0.67(●) spin/unit in 2-methyltetrahydrofuran glass at 2K, where θ is a weak antiferromagnetic term. The theoretical Brillouin curves for S=1/2 (– –) and 4/2 (–···–), and the Brillouin curves simulated for the spin concentration of 0.15(—), 0.42(--), and 0.67 (····) spin/unit.

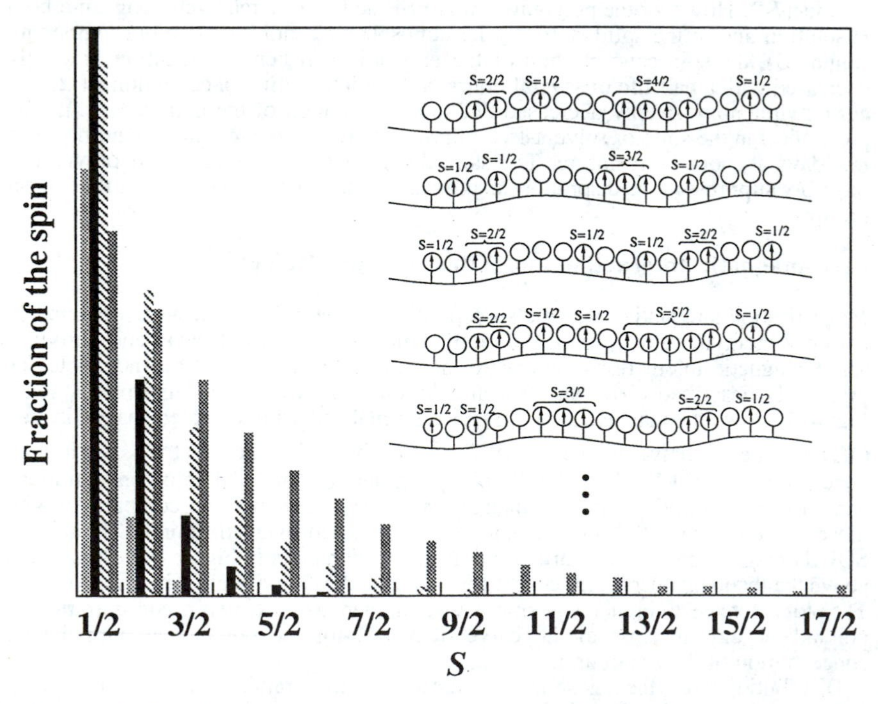

Figure 5. Distribution of the ground spin states simulated under the assumption of random radical generation for the polyradical with a degree of polymerization of 17 and with a spin concentration of 0.2(lightly shadowed box), 0.4(closed box), 0.6(striped box), and 0.8(darkly shadowed box) spin/unit. Insert: radical distribution simulation for a spin concentration of 0.47 spin/unit.

degree of polymerization (DP) of 17 (the experimental magnetization data in Figure 4 are also for the polyradical with DP=17), and with spin concentrations of 0.2, 0.4, 0.6 and 0.8 spin/unit under the assumption that ferromagnetic spin coupling among the unpaired electrons of the radical could occur only among neighboring units (Figure 5). The magnetization curves were estimated by modifying the Brillouin function ($B(X)$),

$$M/M_S = \sum x_S B(X), \quad \sum x_S = 1 \quad (\sum: S \text{ from 0 to 17}) \tag{1}$$

where x_S is the fraction for each S over the total spin amount.

The modified and simulated Bullouin curves for the polyradicals with DP=17 and spin concentrations of 0.15, 0.42 and 0.67 spin/unit are also given in Figure 4. The experimental magnetization plots are shown to be close to the simulated curve for each spin concentration, however, they significantly deviate upward (higher spin state). This result supports a random radical (spin) distribution along the chain, and indicates that ferromagnetic spin coupling occurs even for the radical units having one non-radical phenylenevinylene unit (radical defect). A defect in the radical generation thus is not fatal, and gives a partial but ferromagnetic spin alignment between the side-chain unpaired electrons through the π-conjugated poly(phenylenevinylene) backbone.

Polymer Effect on the Ferromagnetic Spin Alignment

Figure 6 shows plots of the product of molar magnetic susceptibility and T ($\chi_{mol}T$) *vs* T using polyradicals with the same spin concentration of *ca.* 0.6 spin/unit and with different molecular weights (DP=8, 17 and, 20). $\chi_{mol}T$ increases at low temperature from the theoretical value ($\chi_{mol}T$ = 0.375) for S=1/2 which indicates ferromagnetic behavior. The $\chi_{mol}T$ increase at low temperature is enhanced for the polyradical with a higher molecular weight.

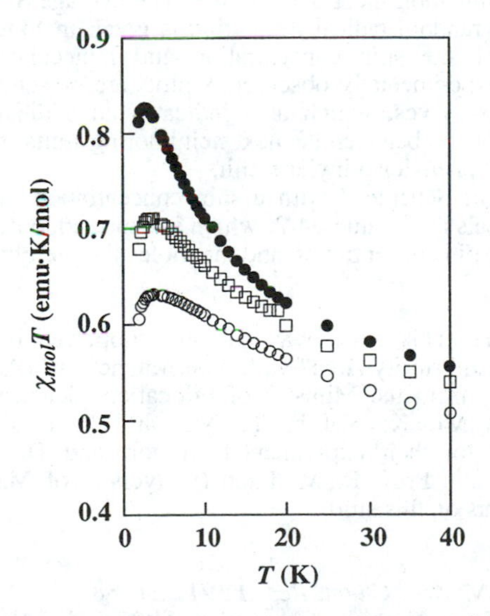

Figure 6. Plots of the products of molar magnetic susceptibility and T ($\chi_{mol}T$) *vs* T for the polyradicals with the degree of polymerization and spin concentration of 8 and 0.60(□), 17 and 0.55(○), and 20 and 0.54(●); DP and spin/unit, respectively.

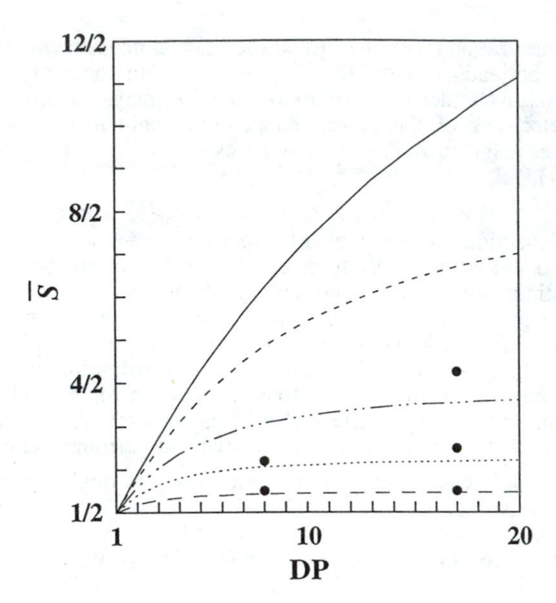

Figure 7. Effect of the degree of polymerization (DP) on the average spin quantum number (S) simulated by the random radical generation for the polyradicals with the spin concentration of 0.2(– –), 0.4(····), 0.6(–···–), 0.8(---), and 0.9(—) spin/unit. Experimentally observed S (●).

This polymer effect on the ferromagnetic interaction is also supported by the simple statistical simulation mentioned above. The average S values estimated from the simulation of a random radical generation is given in Figure 7. The average S increases both with the spin concentration and molecular weight (DP) of the polyradicals. The experimentally observed S plots are presented close to but higher than the simulation curves, which also indicates an additive contribution of the ferromagnetic interaction between the next neighboring units through the conjugated but spacing (defect) phenylenevinylene unit.

The polyradical of Scheme 3 with a spin concentration of 0.67 and molecular weight DP=17 reveals an S value ≥4/2, which is expected to be much enhanced both by increases in the spin concentration and the molecular weight.

Acknowledgment. This work was partially supported by a Grant-in-Aid for Scientific Research in Priority Area "Molecular Magnetism" (Area No.228/04242104 and 228/04242105) from the Ministry of Education, Science and Culture, Japan. The author thanks Mr. K. Katoh, T. Nii, and Prof. E. Tsuchida (Waseda University, Japan) for their experimental support and Dr. T. Kaneko (Niigata University, Japan) and Prof. P. M. Lahti (University of Massachusetts) for their productive comments on this study.

Literature Cited.

(1) Dougherty, D. A., *Acc. Chem. Res*. **1991**, *24*, 88.
(2) Iwamura, H.; Koga, N., *Acc. Chem, Res*. **1993**, *26*, 346.
(3) Rajca, A., *Chem. Rev*. **1994**, *94*, 871.

(4) Nishide, H., *Adv. Mater.* **1995**, *11*, 937.
(5) Utamapaya, S., Kakegawa, H.; Bryant, L.; Rajca, A., *Chem. Mater.* **1993**, *5*, 1053.
(6) Nishide, H.; Kaneko, T.; Nii, T.; Katoh, K.; Tsuchida, E.; Yamaguchi, K., *J. Amer. Chem. Soc.* **1995**, *117*, 548.
(7) Nishide, H.; Kaneko, T.; Toriu, S.; Kuzumaki, Y.; Tsuchida, E., *Bull. Chem. Soc., Jan.* **1996**, *69*, 499.
(8) Lahti, P.M.; Ichimura, A. S., *J. Org. Chem.* **1991**, *56*, 3030.
(9) Murase, I.; Ohnishi, T.; Noguchi, T.; Hirooka, M., *Polym. Commun.* **1984**, *25*, 327; Gagnon, D. R.; Capistran, J. D.; Karsaz, F. E.; Lenz, R. W., *Polym. Bull.* **1984**, *12*, 293.
(10) McDonald, R. N; Campbell, T. W., *J. Amer. Chem. Soc.* **1960**, *82*, 4669; Lapitskii, A.; Makin, S. M.; Berlin, A. A., *Vysokomol. Soedin.* **1967**, *A9*, 1274.
(11) Heck, R. F., *Org. React.* **1982**, *27*, 345; Heitz, W.; Brügging, W.; Freund, L,; Gailberger, M,; Greiner, A.; Jung, H,; Kampschulte, U,; Niessner, N.; Osan, F,; Schmidt, H. W.; Wicker, M., *Makromol. Chem.* **1998**, *189*, 119.
(12) Kaneko, T.; Ito, E.; Seki, K.; Tsuchida, E.; Nishide, H., *Polymer J.* **1996**, *28*, 182.
(13) Cook, C. D.; Norcross, B. E., *J. Amer. Chem. Soc.* **1959**, *81*, 1176.
(14) Becker, H. D., *J. Org. Chem.* **1969**, *34*, 1211.
(15) Altwicker, E. R., *Chem. Rew.* **1967**, *67*, 475.
(16) Nishide, H.; Kaneko, T.; Igarashi, M.; Tsuchida, E.; Yoshioka, N,; Lahti, P. M., *Macromolecules* **1994**, *27*, 3082.

Chapter 17

Very High Spin Polyradicals

Andrzej Rajca

Department of Chemistry, University of Nebraska, P.O. Box 880304, Lincoln, NE 68588–0304

I. Introduction

Synthesis and study of mesoscopic size molecules which are relevant to properties at the interface between the condensed phase and small molecule is one of the interdisciplinary frontiers in organic chemistry (*1*). Recent advances in organic synthesis and characterization allow for preparation and study of large organic molecules with large numbers of "unpaired" electrons. Such molecules address fundamental aspects of electronic structure and are relevant to search for novel materials (*2-6*).

This chapter is focused on macrocyclic high-spin polyradicals. Well-defined molecular mesoscopic-size high-spin polyradicals will test the possibility of maintaining strong magnetic coupling over large distances in organic molecules and will provide novel opportunities for studying magnetic phenomena on nanometer-scale within a single molecule. Such magnetic phenomena in mesoscopic-size systems are of both fundamental and future technological interest (*7-10*).

Macrocyclic polyradicals are building blocks (cross-linking units) for network polyradicals. Development of synthetic routes toward two- and three-dimensional network polyradicals is a key stepping stone to organic ferromagnetism with high Curie temperature (T_C) (*2*).

II. Prerequisites for a Very-High-Spin Polyradical (*2*)

Spin coupling in π-conjugated polyradicals is *primarily* controlled by molecular connectivity (*11,12*), which may be discussed in terms of ferromagnetic coupling units (fCU's), antiferromagnetic coupling units (aCU's), and spin sites (*2,3*). The well-known spin coupling units are 1,3-phenylene (fCU) and 1,4-phenylene (aCU); e.g., the triplet ground state Schlenk Hydrocarbon may be viewed as two triarylmethyl spin sites connected to a fCU and the singlet ground state Thiele Hydrocarbon contains two spin sites connected to an aCU (Figure 1).

The energy difference between the singlet and triplet states (ΔE_{ST}) in a diradical measures the strength of the spin coupling. Selected organic diradicals possess very

0097–6156/96/0644–0258$15.00/0

strong ferromagnetic coupling (e.g., in trimethylenemethane, ΔE_{ST} = 15 kcal/mol) (*13*); for a relatively "spin-diluted" Schlenk Hydrocarbon, an example of 1,3-connected polyarylmethyl diradical, ΔE_{ST} > 1 kcal/mol is estimated (*2*).

3,4'-Biphenylene-based polyarylmethyl diradicals are expected to possess ferromagnetic coupling (triplet ground state) (*14*); their fCU may be viewed as sequential connection of a fCU and an aCU (Figure 2) (*2*). Application of an empirical relationship between spin density distribution and strength of spin coupling predict that strength of ferromagnetic coupling will decrease by factor of 5 - 6, when 1,3-phenylene is replaced with 3,4'-biphenylene in a polyarylmethyl diradical (*2,15*).

The conceptual extension from diradicals to high-spin polyradicals is attained by connecting, in an alternating mode, the fCU's and the spin sites (Figure 3). [In particular, fCU's coordinated with three spin sites may correspond to 1,3,5-phenylyne or 3,5,4'-biphenylyne.] This approach was applied to construct series of polyarylmethyls, from diradicals with two spin sites to polyradicals with 31 spin sites, using either star-branched or dendritic connectivity (*2*). [The Iwamura group used analogous approach to construct linear and star-branched polycarbenes (*16*).]

Preparation of very-high-spin polyradicals may be affected by defects, which we define as those spin sites where an "unpaired" electron has failed to be generated. Probability of having polyradicals with defects increase rapidly with the number of spin sites. Therefore, it is essential to maintain ferromagnetic coupling between the "unpaired" electrons in the presence of defects, if very-high-spin polyradicals are to be attained. In acyclic systems, only a single pathway links any pair of spin sites; therefore, even a single defect at an inner site may interrupt π-conjugation (and ferromagnetic coupling), leading to a mixtures of spin systems with drastically lower than expected spin values (Figure 4) (*17*).

The recently synthesized macrocyclic polyradicals, which possess at least two spin coupling pathways between any pair of spin sites, show great promise in addressing the problem of defects (*18*).

III. Polymacrocyclic Polyradicals

Annelation of macrocylic polyradicals provides polymacrocyclic structures with greater resistance to defects. The resistance to defects may be evaluated using parameter "Q". We define "Q" as percentage of those polyradicals with *all* "unpaired" electrons ferromagnetically coupled; i.e., each of those polyradicals consists of a *single* spin system. Assuming that the yield per site for generation of an "unpaired" electron is 95 %, which is a typical value for non-dendritic polyarylmethyl polyradicals, the calculated values of "Q" decrease rapidly with the number of sites for macrocyclic polyradicals, but very slowly for polymacrocyclic polyradicals. Therefore, polymacrocyclic structures are promising targets for very-high-spin polyradicals (Figure 5).

IV. Elongated Shape Very-High-Spin Molecular Polyradicals

A fascinating feature of mesoscopic polyradicals (>10 spin-coupled electrons) is the possibility of observing magnetic phenomena associated with anisotropy barrier (E_A) in a single molecule (e.g., superparamagnetism). The elongated shape of the polyradical would increase E_A by magnetic dipole-dipole interactions (*19*). Estimates of this effect in polyarylmethyl polyradicals can be found in our previous work (*17*).

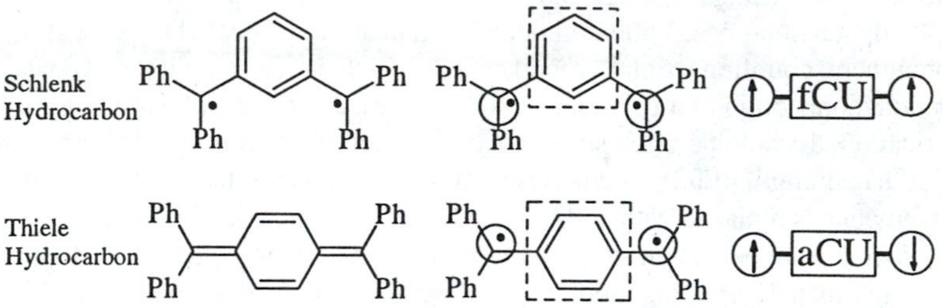

Figure 1. Spin coupling in Schlenk and Thiele Hydrocarbons.

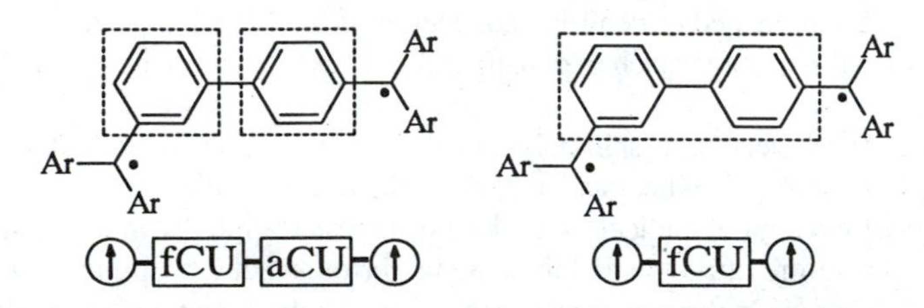

Figure 2. Spin coupling in 3,4'-biphenylene-based diradical.

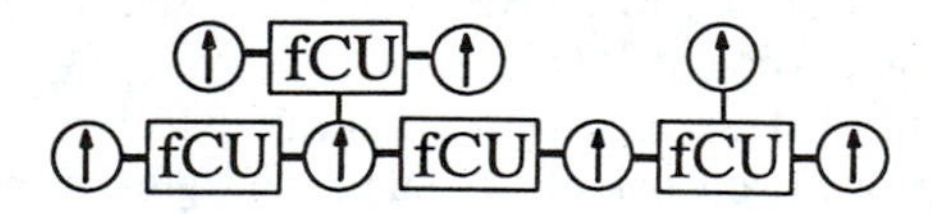

Figure 3. Connectivity of ferromagnetic coupling units and spin sites in a high-spin polyradical.

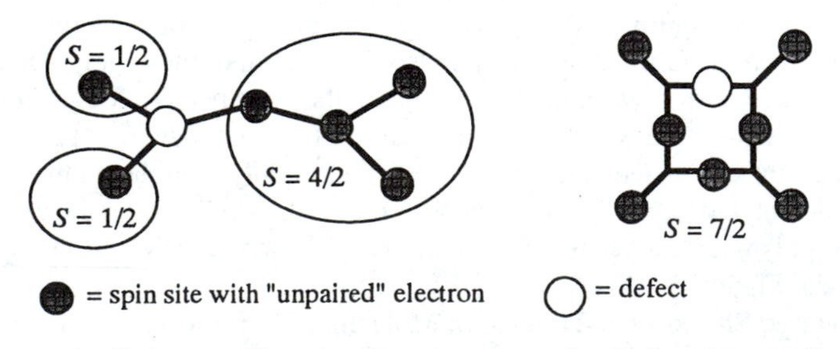

Figure 4. Spin coupling in the presence of a defect in acyclic vs. macrocyclic polyardical.

Rigorously controlled size, which is attained via stepwise convergent organic synthesis, is especially important in search for quantum tunneling phenomena on mesoscopic scale (*8*).

Iwamura and coworkers recently reported an onset of superparamagnetism in an organic molecule, the $S = 9$ nonacarbene (Figure 6), in the liquid helium temperature range, but the origin of the effect is not clear (*16*).

In the proposed elongated polyradicals, the dipolar interactions are more likely to be the dominant contributor to E_A; thus, calculation of E_A would be straightforward, once approximate conformation (shape) of elongated-shape polyradical is known. Because impact of defects on spin coupling has to be minimized, elongated shapes should be implemented with macrocyclic polyradicals. This can be accomplished by either annelating "non-elongated shape" macrocycles (e.g., calix[4]arene rings) into 2-strands or constructing "elongated shape" macrocycles; presumably, the ultimate approach would be to use "elongated shape" macrocycles to build a 2-strand. "Elongated shape" macrocycle, which we are interested in, can be viewed as derived from calix[4]arene ring by replacing 1,3-phenylene fCU's with 3,4'-biphenylene or 3,5,4'-biphenylyne fCU's (Figure 7).

V. Network Polyradicals

Interesting bulk magnetic phenomena such as ferromagnetism, ferrimagnetism, etc., require at least two-dimensional (2D), preferably 3D, spin ensembles. In the last decade, many molecular solid-based magnets, which were relying on *intermolecular* interactions between organic radicals and/or transition metals, were reported (*20*). Well-documented organic ferromagnets (without transition metals) are currently limited to nitroxide-based molecular solids with the Curie temperature, T_C, <4 K (*21*); also, an interesting magnetic behavior of C_{60}-fullerene-based charge-transfer solid was interpreted as soft ferromagnetism with $T_C = 16$ K (*22*). Because *intermolecular* ferromagnetic interactions between organic radicals are very weak, the T_C's in these molecular magnets are low. However, *intramolecular* (through π-conjugated system) ferromagnetic interactions in organic diradicals such as polyarylmethyls can be much stronger, i.e., by factor of 100 or more, than intermolecular interactions. Thus, a network based on *intramolecular* ferromagnetic interactions may be a route to a high-T_C organic ferromagnet.

In 1968, Mataga proposed several polyarylmethyl 2D-network polycarbenes and polyradicals (*23*); the calix[3]arene-based polyradical (Mataga Polymer in Figure 8) is the most promising and relevant to the macrocyclic polyradicals, recently studied in our laboratory (*18*).

In the design of a synthetic route toward network polyradical, insolubility of highly cross-linked networks with high molecular weights should be taken into consideration (*24*). Thus, the cross-linking should be the final step in the synthesis of the precursor to the polyradical. The radical sites could be generated from the precursor by electron transfer reactions (e.g., similar to those used in the preparation of polyarylmethyls), in the manner analogous to the doping of polyacetylenes and similar materials (*25*). This strategy can be implemented by using the "elongated shape" macrocycles to form cross-links (Figure 9).

Interesting alternatives are the isoelectronic boron- and nitrogen-based networks, which consist of stable triarylboron radical anions and triarylamine radical cations

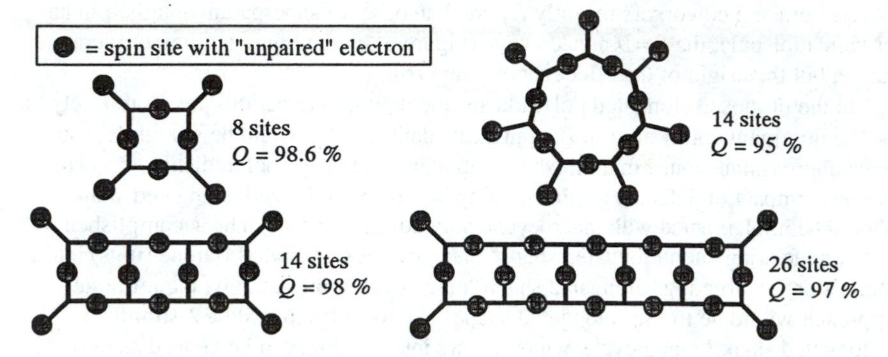

Figure 5. The effect of defects on macrocyclic and annelated macrocyclic polyradicals.

H
3
H
3
3
H
Nonacarbene (S = 9)

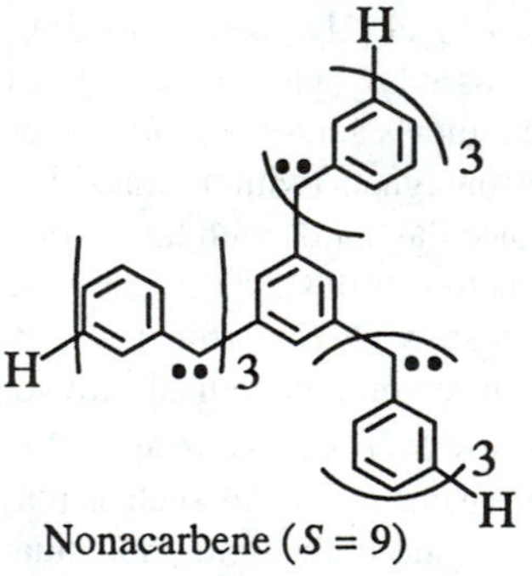

Figure 6. The S = 9 nonacarbene.

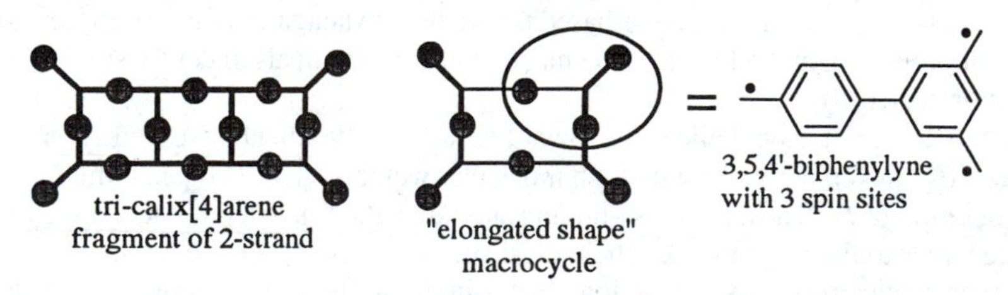

Figure 7. “Elongated shape” macrocyclic polyradicals.

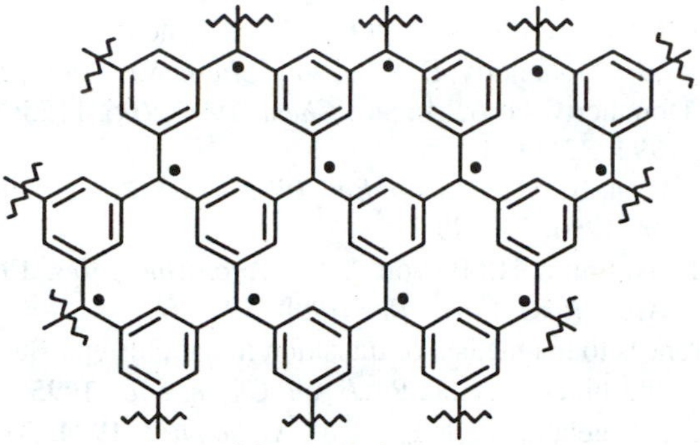

Mataga Polymer

Figure 8. Mataga polymer.

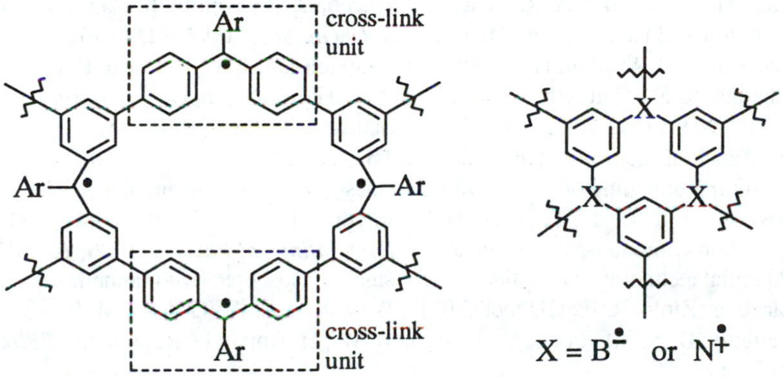

Figure 9. Plausible structures for network polyradicals.

(triarylaminium salts) (*26*). Recently, polyarylaminium di- and triradicals, and the related boron-based diradical dianion were reported (*27*,*28*).

Acknowledgements. I thank my co-workers, named in the references, for their contributions to this field of study. I gratefully acknowledge the Division of Chemistry and Division of Materials Research of National Science Foundation for the support of research on polyarylmethyl polyradicals.

References

1. Lehn, J.-M. *Angew. Chem.* **1990**, *102*, 1347; *Angew. Chem. Int. Ed. Engl.* **1990**, *29*, 1304. Escamilla, G. H.; Newkome, G. R. *Angew. Chem.* **1994**, *106*, 2013; *Angew. Chem. Int. Ed. Engl.* **1994**, *33*, 1937.
2. Rajca, A. *Chem. Rev.* **1994**, *94*, 871.

3. Dougherty, D. A. *Acc. Chem. Res.* **1991**, *24*, 88. Jacobs, J. S.; Shultz, D. A.; Jain, R.; Novak, J.; Dougherty, D. A. *J. Am. Chem. Soc.* **1993**, *115*, 1744. Jacobs, J. S.; Dougherty, D. A. *Angew. Chem.* **1994**, *106*, 1155; *Angew. Chem. Int. Ed. Engl.* **1994**, *33*, 1104.
4. Iwamura, H.; Koga, N. *Acc. Chem. Res.* **1993**, *26*, 346. Iwamura, H. *Adv. Phys. Org. Chem.* **1990**, *26*, 179.
5. Borden, W. T.; Iwamura, H.; Berson, J. A. *Acc. Chem. Res.* **1994**, *27*, 109.
6. Buchachenko, A. L. *Russ. Chem. Rev.* **1990**, *59*, 307.
7. Leading references to magnetism of transition metal clusters: Barra, A. L.; Caneshi, A.; Gatteshi, D.; Sessoli, R. *J. Am. Chem. Soc.* **1995**, *117*, 8855. Billas, I. M. L.; Chatelain, A.; de Heer, W. A. *Science* **1994**, *265*, 1682. Gatteshi, D.; Caneshi, A.; Pardi, L. Sessoli, R. *Science* **1994**, *265*, 1054. Micklitz, W.; McKee, V.; Rardin, R. L.; Pence, L. E.; Papaethymiou, G. C.; Bott, S. G.; Lippard, S. J. *J. Am. Chem. Soc.* **1994**, *116*, 8061. Goldberg, D. P.; Caneshi, A.; Lippard, S. J. *J. Am. Chem. Soc.* **1993**, *115*, 9299. Sessoli, R.; Tsai, H.- L.; Schake, A. R.; Wang, S.; Vincent, J. B.; Folting, K.; Gatteschi, D.; Christou, G.; Hendrickson, D. N. *J. Am. Chem. Soc.* **1993**, *115*, 1804. Christmas, C. A.; Tsai, H.- L.; Pardi, L.; Kesselman, J. M.; Gantzel, P. K.; Chadha, R. K.; Gatteschi, D.; Harvey, D. F.; Hendrickson, D. N. *J. Am. Chem. Soc.* **1993**, *115*, 12483. Caneshi, A.; Gatteshi, D.; Laugier, J.; Rey, P.; Sessoli, R.; Zanchini, C. *J. Am. Chem. Soc.* **1988**, *110*, 2795.
8. Quantum tunnelling of magnetization in mesoscopic-size magnetic particle: Awshalom, D. D.; DiVincenzo, D. P.; Smyth, J. F. *Science* **1992**, *258*, 414.
9. Quantum computers: Feynman, R. P. *Foundations of Physics* **1986**, *16*, 507.
10. Potential technological applications for small mesoscopic size magnetic particles: Ziolo, R. F.; Giannelis, E. P.; Weinstein, B. A.; O'Horo, M. P.; Ganguly, B. N.; Mehrotra, V.; Russell, M. W.; Huffman, D. R. *Science* **1992**, *257*, 219.
11. Berson, J. A. in *The Chemistry of Quinoid Compounds*, Patai, S.; Rappaport, Z., Eds.; Wiley: 1988, Vol. II, Chapter 10.
12. Longuet-Higgins, H. C.; Rector, C. W.; Platt, J. R. *J. Chem. Phys.* **1950**, *18*, 1174. Borden, W. T.; Davidson, E. R. *J. Am. Chem. Soc.* **1977**, *99*, 4587. Ovchinnikov, A. A. *Theor. Chim. Acta* **1978**, *47*, 497.
13. (a) Trimethylenemethane: reviews, Dowd, P. *Acc. Chem. Res.* **1972**, *5*, 242. Berson, J. A. *Acc. Chem. Res.* **1978**, *11*, 446. Borden, W. T. in *Diradicals*; Borden, W. T., Ed.; Wiley: New York, 1982, p. 1; Berson, J. A. *ibid*, p. 151; ESR spectroscopy, Dowd, P. *J. Am. Chem. Soc.* **1966**, *88*, 2587. Baseman, R. J.; Pratt, D. W.; Chow, M.; Dowd, P. *J. Am. Chem. Soc.* **1976**, *98*, 5726; ab initio calculations, Dixon, D. A.; Dunning, T. A.; Eades, R. A.; Kleier, D. A. *J. Am. Chem. Soc.* **1981**, *103*, 2878; tetraradical derivatives, ref. 3: (b) For a recent report on experimental measurement of a large ΔE_{ST} in a triplet ground state diradical, see: Khan, M. I.; Goodman, J. L. *J. Am. Chem. Soc.* **1994**, *116*, 10342.
14. Topologically related nitroxides were reported: Kanno, F.; Inoue, K.; Koga, N.; Iwamura, H. *J. Phys. Chem.* **1993**, *97*, 13267.

15. ENDOR studies triarylmethyl radicals: Maki, A. H.; Allendoerfer, R. D.; Danner, J. C.; Keys, R. T. *J. Am. Chem. Soc.* **1968**, *90*, 4225.
16. Nonacarbene: Nakamura, N.; Inoue, K.; Iwamura, H. *Angew. Chem. Int. Ed. Engl.* **1993**, *32*, 872.
17. Dendritic polyradicals: Rajca, A.; Utamapanya, S. *J. Am. Chem. Soc.* **1993**, *115*, 10688.
18. Macrocyclic polyradicals: Rajca, A.; Rajca, S.; Padmakumar, R. *Angew. Chem.* **1994**, *106*, 2193; *Angew. Chem. Int. Ed. Engl.* **1994**, *33*, 2091. Rajca, A.; Rajca, S.; Desai, S. R. *J. Am. Chem. Soc.* **1995**, *117*, 806.
19. Magnetic dipole-dipole interactions: Osborn, J. A. *Phys. Rev.* **1945**, *67*, 351. Carey, R.; Isaac, E. D. *Magnetic Domains an Techniques for their Observation*; Academic: New York, 1966.
20. Leading reference to molecular solid-based magnets: Miller, J. S.; Epstein, A. J. *Angew. Chem. Int. Ed. Engl.* **1994**, *33*, 385.
21. (a) Tamura, M.; Nakazawa, Y.; Shiomi, D.; Nozawa, K.; Hosokoshi, Y.; Ishikawa, M.; Takahashi, M.; Kinoshita, M. *Chem. Phys. Lett.* **1991**, *186*, 401. Chiarelli, R.; Novak, A.; Rassat, J. L. *Nature* **1993**, *363*, 147. Chiarelli, R.; Rassat, A.; Rey, P. *Chem. Commun.* **1992**, 1081. (b) Example for exceptionally strong intermolecular ferromagnetic coupling: Inoue, K.; Iwamura, H. *Chem. Phys. Lett.* **1993**, *207*, 551.
22. Allemand, P.-M.; Khemani, K. C.; Koch, A.; Wudl, F.; Holczer, K.; Donovan, S.; Gruner, G.; Thompson, J. D. *Science* **1991**, *253*, 301. Tanaka, K.; Zakhidov, A. A.; Yoshizawa, K.; Okahara, K.; Yamabe, T.; Yakushi, K.; Kikuchi, K.; Suzuki, S.; Ikemoto, I.; Achiba, Y. *Phys. Lett. A* **1992**, *164*, 221.
23. Mataga, N. *Theor. Chim. Acta* **1968**, *10*, 372.
24. Network polymers: (a) carbyne-based, Visscher, G. T.; Bianconi, P. A. *J. Am. Chem. Soc.* **1994**, *116*, 1805; (b) silylyne-based, Bianconi, P. A.; Weidman, T. W. *J. Am. Chem. Soc.* **1988**, *110*, 2342; Bianconi, P. A.; Schilling, F. C.; Weidman, T. W. *Macromolecules* **1989**, *22*, 1697.
25. Doping of polyacetylenes: Chung, T.- C.; Feldblum, A.; Heeger, A. J.; MacDiarmid, A. G. *J. Chem. Phys.* **1981**, *74*, 5504. *Handbook of Conducting Polymers*, Skotheim, T. A., Ed.; Marcel Dekker: New York, 1986.
26. Forrester, A. R.; Hay, J. M.; Thomson, R. H. *Organic Chemistry of Free Radicals*, Academic Press: New York, 1968; Chapters 6 and 8.
27. Polyarylaminium di- and triradicals: Stickley, K. R.; Blackstock, S. C. *J. Am. Chem. Soc.* **1994**, *116*, 11576.
28. Rajca, A.; Rajca, S.; Desai, S. R. *Chem. Commun.* **1995**, 1957.

Chapter 18

Assembly of Imino Nitroxides with Cu(I) Halides

H. Oshio and T. Ito

Department of Chemistry, Faculty of Science, Tohoku University, Aoba-ku, Sendai 980–77, Japan

In this article, a ferromagnetic interaction of imino nitroxides through a Cu(I) ion and cubane and 2-D assemblies of imino nitroxies with Cu(I) halides are described.

There has been increasing interest in molecular assemblies which have macroscopic properties like ferromagnetism. Since the first molecular based ferromagnets [Fe-$(C_5H_5)_2$][TCNE] (1) and [MnCu(pba)$(H_2O)_3$]$\cdot 2H_2O$ (2) in 1986 have been reported, several ferro- and ferrimagnets were reported. Especially, combinations of metal complexes and nitroxides have been proven to be good components for building such magnetic materials. For example, [M$(hfac)_2$][NITR] (M = Mn(II), Ni(II), hfac = hexafluoroacetylacetone, and NITR = nitronyl nitroxide derivertives) (3), [polynitroxide]$_2$[Mn$(hfac)_2$]$_3$ (4), and (radical)$_2$Mn$_2$[Cu(opba)]$_3$(DMSO)$_2\cdot 2H_2O$ (5) have been reported to show spontaneous magnetization at 10, 3.4, 22.5 K, respectively. A key point for such macroscopic properties in solids is to have strong intermolecular interactions.

Diamagnetic metal ions have been believed not to mediate magnetic interactions (6). Some diamagnetic metal complexes with organic radicals as ligands, however, have shown that ferromagnetic interactions were operative through the diamagnetic metal ions. Pseudo-octahedral coordination geometry provides for an orthogonal coordination of semiquinones. [M(III)(3,6-DBSQ)$_3$] (M = Al and Ga, and 3,6-DBSQ = 3,6-di-tert-butylsemiquinone) (7) and [Ga(III)(3,5-dtbsq)$_3$] (3,5-dtbsq = 3,5-di-tert-butyl-1,2-benzosemiquinone) (8) showed weak ferromagnetic interactions (J = 6.2, 8.6, and 7.8 cm^{-1} where $H = -2J\Sigma S_i \cdot S_j$) between semiquinones, while [M(IV)(Cat-N-SQ)$_2$] (M = Ti, Ge, and Sn) (Cat-N-SQ = tridentate Schiff base biquinone) (9) were characterized by a triplet ground state with the exchange coupling constants of J = -56, -27, and -23 cm^{-1} ($H = J\Sigma S_1 \cdot S_2$), respectively. If the certain conditions are fulfilled, these diamagnetic metal ions are considered to be particularly suitable for linking organic radicals. We describe here that a tetrahedral coordiantion of Cu(I) ion can be utilized to have the orthogonal arrangement of bidentate imino nitroxides and this leads to the strong ferromagnetic interaction. On the other hand, Cu(I) halides have wide variety in structures. These molecules form discrete geometries of varied nuclearities or polymeric extended systems (10). If organic radicals are introduced into the

0097–6156/96/0644–0266$15.00/0

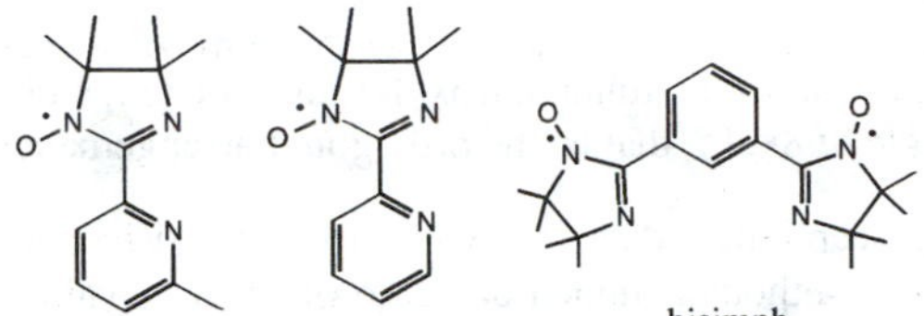

halocuprate cluster or network, interesting magnetic materials might be obtained. We also describe the strucutres and mangetic properties of imin nitroxyl networks assembled by Cu(I) halides. Imino nitroxides used in this study are depicted below.

Ferromagnetic Interaction in [Cu(I)(immepy)$_2$](PF$_6$)

Magnetic interactions between paramagnetic centers through diamagnetic metal ions are negligibly small or weakly antiferromagnetic. However, some diamagnetic metal ions have been reported to mediate magnetic interactions between coordinated radicals. (6 - 9) The magnitude of magnetic interactions through diamagnetic ions depend strongly on energy level of the dπ orbitals. Which diamagnetic metal complex provides the appropriate symmetry and orbital energy to propagate ferromagnetic interaction? Ab-initio molecular orbital calculations of the divalent metal oxides have proven that among the first row transition metal ions, d-orbital energy of the Cu(II) ion is the closest to the oxygen p-orbital, that is, the Cu(I) ion is also expected to have the strongest electronic interactions with organic molecules among the first raw transitin metal ions. In addition, Cu(I) ions are known to favor a tetrahedral coordination geometry (11), which is suitable for the orthogonal arrangement of two bidentate ligands. Thus, it is anticipated that coordination of the two bidentate radical ligands to the Cu(I) ion would lead to a ferromagnetic interaction between the radicals. In this section, we describe the fairly strong ferromagnetic interactions of imino nitroxides in [Cu(I)(immepy)$_2$](PF$_6$) (12).

Cu(I) complex with imino nitroxide as a ligand, [Cu(I)(immepy)$_2$]$^+$ (Figure 1), has a C_2 axis in the molecule and the Cu(I) ion is coordinated by the crystallographically equivalent immepy ligands acting as bidentate ligands. The coordination geometry about Cu(I) ion is pseudotetrahedral with the four coordination sites being occupied by four nitrogen atoms. The Cu-N(imino nitroxide) bond is slightly shorter (1.953(5) Å) than the Cu-N(pyridine) bond (2.081(6) Å).

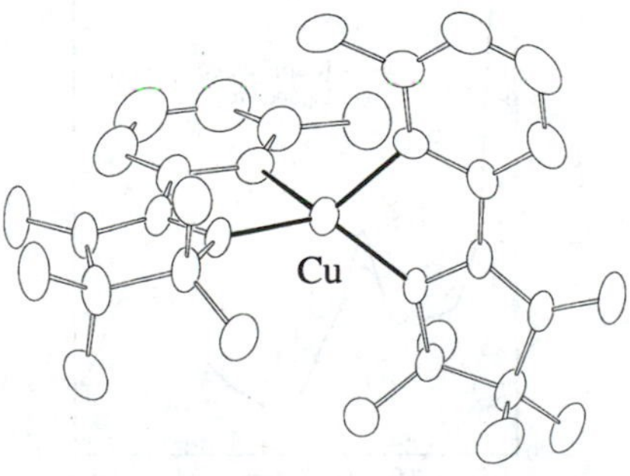

Figure 1. ORTEP drawing of [Cu(I)(immepy)$_2$]$^+$.

The two radical planes (imino nitroxyl plane), where the magnetic orbital (π^* orbital) of the radical is perpendicular to the imino nitroxyl plane, are perpendicular to each other with the dihedral angle of 88.7 °, that is, the orthogonal arrangement of the radicals was obtained.

The magnetic susceptibility data for $[Cu(I)(immepy)_2](PF_6)$ (Figure 2) show a ferromagnetic behavior, although sudden decrease of the χ_mT values below 20 K was observed. The least squares fitting for the data above 20 K by using the Bleaney-Bowers equation (H = -2JS1.S2) (13) gives an exchange coupling constant 2J value of 103(2) cm^{-1} with g value being 1.977(3). The sudden decrease below 20 K is due to either intermolecular antiferromagnetic interaction or zero-field splitting of the triplet state.

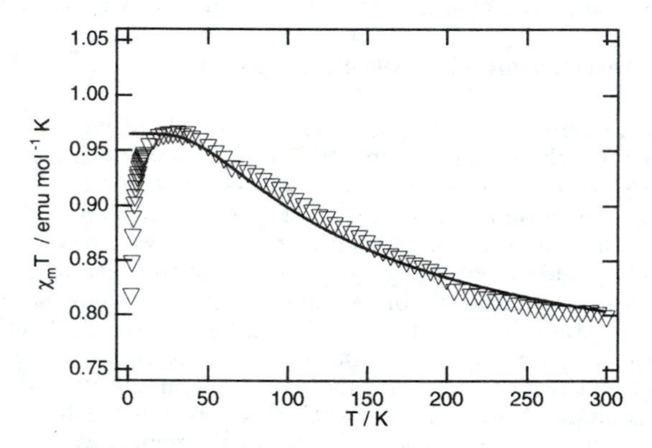

Figure 2. χ_mT versus T plot of $[Cu(I)(immepy)_2](PF_6)$.

$[Cu(I)(immepy)_2](PF_6)$ shows the fairly strong ferromagnetic interaction. A uv-visible spectrum may give some insight to interpret the magnetic property of the complex. The uv-visible absorption spectra of $[Cu(I)(immepy)_2](PF_6)$ along with the ligand of immepy show interesting feature (Figure 3). In acetonitrile solution, $[Cu(I)(immepy)_2](PF_6)$ shows intense absorption bands at 766 nm ($\varepsilon = 5000\ M^{-1}\ cm^{-1}$) and 464 nm ($\varepsilon = 6300\ M^{-1}\ cm^{-1}$) with a shoulder band at 510 nm.

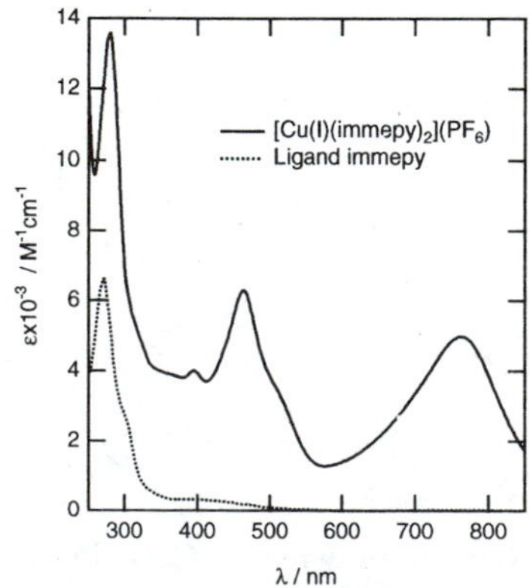

Figure 3. UV-visible spectra of ligand immepy (dotted line) and $[Cu(I)(immepy)_2](PF_6)$ (solid line) in acetonitrile.

A PM3 molecular orbital calculation for the ligand immepy (where the imino nitroxide fragment and the pyridine ring are coplanar) shows that the singly occupied molecular orbital (SOMO) is mainly composed of the imino nitroxide fragment, while the next lowest unoccupied orbital (NLUMO) is composed of the whole π system. Both the SOMO and the NLUMO are of proper symmetry to overlap with the d_{xz} and d_{yz} orbitals. By analogy of Cu(I)-diimine complex, $[Cu(I)(TET)]^+$ (TET = 2,2'-bis(6-(2,2'-bipyridyl)biphenyl) (14), and Cu(I)-o-semiquinone complex, $[Cu(I)_2\{\mu\text{-}N_2[CO(OBu^t)]_2[\mu\text{-}Ph_2P(CH_2)_6PPh_2\}_2]BPh_4$ (15), the lower energy band (766 nm) can be assigned to an $e(d_{xz},d_{yz}) \rightarrow$ SOMO transition and the shoulder at 510 nm to $b_2(d_{xy}) \rightarrow$ NLUMO, while the higher energy band (464 nm) to an $e(d_{xz},d_{yz}) \rightarrow$ NLUMO. (Scheme I).

The strong ferromagnetic interaction observed in this complex can be attributed to the charge transfer interaction, a strong absorption band at 766 nm which corresponds to electron transfer from the $e(d_{xz}, d_{yz})$ orbital to the SOMO (radical). The π-back donation, which mixes the $e(d_{xz}, d_{yz})$ orbital and SOMO, induces a large spin delocalization of the each coordinated imino nitroxide onto the Cu(I) ion; the resultant spin on the Cu(I) ion leading to the strong ferromagnetic interactions.

Scheme I

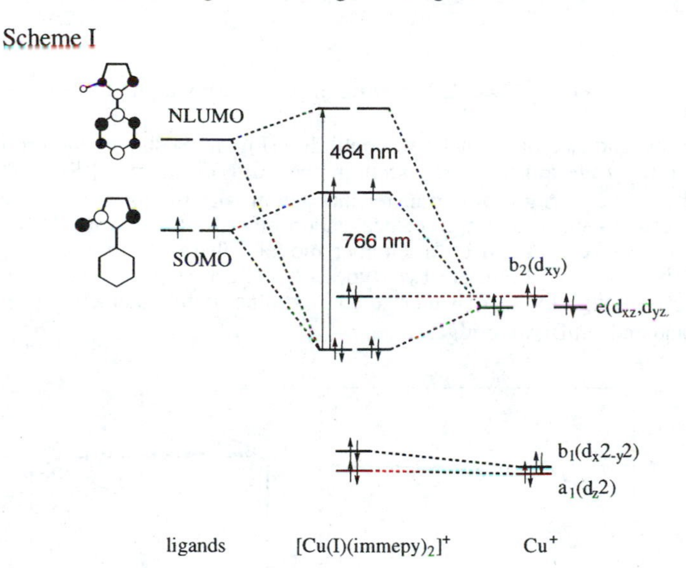

Assembles of Iminonitroxides with Cu(I) Halides

Radical Cubane of $[Cu_4Br_4(bisimph)_2]$. The reaction of biradical, bisimph, with $[Cu(I)(CH_3CN)_4](PF_6)$ and NaBr in methanol gave a bromo-bridged tetranuclear Cu(I) complex, $[Cu_4Br_4(bisimph)_2]$ (Figure 4). Each Cu(I) ion in the complex has a coordinated imino nitroxide. That is, the four imino nitroxides are assembled to form a radical cubane structure. The Cu_4Br_4 core has distorted cubane structure with Cu-Br bond lengths of 2.489(1) - 2.711(1) Å. Coordination geometry about the Cu(I) ion is a distorted tetrahedron with one nitrogen atom of the imino nitroxide and three bromide ions. The Cu-N distances are in the range of 1.965(5) - 1.988(5) Å.

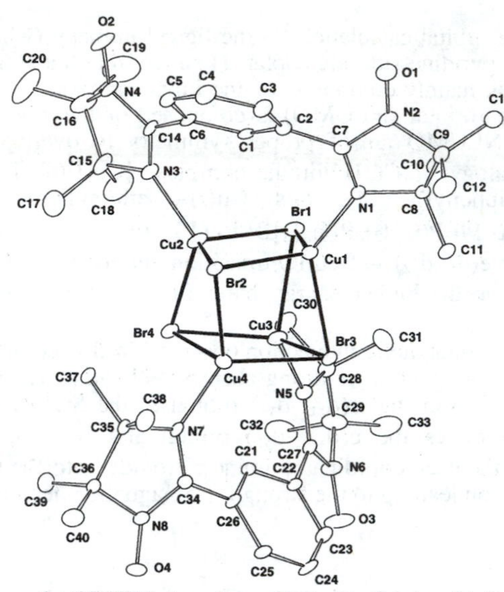

Figure 4. ORTEP drawing of $[Cu_4Br_4(bisimph)_2]$.

Temperature dependence of magnetic susceptibility (Figure 5) shows antiferromagnetic behavior for the four radicals. As seen in the $[Cu(I)(immepy)_2](PF_6)$, the imino nitroxide shows the strong charge transfer interaciton with the dπ orbital of the Cu(I) ion, which leads to substantial spin delocalization onto the Cu(I) ion. Therefore, the magnetic data can be analyzed by a tetramer model, where the four spins locate on cornors of the cubane. There are two types of interacting path ways. One is the interaction (J_2) through the $CuBr_2Cu$ brige and the other is the interaction (J_1) through the intraligand and $CuBr_2Cu$ bridge.

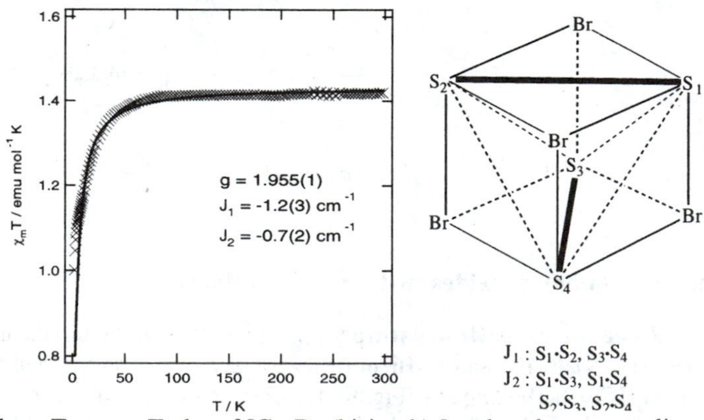

Figure 5. $\chi_m T$ versus T plot of $[Cu_4Br_4(bisimph)_2]$ and exchange coupling scheme.

The coupling scheme is depicted in Figure 6 and the spin Hamiltonian can be expressed as follows,

$$H = -2J_1(S_1.S_2+S_3{\cdot}S_4)-2J_2(S_1{\cdot}S_3+S_1{\cdot}S_4+S_2{\cdot}S_3+S_2{\cdot}S_4)$$

The best fit parameter for J_1, J_2 and g are -1.2(3), 0.7(2) cm^{-1} and 1.955(1), respectively.

Two dimensional network of Iminonitroxides with $(CuI)_2$ Unit.

$[Cu(\mu\text{-}I)(impy)]_2$ and $[Cu(\mu\text{-}I)(immepy)]_2$. Complexes of $[Cu(\mu\text{-}I)\text{-}(impy)]_2$ and $[Cu(\mu\text{-}I)(immepy)]_2$ crystallize in the orthorhomic space group $P2_12_12_1$ and the monoclinic space group $P2_1/c$, respectively. The structures of the complex molecules consist of discrete dimers (Figure 6), where two Cu(I) ions are bridged by two iodide ion.

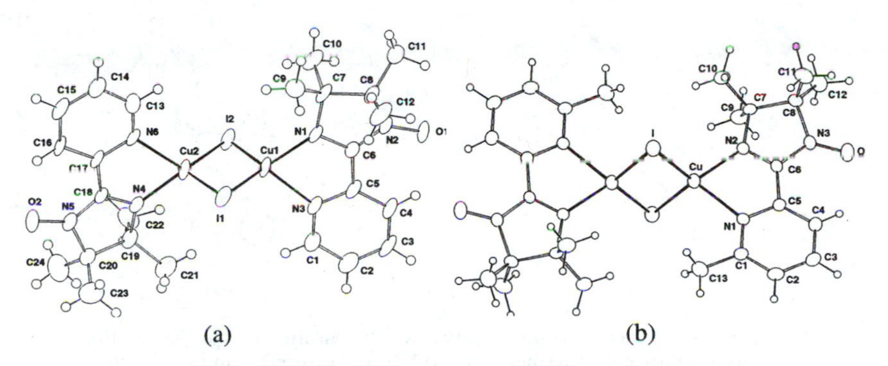

Figure 6. ORTEP drawings of (a) $[Cu(\mu\text{-}I)(impy)]_2$ and (b) $[Cu(\mu\text{-}I)\text{-}(immepy)]_2$.

Coordination geometries about the Cu(I) ion in these complexes are very similar to each other, where distorted tetrahedral coordination sites are occupied with two nitrogen atoms and two iodide ions. Cu-N(pyridine), Cu-N (imino nitroxide), and Cu-I distances are in the range of 2.162(9) - 2.173(3) Å, 2.005(9) - 2.030(3) Å, and 2.571(2) - 2.631(2) Å, respectively. Coordinating ligands in the adjacent molecules are stacked to form one dimensional chain and the chains are linked by CuI_2Cu units, that is, forming a two dimensional network (Figure 7).

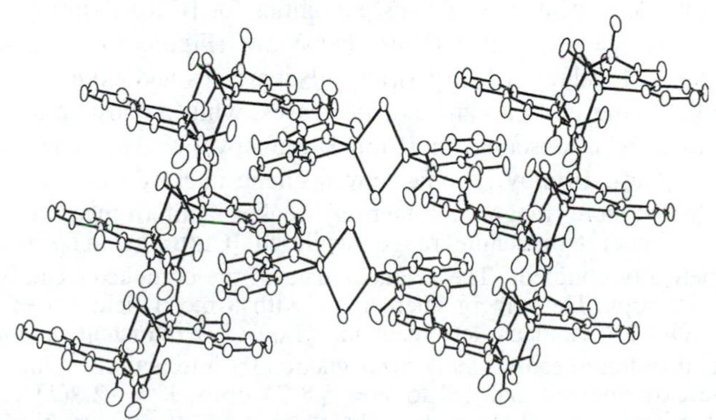

Figure 7. 2D network of imino nitroxide in $[Cu(\mu\text{-}I)(impy)]_2$.

The adjacent iminonitroxides in the two complexes, however, have subtly different stacking modes. Figure 8 shows the stacking arrangements in views parallel and perpendicular to the iminonitroxide fragments.

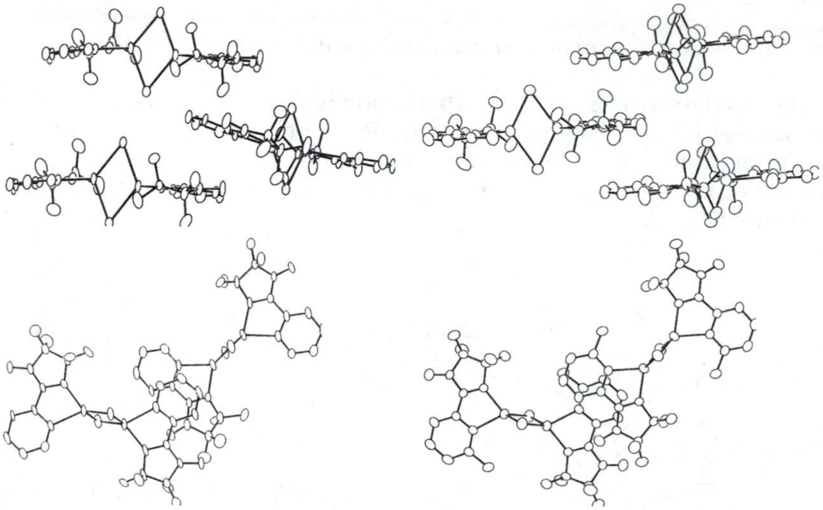

(a) (b)

Figure 8 The stacking arrangements in views parallel and perpendicular to the imino nitroxide fragments of (a) $[Cu(\mu\text{-}I)(impy)]_2$ and (b) $[Cu(\mu\text{-}I)\text{-}(immepy)]_2$.

There are two distinct differences in the stacking mode. (i) $[Cu(\mu\text{-}I)(impy)]_2$ has closer intermolecular distance (O1···C18' = 3.33(1) Å) between the oxygen atom (O1) of the nitroxide and the sp^2 carbon atom of the next imino nitroxide moiety than that (O1···C6' = 3.751(8) Å) of $[Cu(\mu\text{-}I)(immepy)]_2$. (ii) The conjugated imino-nitroxyl fragments of the two adjacent molecules in $[Cu(\mu\text{-}I)(impy)]_2$ are tilted to each other (two adjacent imino-nitroxyl planes make an angle of 22.0(8)°), while those of the $[Cu(\mu\text{-}I)\text{-}(immepy)]_2$ are stacked with almost parallel arrangements (a dihedral angle of two adjacent imino-nitroxyl groups is only 6.8(4)°).

Temperature dependent magnetic susceptibilities for $[Cu(\mu\text{-}I)(impy)]_2$ and $[Cu(\mu\text{-}I)(immepy)]_2$ have shown quite different behaviors (Figure 9). On lowering the temperature, the χ_mT value for $[Cu(\mu\text{-}I)(impy)]_2$ increases and exhibits a maximum at 16 K (χ_mT = 0.80 emu mol^{-1} K) and then decreases, while a constant decrease of χ_mT value down to 2 K is observed in $[Cu(\mu\text{-}I)(immepy)]_2$. The magnetic behaviors suggest that in $[Cu(\mu\text{-}I)(impy)]_2$ some ferromagnetic interaction is predominant at a intermediate temperature range and then a weaker antiferromagnetic coupling is involved at the lower temperature range, while in $[Cu(\mu\text{-}I)(immepy)]_2$ radicals are antiferromagnetically coupled. The magnetic data of the complexes can be interpreted by assuming isotropic Heisenberg interactions with a mean field correction, that is, ferromagnetic (J_F) (16) and antiferromagnetic (J_{AF}) (17) intrachain interactions with weak interchain (intramolecular) antiferromagnetic (J') interactions. The least squares fittings of the experimental data led to J_F = 5.8(2) cm^{-1}, J' = -2.2(1) cm^{-1} and g = 1.94(1) for $[Cu(\mu\text{-}I)(impy)]_2$, and J_{AF} = -0.68(2) cm^{-1}, J' = -0.40(4) cm^{-1} and g =

1.91(1) for [Cu(μ-I)(immepy)]$_2$. The estimated interchain interactions are rather large, which might not satisfy the mean field corrections. However, the fittings support the hypothesis of the ferro and antiferromagnetic intrachain interations in [Cu(μ-I)(impy)]$_2$ and [Cu(μ-I)-(immepy)]$_2$, respectively.

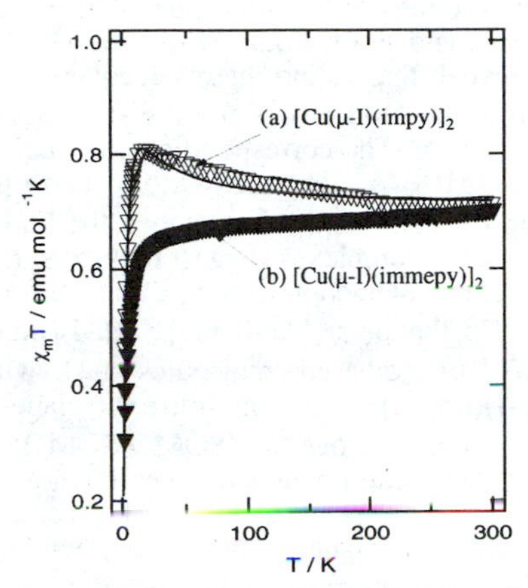

Figure 9. χ_mT versus T plot of (a) [Cu(μ-I)(impy)]$_2$ and (b) [Cu(μ-I)-(immepy)]$_2$.

The sign and magnitude of the intermolecular magnetic interactions depend strongly on a relative arrangement of adjacent N-O groups, and the correlation between the magnetic interactions and geometrical parameters have been discussed (18). There are two types of stacking modes which lead to the ferrromagnetic and antiferromagnetic intrachain interactions, respectively (Scheme II). (i) Ferromagnetic stacking mode (McConnell mechanism (19)): Spin polarization induces opposite or negative spin density on a molecule. If molecules overlap with opposite sign of the spin density, the intermolecular ferromagnetic interaction would be expected. (ii) Antiferromagnetic stacking mode (SOMO-SOMO overlap): When two SOMOs overlap, bonding and antibonding orbitals are formed. Two electrons from the SOMOs locate on the bonding MO, and the singlet state will be the ground state, that is, SOMO-SOMO overlap stabilizes the antiferromagnetic state.

Scheme II

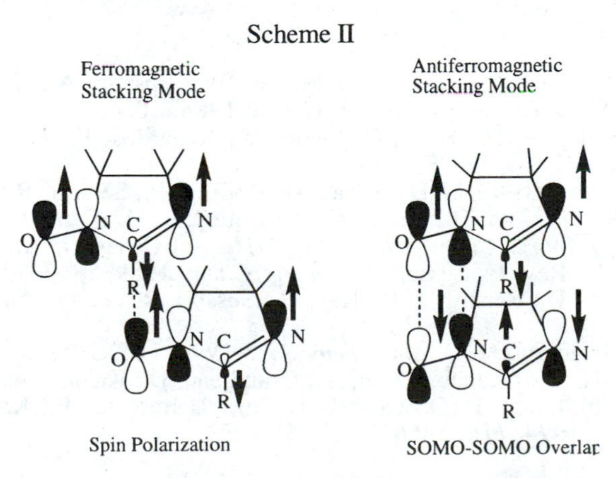

The spin density distributions in some nitronyl nitroxides have been determined by the polarized neutron diffraction study and MO calculation (20). The positive spins are populated over the N-O groups, while the large negative spin density appears on the sp^2 carbon atom bridging two N-O groups. This indicates large spin-polarization effect on the nitronyl nitroxide. The electronic structure of the pyridyl-imino nitroxide is considered to be very similar to the nitronyl nitroxide. The positive spin density locates on both imino-nitrogen atom and N-O group, while the negative spin on the carbon atom. Structural analysis reveals that a short intermolecular contact in $[Cu(\mu\text{-}I)(impy)]_2$ involves the oxygen atom of the N-O group and the sp^2 carbon atom of the adjacent molecule (O1···C18' = 3.33(1)Å). The corresponding intermolecular contact (O1···C6') in $[Cu(\mu\text{-}I)(immepy)]_2$ is 3.751(8) Å. These two atoms carry the opposite sign of the spin which alternate along the stack and this matches the McConnell's criteria. The observed contact distances in the complexes suggest that the spin polarization leading to the intermolecular ferromagnetic interaction is more effective for $[Cu(\mu\text{-}I)(impy)]_2$ than for $[Cu(\mu\text{-}I)(immepy)]_2$. On the other hand, in $[Cu(\mu\text{-}I)(immepy)]_2$ the conjugated imino-nitroxyl fragments of two adjacent molecules stack with a parallel alignment (6.8(4)°), while in $[Cu(\mu\text{-}I)(impy)]_2$ two imino-nitroxyl planes tilt toward each other with an angle of 22.0(8)°. A σ-type overlap (SOMO-SOMO overlap) of the nitronyl nitroxide π^* orbital leads to the antiferromagnetic interaction and this will be maximum when the adjacent N-O groups are parallel (19). The resulting overlap in $[Cu(\mu\text{-}I)(immepy)]_2$ favors the antiferromagnetic interaction, while the tilted stacking in $[Cu(\mu\text{-}I)(impy)]_2$ diminishes the antiferromagnetic contribution. As a result, in spite of the fact that the spin polarization contributes to the stabilization of the intermolecular ferromagnetic interaction, the intrachain magnetic interaction for $[Cu(\mu\text{-}I)(impy)]_2$ is ferromagnetic and that for $[Cu(\mu\text{-}I)(immepy)]_2$ is antiferromagnetic.

Conclusion

We described here that the copper(I) ion provides the orthogonal arrangement for the two coordinated imino nitroxides and this leads to the fairly strong intramolecular ferromagnetic interaction between the imino nitroxides. We also focused on the validity of the halocuprate unit to increase dimensionality or nuclearity of the imino nitroxides. Imino nitroxides in the complexes studied here have the cubane structure or two dimensional structures in which one dimensional iminonitroxide chains are linked by the $Cu\text{-}I_2\text{-}Cu$ unit. It should be concluded that the magnetic interaction of the iminonitroxides through Cu(I) halide bridges is not strong.

References

1. Miller, J. S.; Calabrese, J. C.; Bigelow, R. W.; Epstein, A.; J. Zhang, R. W.; Reiff, W. M. J. *Chem. Soc. Chem. Comm.* **1986**, 1026.
2. Pei, Y.; Verdaguer, M.; Kahn, O.; Sletten, J.; Renard, J. P. *J. Am. Chem. Soc.* **1986**, *108*, 7428
3. a) Caneschi, A.; Gatteschi, D.; Renard, J. P.; Rey, P.; Sessoli, R. *J. Am. Chem. Soc.* **1989**, *111*, 785. b) Caneschi, A.; Gatteschi, D.; Renard, J. P.; Rey, P.; Sessoli, R. *Inorg. Chem.* **1989**, *28*, 1976. c) Caneschi, A.; Gatteschi, D.; Renard, J. P.; Rey, P.; Sessoli, R. *Inorg. Chem.* **1989**, *28*, 3314. d) Caneschi, A.; Gatteschi, D.; Renard, J. P.; Rey, P.; Sessoli, R. *Inorg. Chem.* **1989**, *28*, 2940.
4. Inoue, K; Iwamura, H. *J. Am. Chem. Soc.* **1994**, *116*, 3173.
5. a) Stumpf, H. O.; Ouahab, L. Pei, Y.; Grandjean, D.; Kahn, O. *Science*, **1993**, *261*, 447. b)Stumpf, H. O.; Ouahab, L. Pei, Y.; Bergerat, P.; Kahn, O. *J. Am. Chem. Soc.* **1994**, *116*, 3866.

6. a) Chaudhuri, P.; Winter, M.; Della Védova, B. B. C.; Bill, E.; Trautwein, A.; Gehring S.; Fleischhauer, P.; Nuber, B.; Weis, J. *Inorg. Chem.* **1991**, *30*, 2148. b) Chaudhuri, P.l Winter, M.; Fleischhauer, P.; Haase, W.; Flörke, U.; Haupt, H. J. *J. Chem. Soc., Chem. Comm.* **1990**, 1728.
7. Lange, C. W.; Conklin, B. J.; Pierpont, C. G. *Inorg. Chem.* **1994**, *33*, 1276.
8. (a) Adams, D. M.; Rheingold, A. L.; Dei, A.; Hendrickson, D. N. *Angew. Chem. Int. Ed. Engl.* **1993**, *32*, 391. (b) Ozarowski, A.; Mcgarvey, B. R.; El-Hadad, A.; Tian, Z.; Tuck, D. G.; Krovich, D. J.; DeFtis, G. C., *Inorg. Chem.* **1993**, *32*, 841.
9. Bruni, S.; Caneschi, A.; Cariati, F.; Delfs, C.; Dei, A.; Gatteschi, D. *J. Am. Chem. Soc.* **1994**, *116*, 1388.
10. a) Subramanian, L.; Hoffmann, R. *Inorg. Chem.* **1993**, *31*, 1021. b) Vitale, M.; Ryu, C. K.; Palke, W. E.; Ford, P. C. *Inorg. Chem.* **1994**, *33*, 561. c) Raston, C. L.; White, A. H.; White, A. H. *J. Chem. Soc. Dalton Trans.* **1976**, 2153. d) Andersson, S.; Jagner, S. *Acta Chem. Scand., A* **1986**, *40*, 177. e) Jagner, S.; Helgesson, G. *Adv. Inorg. Chem.* **1991**, *37*, 1. f) Bowmaker, G. A. *Adv. Spectrosc.* **1987**, *14*, 1. g) Bigalke, K. P.; Hans, A.; Hartl, H. *Z. Anorg. Allg. Chem.* **1988**, *563*, 96. h) Hoyer, M.; Hartl, H. *Z. Anorg. Allg. Chem.* **1990**, *587*, 23.
11. Cotton, F. A.; Wilkinson, G. *Advanced Inorganic Chemistry 5th edition*, Wiley, New York, 1988.
12. Oshio, H.; Watanabe, T.; Ohto, A.; Ito, T; Nagashima, U. *Angew. Chem. Int. Ed. Engl.* **1994**, *33*, 670.
13. Bleaney, B.;. Bowers, K. D. *Proc. R. Soc. London, Ser. A* , **1952**, *214*, 451.
14. Müller, E.; Piguet, C.; Bernardinelli, G.; Williams, A. F. *Inorg. Chem.* **1988**, *27*, 849.
15. Moscherosch, M.; Field, J. S.; Kaim, W.; Kohlmann, S.; Krejcik, M. *J. Chem. Soc. Dalton Trans*, **1993**, 211.
16. Baker, G. A.; Rushbrooke, G. S.; Gilbert, H. E. *Phys. Rev.* **1964**, *135*, A1272.
17. Bonner, J. C.; Fisher, M. E. *Phys. Rev.* **1964,** *135*, A640.
18. a) Caneschi, A.; Ferrara, F.; Gatteschi, D.; Rey, P.; Sessoli, R.; *Inorg. Chem.* **1990**, *29*, 1756. b) Panthou, F. L.; Luneau, D.; Laugier, J.; Rey, P. *J. Am. Chem. Soc.* **1993**, *115*, 9095.
19. McConnel, H. M. *J. Chem. Phys.* **1963**, *39*, 1910.
20. a) Zheludev, A.; Barone, V.; Bonnet, M.; Delley, B.; Grand, A.; Ressouche, E.; Rey, P.; Subra, R.; Schweizer, J. *J. Am. Chem. Soc.* **1994**, 116, 2019. b) Yamaguchi, K.; Okumura, M.; Maki, J.; Noro, T.; Namimoto, H.; Nakano, M.; Fueno, *Chem. Phys. Lett.* **1992**, *190*, 353. c) Yamaguchi, K.; Okumura, M.; Nakano, *Chem. Phys. Lett.* **1992**, *191*, 237.

Chapter 19

Weak-Ferromagnetism and Ferromagnetism in Tetrafluorotetracyanoquinodimethanide Salts

Toyonari Sugimoto[1], Kazumasa Ueda[1], Nobuko Kanehisa[2], Yasushi Kai[2], Motoo Shiro[3], Nobuyoshi Hosoito[4], Naoya Takeda[5], and Masayasu Ishikawa[5]

[1]Research Institute for Advanced Science and Technology, University of Osaka Prefecture, Sakai, Osaka 593, Japan
[2]Department of Applied Chemistry, Osaka University, Suita, Osaka 565, Japan
[3]Rigaku Corporation, Akishima, Tokyo 196, Japan
[4]Institute for Chemical Research, Kyoto University, Uji, Kyoto 611, Japan
[5]Institute for Solid State Physics, University of Tokyo, Roppongi, Tokyo 106, Japan

Several salts of the radical anion of tetrafluorotetracyanoquinodimethane (TCNQF4) (tetrafluorotetracyanoquinodimethanide, $TCNQF4^{-\cdot}$) exhibited very unique magnetisms, i.e., weak-ferromagnetism and ferromagnetism. The weak-ferro-magnetism appeared below 12 K in the Li^+ salt of $TCNQF4^{-\cdot}$. On the other hand, the ferromagnetism was observed in the charge-transfer (CT) complexes of $TCNQF4^{-\cdot}$ with pyridinium-substituted imidazolin-1-oxyls as well as in the $N(CH_3)_4^+$ salt of $TCNQF4^{-\cdot}$ with a half molecule of TCNQF4. The ferromagnetic phase-transition temperature (Curie temperature, T_C) was in the range of 0.4 to 0.55 K in the former salt, which can be declared as a first purely organic ferromagnet based on a well-characterized CT complex. Very interestingly, the latter $TCNQF4/TCNQF4^{-\cdot}$ mixed salt revealed a remarkable high T_C close to room temperature.

Tetracyanoquinodimethane (TCNQ) (*1*) and its tetrafluoro-substituted derivative (TCNQF4) (*2*) are well known as common electron acceptors in the formation of charge-transfer (CT) complexes. Most notably, the first synthetic metal utilized TCNQ as the organic acceptor (*3,4*). This discovery has triggered off a great advance in electrical conduction in organic materials, and at last led to a first organic superconductor in 1980 (*5*).

0097–6156/96/0644–0276$15.00/0

Ferromagnetism is another target to be acieved in organic materials. Since the first discovery of ferromagnetic interaction in a galvinoxyl crystals (*6*), similar phenomena have also been recognized in almost 20 organic radical crystals (*7*). Very recently, a purely organic ferromagnet has been synthesized in a *p*-nitrophenyl nitronyl nitroxide crystal (*8*) and successively in several nitroxide crystals (*9-12*). However, the highest Tc value is only 1.48 K (*9*). In addition, there has so far been almost no progress in the achievement of ferromagnetism even in purely organic CT complexes, which have the advantage of different compositions between donors and acceptors, high stability, strong spin-spin coupling by aid of Coulombic interaction between the positive and negative charges, and coexistence of local spins and conduction electrons interacting with each other. This article presents weak-ferromagnetism in the Li^+ salt of the radical anion of TCNQF4 ($TCNQF4^{-\cdot}$) and ferromagnetism in the CT complexes of $TCNQF4^{-\cdot}$ with pyridinium-substituted imidazolin-1-oxyls as well as in the tetramethylammonium salt of $TCNQF4/TCNQF4^{-\cdot}$. These results bring an expectation that TCNQF4 (and also TCNQ in future) might make a great contribution to the development of high T_c purely organic purely ferromagnets.

Weak-Ferromagnetism in the Li^+ salt of $TCNQF4^{-\cdot}$ (*13*)

Prior to studying the magnetic properties of the CT complexes with $TCNQF4^{-\cdot}$, the spin-spin interaction between $TCNQF4^{-\cdot}$ molecules was investigated by using the Li^+, Na^+, K^+, Rb^+, Cs^+ and $N(CH3)_4^+$ salts Previous works indicate that in salts other than $Li^+ \cdot TCNQF4^{-\cdot}$ strong antiferromagnetic interaction occurs as a result of dimer formation between $TCNQF4^{-\cdot}$ molecules, but in the Li^+ salt the temperature dependence of paramagnetic susceptibility (χ_p) follows the Curie law in the temperature range 110 to 300 K, indicating no significant spin interaction between $TCNQF4^{-\cdot}$ molecules (*14*). It is of much interest to examine in the spin-spin interaction in the lower temperature region (ca. 2 K) in the Li^+ salt, where the dimer formation might also occur bringing about disappearance of the magnetic moment. However, contrary to this expectation weak-ferromagnetism was observed below 12 K. This magnetic phenomenon is well known in several inorganic systems involving α-F_2O_3 (*15*) and NiF_2 (*16-18*), but there has been no report in purely organic systems until this discovery. At almost the same time weak-ferromagnetism was also recognized in two organic radical crystals of 1,3,5-triphenylverdazyl (*19*) and 1,3,5-triphenyl-6-oxoverdazyl (*20*).

The temperature dependences of the χ_p values in the Li^+ and Na^+ salts of $TCNQF4^{-\cdot}$ are shown in Figure 1. The magnetic susceptibility was measured in the temperature range of 2 to 300 K at applied magnetic field of 500 Oe by using a SQUID magnetometer, and the χ_p value was obtained after subtracting the diamagnetic contribution calculated by Pascal's method from the observed value. As seen in Figure 1, $Na^+ \cdot TCNQF4^{-\cdot}$ has almost no contribution to χ_p as result of strong antiferromagnetic interaction between the $TCNQF4^{-\cdot}$ radical pairs. A similar observation was also made for the K^+, Rb^+, Cs^+ and $N(CH3)_4^+$ salts. The energy difference between the singlet and thermally-accessible triplet states was estimated to be greater than 1500 K based on the analysis by use of a Bleaney-Bowers equation (*21*).

R = H: TCNQ
R = F: $TCNQF_4$

R = H: TCNQ-.
R = F: $TCNQF_4^{\cdot -}$

Formula 1

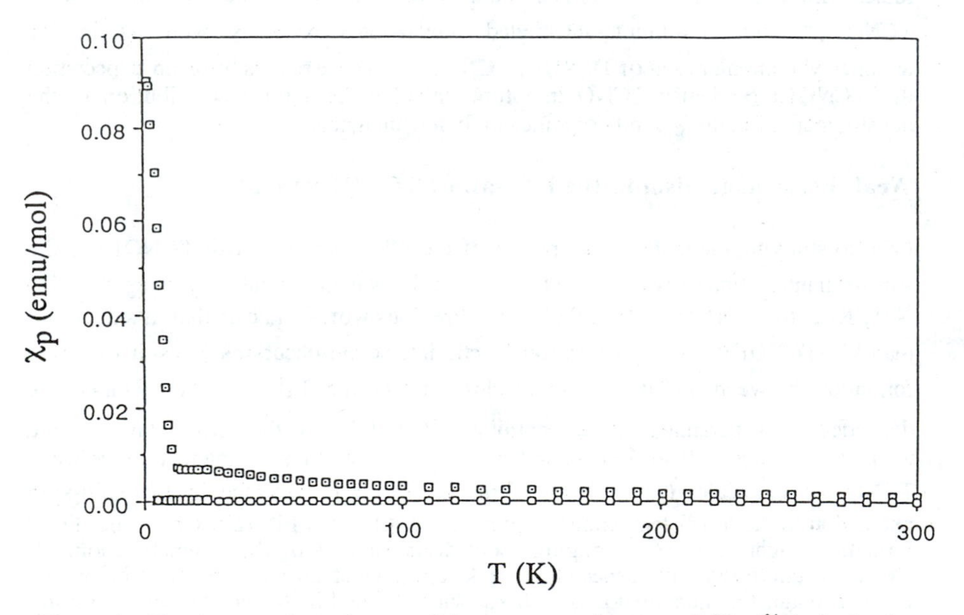

Figure 1. Temperature dependences of the paramagnetic susceptibility (χ_p) obtained by subtracting the diamagnetic contribution from the magnetic susceptibility measured at applied magnetic field of 500 Oe in $Li^+ \cdot TCNQF4^{\cdot -}$ (⊡) and $Na^+ \cdot TCNQF4^{\cdot -}$ (□). Reprinted from T. Sugimoto, M. Tsujii, H. Matsuura, N. Hosoito/Weak ferromanetism below 12 K in a lithium tetrafluorotetracyanoquinodimethanide salt, 1995, pp. 183 - 186, vol. 235, with kind permission from Elsevier Science - NL, Sara Burgerhartstraat 25, 1055 KV Amsterdam, The Netherlands (reference 13).

Contrary to the Na^+ salt, for $Li^+ \cdot TCNQF4^{\cdot -}$ χ_p increases as the temperature is lowered. The temperature dependence of χ_p can be well-expressed by the Curie-Weiss law above 50 K with a Curie constant of 0.36±0.01 emu·K/mol and a negative asymptotic Curie temperature of -30±2 K (see Figure 2(a)). The radical concentration, as shown from the Curie constant, corresponds to one S=1/2 spin per molecule. The χ_p shows a plateau around 25 K, and a sharp increase in χ_p is observed near 12 K and continues to about 5 K. There was no increase in χ_p below 5 K. The anomalous magnetic behavior below 12 K can be more easily visualized in Figure 2(b), where the product of χ_p and temperature ($\chi_p \cdot T$) is plotted against temperature. The $\chi_p \cdot T$ behavior was reproducible for a fresh sample of $Li^+ \cdot TCNQF4^{-\bullet}$. However, enhancement of χ_p slightly diminished when the samples were exposed to air for some time.

A field-cooled magnetization experiment obtained at an applied field of 3 Oe is shown in Figure 3. The curve gradually increases below 12 K with decreasing temperature. At 2 K the applied field was switched off and the remanence was measured with increasing temperature. The temperature dependence of the remanence is similar to that of the field-cooled magnetization. The remanence disappeared at about 10 K. From these results a magnetic phase-transition of $Li^+ \cdot TCNQF4^{-\bullet}$ is expected at around 12 K. Figure 4 shows the magnetization curves at 2,5 and 10 K, respectively. At 10 K the remanence was not observed and the magnetization was almost proportional to the magnetic field up to 55 kOe, the upper limit of the measurement. In contrast, the non-zero residual magnetization appeared at 2 and 5 K. The magnitudes of the residual magnetizations are 50 at 2 K and 30 emu/mol at 5 K. At these temperatures the magnetization curves seem to consist of two parts. The magnetization rapidly increases in the lower magnetic field, and then linearly increases in the higher magnetic field.

Among the $TCNQF4^{-\bullet}$ salts used in the present study only $Li^+ \cdot TCNQF4^{-\bullet}$ has a magnetic moment coresponding to an S=1/2 spin entity. In the Li^+ salt the spins interact antiferromagnetically, as is indicated from the negative asymptotic Curie temperature (-30±2 K) as well as the decrease in $\chi_p \cdot T$ with a lowering temperature. The plateau in χ_p observed around 25 K corresponds to the asymptotic Curie temperature. This suggests an onset of antiferromagnetic ordering, though it is not conclusive for the present. On the other hand, Figures 1-3 indicate definitive evidence of a magnetic phase transition around 12 K. Judging from the negative asymptotic Curie temperature, the magnetic ordering is antiferromagnetic. In antiferromagnets the magnetization is usually proportional to the applied magnetic field. Nevertheless, the Li^+ salt has non-zero magnetization at 2 and 5 K, as is shown in Figure 4. This means that the cancellation of the magnetization by sub-lattice moments is not perfect. A possible reason of this occurrence is canting of sub-lattice moments (*22, 23*). The magnetization curve measurement in the lower field at 2 and 5 K can be reasoanbly interpreted as a magnetization process of the unbalanced moments. Assuming S=1/2 moment for each $TCNQF4^{-\bullet}$ molecule, the estimated canting angle at 2 K is about 0.7°.

To understand the origin of weak ferromagnetism in $Li^+ \cdot TCNQF4^{-\bullet}$, information on the crystal structure is of critical importance. The crystal structure of the Rb^+ salt of $TCNQ^{-\bullet}$ is known, and shows tight pairs of $TCNQ^{-\bullet}$ molecules forming one-dimensional stacks (*24*). Judging from the similar magnetic behavior in the other $TCNQF4^{-\bullet}$ salts, tight pairs of $TCNQF4^{-\bullet}$ molecules are conceivable for $Li^+ \cdot TCNQF4^{-\bullet}$.

Ferromagnetism in Pyridinium-Substituted Imidazolin-1-oxyl/ $TCNQF4^{-\bullet}$ Salts (*25, 26*)

The CT complexes of a series of 4,4,5,5-tetramethylimidazolin-1-oxyls substituted with 4-[*N*-alkyl (R) pyridinium] groups ($\mathbf{1}(R)^{+\bullet}$: R=Me, Et, *n*-Pr) at the 2-position

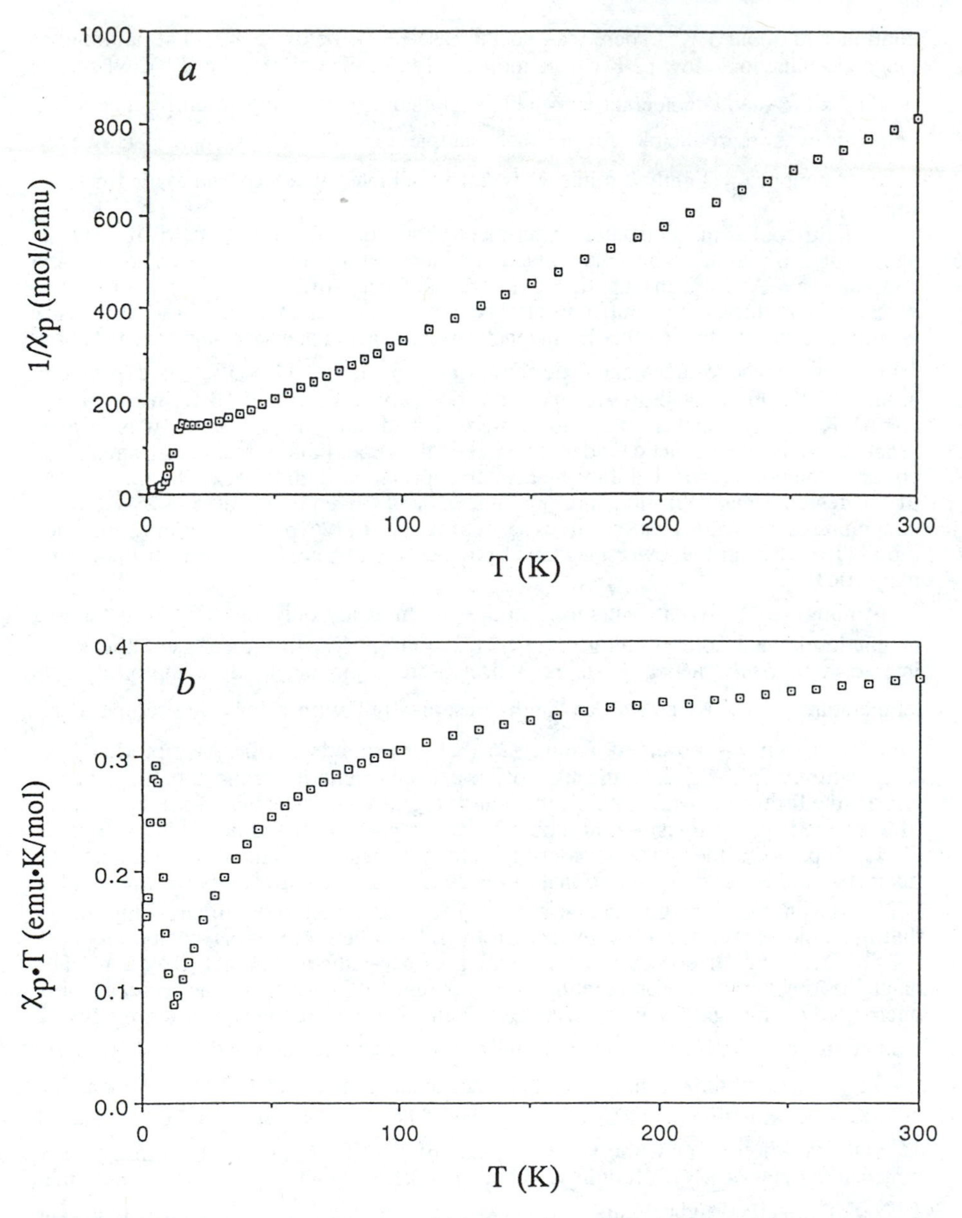

Figure 2. Temperature dependences of (*a*) the reciprocal χ_p ($1/\chi_p$) and (*b*) the product of χ_p and temperature ($\chi_p \cdot T$) in $Li^+ \cdot TCNQF_4^{-\cdot}$. Reprinted from T. Sugimoto, M. Tsujii, H. Matsuura, N. Hosoito/Weak ferromanetism below 12 K in a lithium tetrafluorotetracyanoquinodimethanide salt, 1995, pp. 183 - 186, vol. 235, with kind permission from Elsevier Science - NL, Sara Burgerhartstraat 25, 1055 KV Amsterdam, The Netherlands (reference 13).

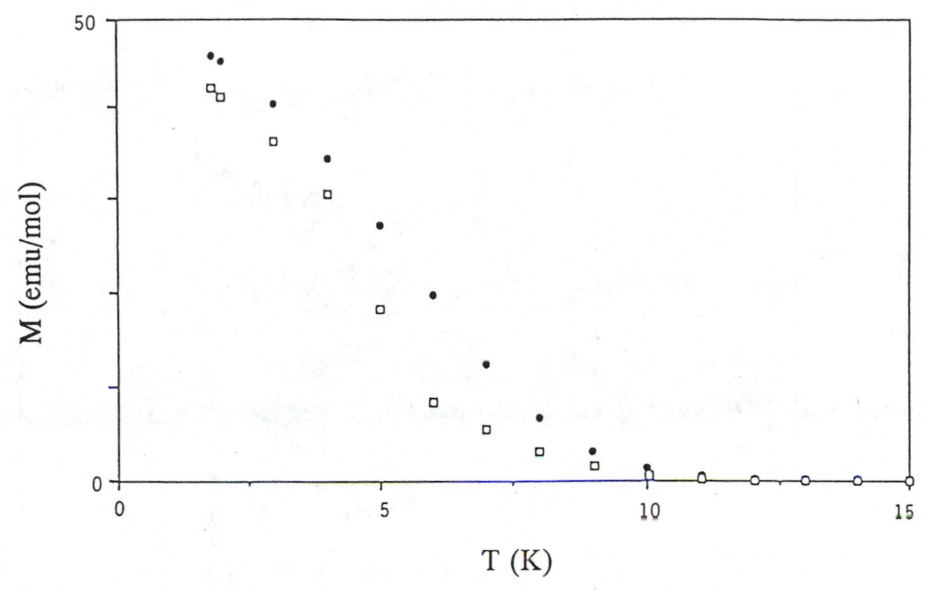

Figure 3. Temperature dependences of the field-cooled magnetization measured at the applied magnetic field of 3 Oe (●) and of the residual magnetization (□). The residual magnetization was measured with increasing temperature after the field-cooling. Reprinted from T. Sugimoto, M. Tsujii, H. Matsuura, N. Hosoito/Weak ferromanetism below 12 K in a lithium tetrafluorotetracyanoquinodimethanide salt, 1995, pp. 183 - 186, vol. 235, with kind permission from Elsevier Science - NL, Sara Burgerhartstraat 25, 1055 KV Amsterdam, The Netherlands (reference 13).

with $TCNQF4^{-\cdot}$ were preapred by mixing aqueous solutions of the iodide salt of **1**$(R)^{+\cdot}$ and of the lithium salt of $TCNQF4^{-\cdot}$ in equal concentrations at room temperature (*27*). All CT complexes of **1**$(R)^{+\cdot}$ with $TCNQF4^{-\cdot}$ were colored blue and blackish brown. The elemental analyses showed a 1:1 composition of the radical cation and the radical anion for all CT complexes. Of the CT complexes only **1**$(Me)^{+\cdot}\cdot TCNQF4^{-\cdot}$ was obtained by recrystallization from acetone/ petroleum ether as single crystals, whose crystal structure analysis was successfully performed. The crystal structure is shown in Figure 5a. The crystal has an alternating stacking of **1**$(Me)^{+\cdot}$ and $TCNQF4^{-\cdot}$ layers along the *b* axis. Within each $TCNQF4^{-\cdot}$ layer the two neighboring $TCNQF4^{-\cdot}$ molecules form a tight dimer structure as evidenced from the closer face-to-face contact (3.18 Å) than a normal π-cloud thickness (3.54 Å) as seen between the dimer units related by an inversion center. On the other hand, within each **1**$(Me)^{+\cdot}$ layer one NO group is connected with the NO group of a neighbor along the *a* and *c* axes (see Figures 5b and 5c) As a result, each of the two-dimensional spin networks are completed within the *a* - *c* plane and furthermore extended to the *b* axis through the interaction with the neighboring spin networks by aid of the neighboring

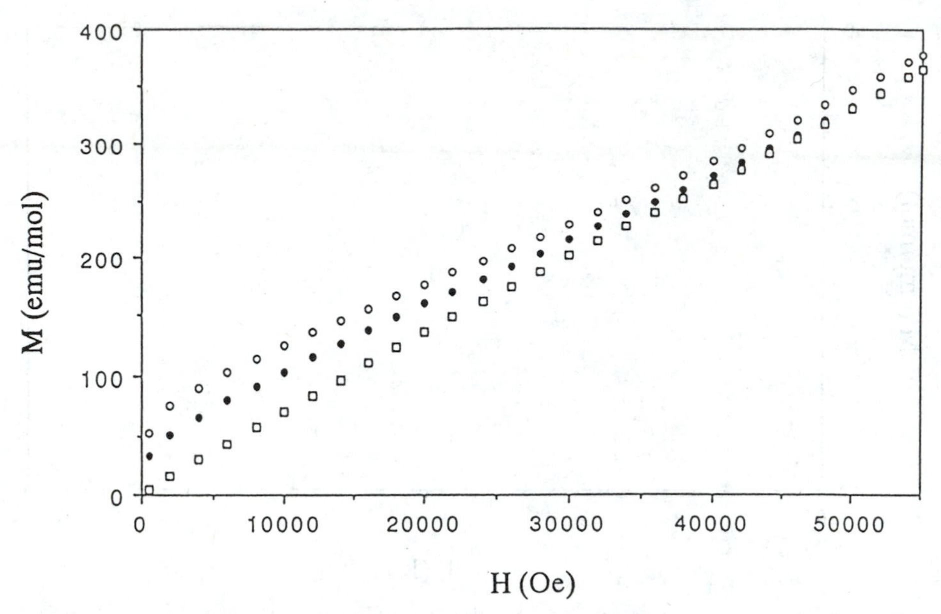

Figure 4. Magnetization (M) as a function of magnetic field (H) in the range of 0 - 55 kOe at 2 (○), 5 (●) and 10 K (□) in $Li^{+}\cdot TCNQF4^{-\cdot}$. Reprinted from T. Sugimoto, M. Tsujii, H. Matsuura, N. Hosoito/Weak ferromanetism below 12 K in a lithium tetrafluorotetracyanoquinodimethanide salt, 1995, pp. 183 - 186, vol. 235, with kind permission from Elsevier Science - NL, Sara Burgerhartstraat 25, 1055 KV Amsterdam, The Netherlands (reference 13).

N
N−R
N
O•

1$(R)^{+\cdot}$ (R = Me, Et, *n*-Pr)

Formula 2

$TCNQF4^{-\cdot}$ molecule (see Figure 5c). It should be noted that the oxygen atom of NO group is disordered. The preferential arrangement of oxygen atom is that as shown in Figure 5a, and the ratio is 83.7 : 16.3. This fact must be taken into consideration in understanding the magnetic properties of this CT complex as well as the others.

The situation of spin-spin interaction in the CT complexes of **1**$(R)^{+\cdot}\cdot TCNQF4^{-\cdot}$ was shown by the magnetic measurement using a SQUID magnetometer (MPMS, Quantum Design) under an applied magnetic field of 500 Oe in the temperature range of

5 to 300 K. The molar paramagnetic susceptibility (χ_p) at each temperature was obtained by correcting the diamagnetic contribution calculated from Pascal's constants The temperature dependences of the product of χ_p and temperature ($\chi_p \cdot T$) are shown in Figure 6 for the CT complexes of $\mathbf{1}(R)^{+\cdot}$ with $TCNQF_4^{-\cdot}$. As is seen from the figure, the following characteristics can be pointed out. (1) For the CT complexes of

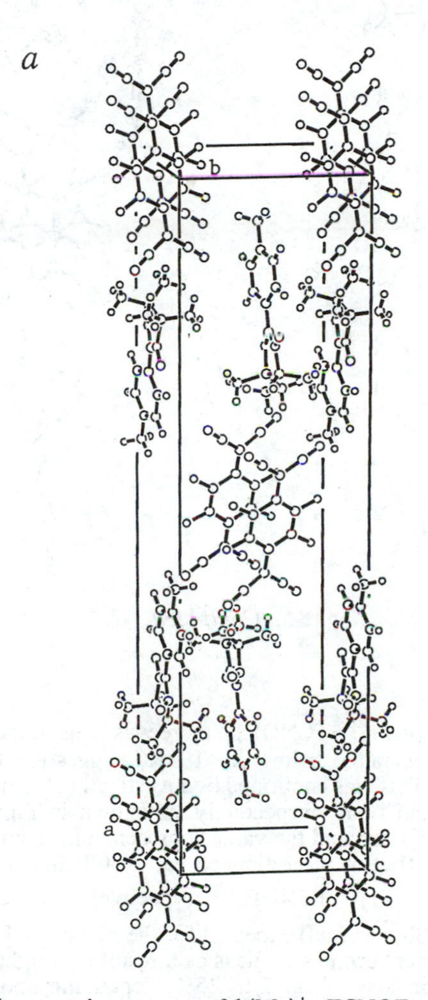

Figure 5. The crystal structures of $\mathbf{1}(Me)^{+\cdot}\cdot TCNQF_4^{-\cdot}$: (a) a whole view (reproduced with permission from reference 26); (b) a view projected down to the *c* axis; (c) a view projected down to the *a* xis.

Continued on next page

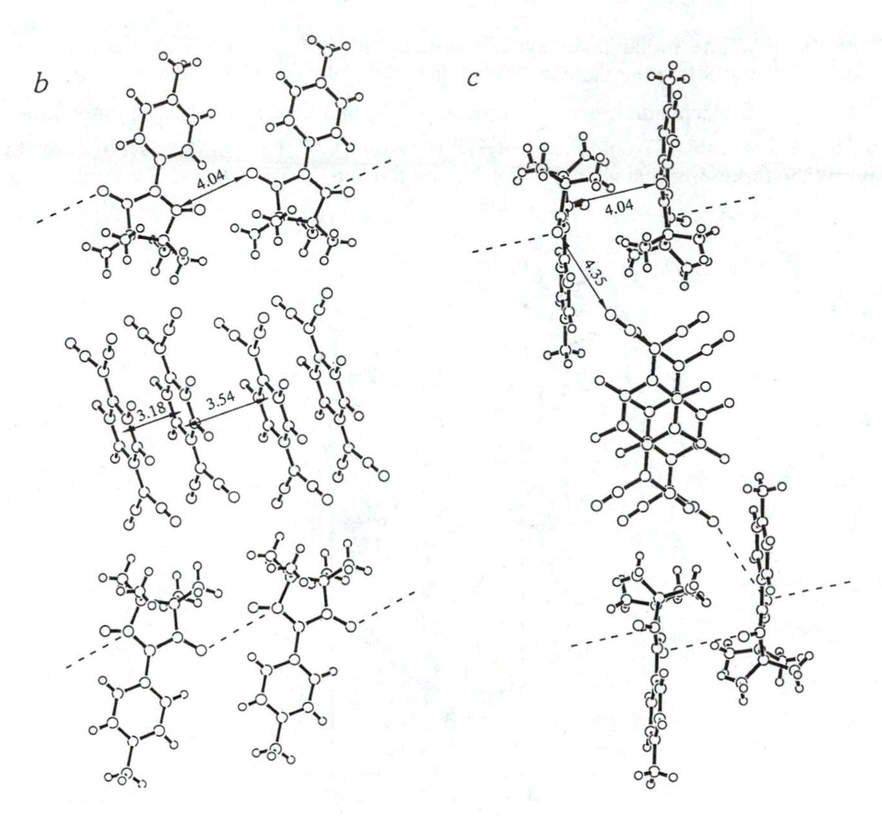

Figure 5. *Continued.*

1(Me)$^{+\cdot}$·TCNQF4$^{-\cdot}$ and **1**(*n*-Pr)$^{+\cdot}$·TCNQF4$^{-\cdot}$ there was a steep increase in the $\chi_p \cdot T$ value with a lowering temperature below ca. 10 K, suggesting the dominance of ferromagnetic interaction. This was confirmed by investigating the field dependence of the magnetization at 2, 5 and 10 K, respectively, as shown in Figure 7. Thus, the lower the temperature, the more rapid the value approached the saturation value. The values of saturation magnetization are estimated ca. 4,000 and 5,200 emu/mol for **1**(Me)$^{+\cdot}$·TCNQF4$^{-\cdot}$ and **1**(*n*-Pr)$^{+\cdot}$·TCNQF4$^{-\cdot}$, respectively. These correspond to 72 and 94% of the value (5,580 emu/mol) expected for the surviving **1**(R)$^{+\cdot}$ spin entity, since the radical anion partner becomes spinless as a result of a tight dimer formation. Accordingly, there is a spin loss of ca. 5 to 25%, depending upon the kind of CT complex. The disorder in the arrangement of NO groups in the crystal is a conceivable cause, as seen in **1**(Me)$^{+\cdot}$·TCNQF4$^{-\cdot}$. For this CT complex the O atom of NO group is arranged at both N atoms on the imidazoline ring and the ratio is 83.7:16.3. Assumed that the preferred arrangement is reponsible for a ferromagnetic interaction between the two neighboring spins, the amount involved in the interaction might be

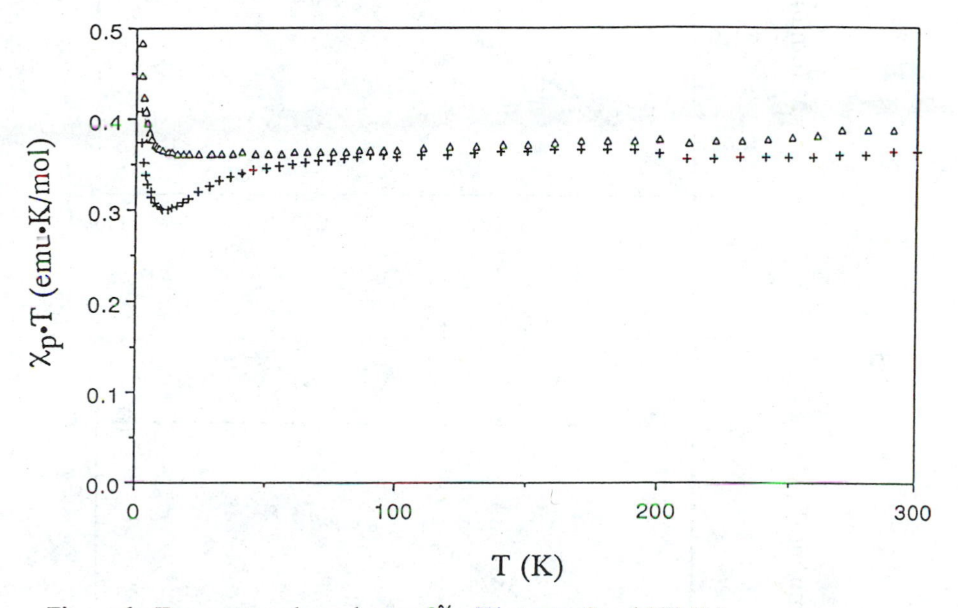

Figure 6. Temperature dependence of $\chi_p \cdot T$ between 2 and 300 K for **1**(Me)$^{+\cdot}\cdot$TCNQF4$^{-\cdot}$ (+) and **1**(*n*-Pr)$^{+\cdot}\cdot$TCNQF4$^{-\cdot}$(Δ). The magnetic susceptibility measurement was carried out under applied magnetic field of 500 Oe. Reproduced with permission from reference 26. Copyright 1995. Gordon and Reach Publishers.

a

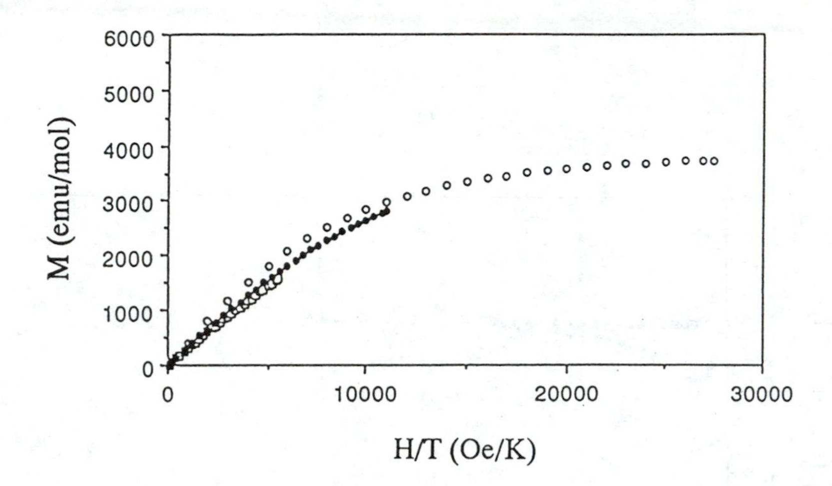

b

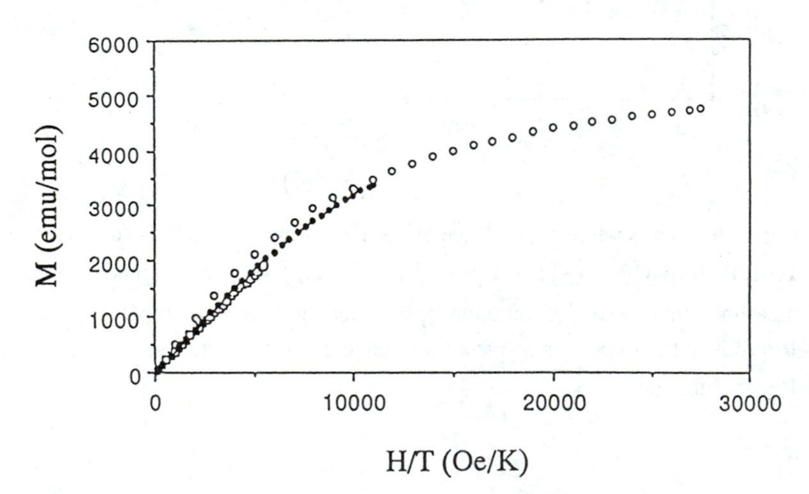

Figure 7 Field dependence of the magnetization (M) at 2(○), 5 (●) and 10 K (□), respectively, for (*a*) **1**(Me)$^{+\cdot}$·TCNQF4$^{-\cdot}$ and (*b*) **1**(*n*-Pr)$^{+\cdot}$·TCNQF4$^{-\cdot}$. Reproduced with permission from reference 26. Copyright 1995. Gordon and Reach Publishers.

reduced to $(0.837)^2 \times 100 = 70\%$, which is well consistent with the experimental result (ca. 72%). The degree of disorder of NO group is smaller for the other CT complexes according to this analysis. The crystal structure analysis of the two remaining CT complexes is by all means necessary to confirm this argument. Furthermore, it remains to be elucidated how the arrangement of NO group in the crystal is different between **1**$(Me)^{+\bullet}\cdot TCNQF_4^{-\bullet}$ and **1**$(n\text{-}Pr)^{+\bullet}\cdot TCNQF_4^{-\bullet}$, and the others which exhibited weak antiferromagnetic interaction. (2) For all CT complexes the $\chi_p \cdot T$ value tends to increase with an increasing temperature in the high temperature region. The magnitude of slope is different for each CT complex. This behavior can be explained by considering the presence of a thermally-accessible triplet of the $TCNQF_4^{-\bullet}$ spin pair in the singlet ground state, judging from the crystal structure of **1**$(Me)^{+\bullet}\cdot TCNQF_4^{-\bullet}$.

By the magnetic measurements at temperatures lower than 5 K it was found that the CT complexes exhibited a ferromagnetic phase transition between 0.4 and 0.55 K. The ac magnetic susceptibility (χ_{ac}) was measured down to ca. 40 mK in a ^3He - ^4He dilution refrigerator at an ac magnetic field of ca. 4 μT (127 Hz). The plot of χ_{ac} against temperature is shown in Figure 8, in which peak maxima albeit in a slightly round shape appear at 0.22 and 0.40 K for **1**$(Me)^{+\bullet}\cdot TCNQF_4^{-\bullet}$ and **1**$(n\text{-}Pr)^{+\bullet}\cdot TCNQF_4^{-\bullet}$, respectively. These peak temperatures should not, however, be assigned as the ferromagnetic phase transition temperatures (T_c's). Rather T_c can be defined in the present study as a temperature where χ_{ac} starts to rapidly increase. There is a slight difference (0.1 - 0.2 K) between the T_c and the peak temperature. When the M - H curves were measured above and below T_c, a typical ferromagnetic hysteresis was observed below T_c (see Figure 9). Both coercive forces ($\leq$ 20 Oe) and saturation fields are low for the three CT complexes like the other organic ferromagnets recently discovered (*8-12*). It should be noted that the estimated saturation moments at the lowest temperature were ca. 0.5 μ_B/formula unit (f.u.) or less than 1 μ_B/f.u. expected for one S = 1/2 spin entity. The main cause of this reduction of the magnetic moment might be attributed to magnetic fluctuations resulting from the low-dimensional character of these CT complexes.

The present CT complexes are another example of non-metal CT complex-based ferromagnets. The first one has been discovered in a fullerene/tetrakis-(dimethylamino)ethylene CT complex (T_c = 16.1 K) (28), whose magnetic details, however, still now remain to be eleucidated because of insufficient structural characterization due to the extreme instability in air. In contrast, the present CT complex-based organic ferromagnets are quite stable in air and have a definitive 1:1 compostion of the radical cation and the radical anion. Through theoretical consideration in combination with crystal structure data, the origin of this ferromagnetism will be pursued. When this problem is resolved, the information will serve as a guiding principle for the synthesis of CT complex-based organic ferromagnets.

Room Temperature Ferromagnetism in a Mixed $TCNQF_4/TCNQF_4^{-\bullet}$ Salt (29)

Further recrystallization of $N(CH_3)_4^+\cdot TCNQF_4^{-\bullet}$ in the presence of a half molar amount of $TCNQF_4$ gave a new compound with a formula of $\{N(CH_3)_4^+\cdot TCNQF_4^{-\bullet}\}\cdot 1/2(TCNQF_4)$. This crystal is blue and quite stable in air for a long period of

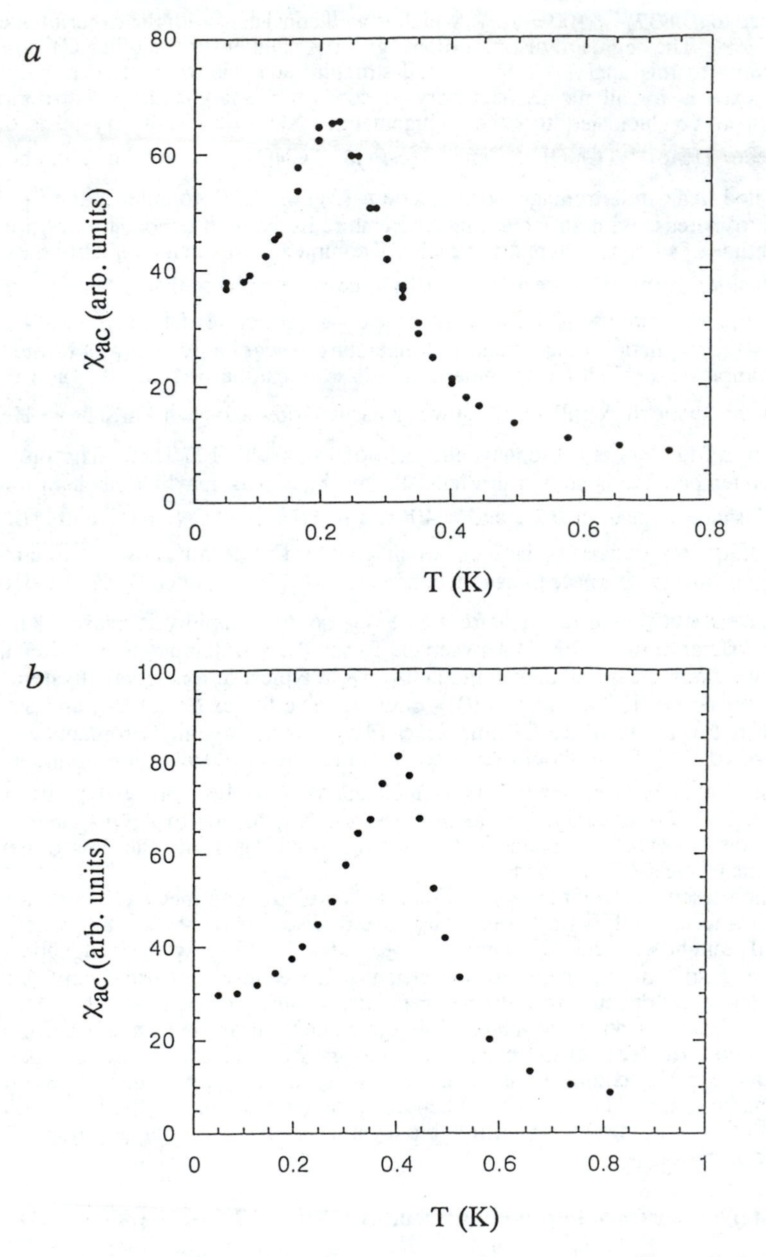

Figure 8 Temperature dependence of the ac magnetic susceptibility (χ_{ac}) below 1 K for (*a*) **1**(Me)$^{+\cdot}$·TCNQF4$^{-\cdot}$ and (*b*) **1**(*n*-Pr)$^{+\cdot}$·TCNQF4$^{-\cdot}$. Reproduced with permission from reference 26.

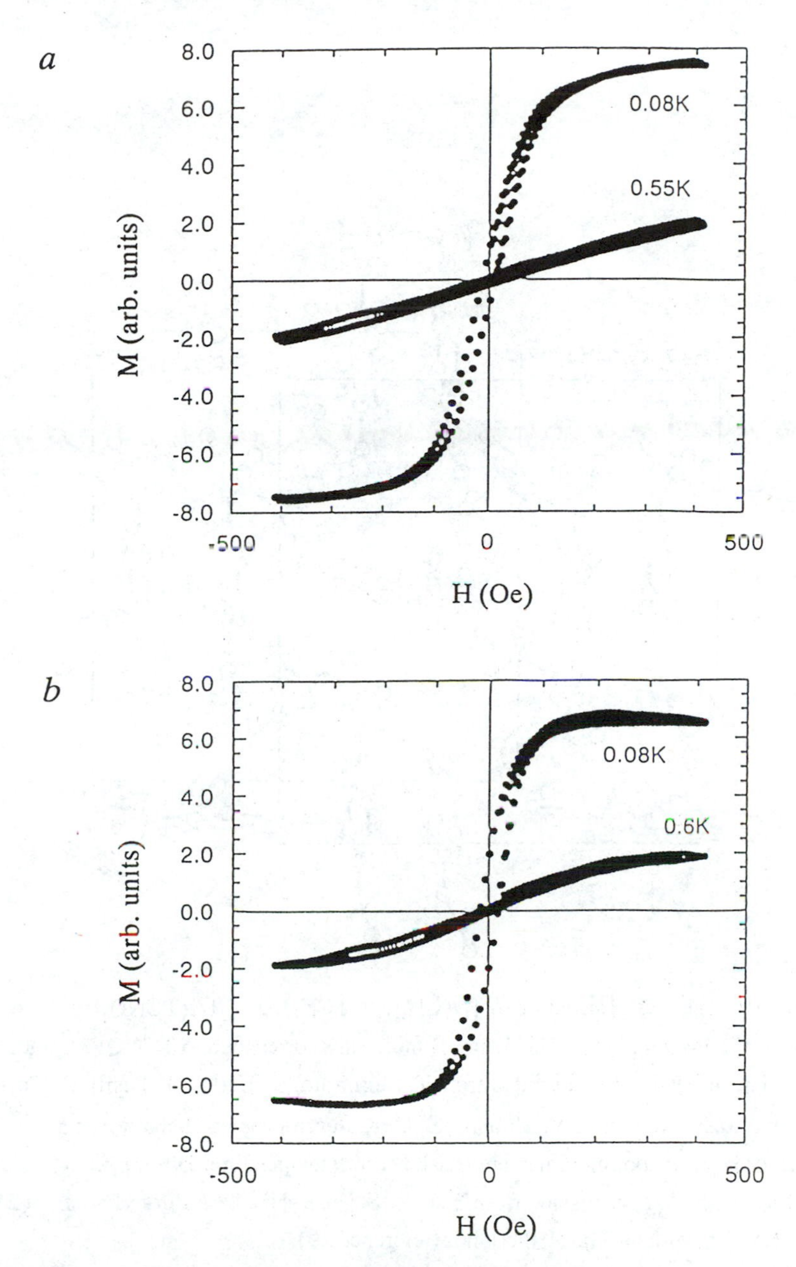

Figure 9 Hysteresis magnetization front and rear the ferromagnetic phase transition for (*a*) **1**(Me)$^{+\cdot}$·TCNQF4$^{-\cdot}$ and (*b*) **1**(*n*-Pr)$^{+\cdot}$·TCNQF4$^{-\cdot}$. The transition temperatures (T_C's) were determined 0.4 to 0.55 K from the experiments of χ_{ac} vs. T. Reproduced with permission from reference 26.

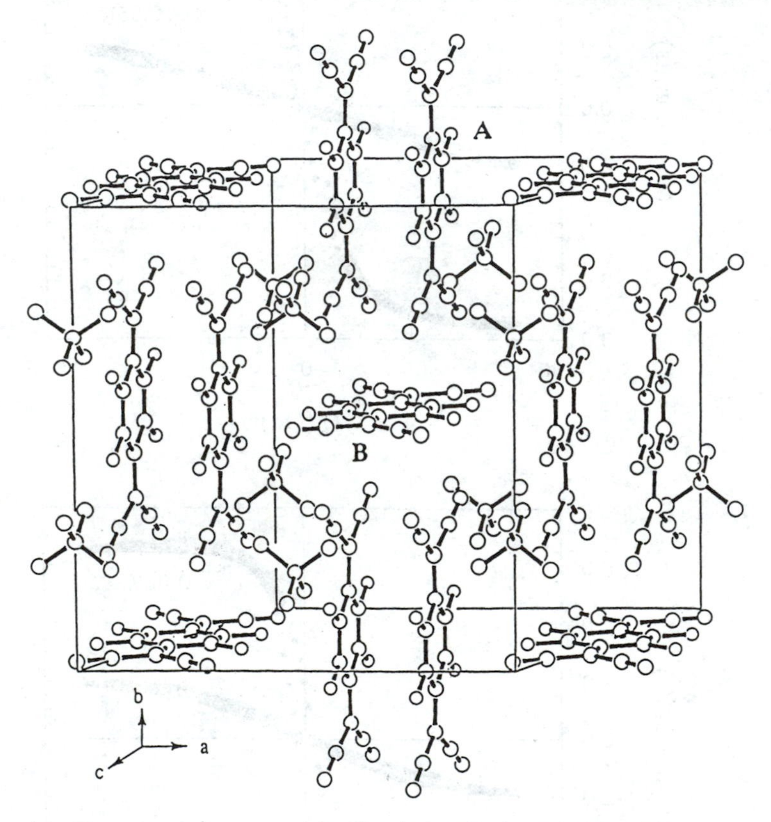

Figure 10. The crystal structure of $\{N(CH_3)_4^+ \cdot TCNQF_4^{-\cdot}\}$ 1/2($TCNQF_4$) projected on the *a-b* plane. The **A** and **B** molecules correspond to $TCNQF_4^{-\cdot}$ and $TCNQF_4$, respectively. Reprinted from T. Sugimoto, K. Ueda, M. Tusjii, H. Fujita, N. Hosoito, N. Kanehisa, Y. Shibamoto, Y. Kai/Ferromagnetic behavior of a tetrafluorotetracyanoquinodimethanide salt at room temperature, 1996, pp. 304 - 308, vol. 249, with kind permission from Elsevier Science -NL, Sara Burgerhartstraat 25, 1055 KV Amsterdam, The Netherlands (reference 29).

time.The crystal structure determined by X-ray analysis is shown in Figure 10. The crystal is composed of alternating $TCNQF_4/TCNQF_4^{-\cdot}$ and $N(CH_3)_4^+$ layers along the *c* axis. The $TCNQF_4/TCNQF_4^{-\cdot}$ layer adopts a unique packing structure, in which each pair of two $TCNQF_4^{-\cdot}$ molecules (denoted as **A** in Figure 10) is arranged perpendicularly to each of one $TCNQF_4$ molecule (denoted as **B**). The **A** and **B** pairs can be assigned to $TCNQF_4^{-\cdot}$ and $TCNQF_4$, respectively, in view of the bond lengths of the two molecules. For the **A** pair bond alternation is very small, suggesting large delocalization of π electrons over the entire molecule. On the other hand, the **B** pair has significant alternation of single and double carbon-carbon bonds. The two neighboring $TCNQF_4^{-\cdot}$ molecules essentially form a tight dimer, judging from the very short contact distance (3.11 Å) as compared with that of a normal π cloud (3.54 Å). Each dimer has two different strong contacts with the neighboring **B** molecule. Along the *a* axis each of **A** molecules in the dimer contacts with one **B** molecule in such a manner that their molecular planes are completely perpendicular to each other (the dihedral angle: 90.08°), and the difluoroethylene bond axis of a **B** molecule passes through the center of the long bisecting axis of an **A** molecule. The closest contact occurs between the difluoroethylene carbons of an **A** molecule and the fluorine atoms in the **B** molecule, and the distances are 3.03 and 3.45 Å. On the other hand, along the *b* axis both **A** molecules in the dimer contact one **B** molecule, and the molecular planes of **A** and **B** molecules are also perpendicular to each other (the dihedral angle is 89.92°). In this case the long bisecting axis of an **A** molecule passes through one carbon atom of the difluoroethylene bond in a **B** molecule. Accordingly, there is a closest contact between the nitrogen atom of cyano group in an **A** molecule and the difluoroethylene carbons in a **B** molecule, and the distance is 2.97 Å. The result is a two-dimensional network between $TCNQF_4$ and $TCNQF_4^{-\cdot}$ molecules achieved in each $TCNQF_4/TCNQF_4^{-\cdot}$ layer. Each $N(CH_3)_4^+$ ion is located near the midpoint in the cavity surrounded by eight cyano groups of four $TCNQF_4^{-\cdot}$ molecules and two cyano groups of two $TCNQF_4$ molecules. Such an arrangement of $N(CH_3)_4^+$ ion is responsible for stabilization of this unique crystal structure by aid of Coulombic interaction between the positively-charged $N(CH_3)_4^+$ ion and the negatively-charged cyano groups. By intervention of the $N(CH_3)_4^+$ layer, the two neighboring $TCNQF_4/TCNQF_4^{-\cdot}$ layers are separated by 9.58 Å (closest contact).

The magnetism of this radical anion salt was investigated using a SQUID magnetometer. Figure 11 shows the temperature dependence of paramagnetic susceptibility (χ_p) under an applied field of 50 Oe between 5 and 300 K. In this temperature range there is no change in χ_p and the value is about 0.012 emu/mol, which is larger by about 10 times than that of non-interacting S=1/2 spins per formula unit. These results suggest that saturation in χ_p is already accomplished even at room temperature. This was indeed confirmed by magnetization vs. applied field measurements. The M - H curve is plotted in Figure 12 (a), where the χ_{dia} contributions from both a sample holder and the sample are subtracted. The magnetization curve is S-shaped even at room temperature. In addition, hysteresis with a coercive field of about 25 Oe is observed (see Figure 12 (b)). The magnetization curve at 5 K is almost the same as that at 300 K, although the coercive field is slightly larger (28 Oe). The saturation magnetization (M_S) at room temperature

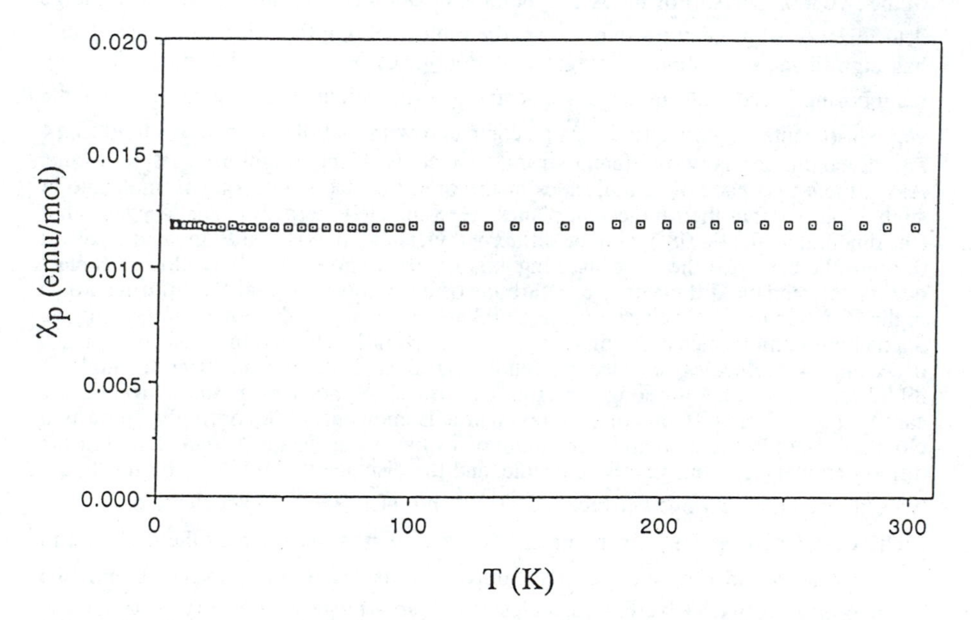

Figure11. Temperature dependence of paramagnetic susceptibility (χ_p) between 5 and 300 K. The measurement was performed by using the microcrystalline sample of $\{N(CH_3)_4^+ \cdot TCNQF_4^{-\cdot}\} \cdot 1/2(TCNQF_4)$ under applied magnetic field of 50 Oe. Reprinted from T. Sugimoto, K. Ueda, M. Tusjii, H. Fujita, N. Hosoito, N. Kanehisa, Y. Shibamoto, Y. Kai/Ferromagnetic behavior of a tetrafluorotetracyanoquinodimethanide salt at room temperature, 1996, pp. 304 - 308, vol. 249, with kind permission from Elsevier Science -NL, Sara Burgerhartstraat 25, 1055 KV Amsterdam, The Netherlands (reference 29).

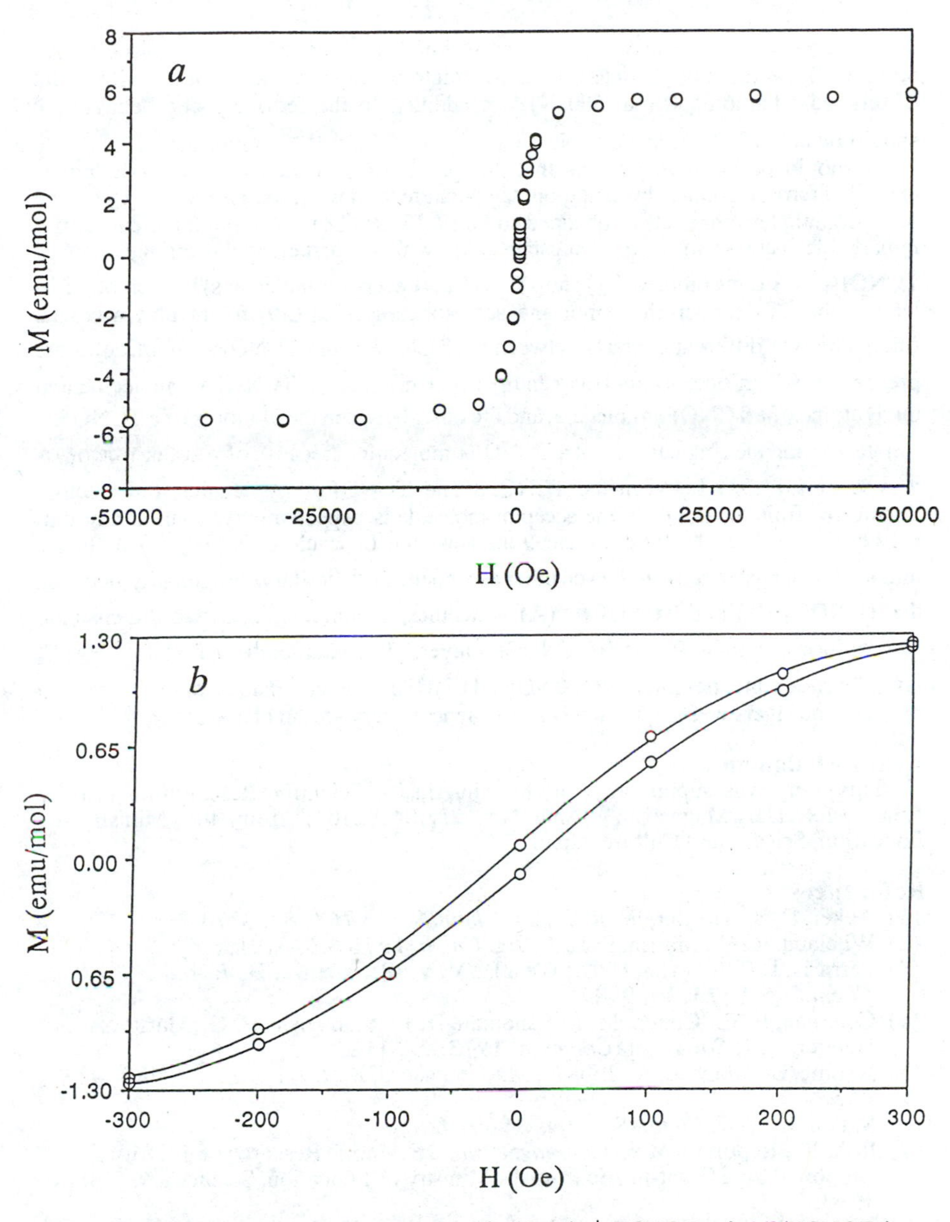

Figure 12 Hysteresis (M vs. H) curves of $\{N(CH3)4^{+}\cdot TCNQF4^{-\cdot}\}$ 1/2(TCNQF4) at room temperature in the magnetic field ranges of (*a*) ±50 kOe and (*b*) ±300 Oe. Reprinted from T. Sugimoto, K. Ueda, M. Tusjii, H. Fujita, N. Hosoito, N. Kanehisa, Y. Shibamoto, Y. Kai/Ferromagnetic behavior of a tetrafluorotetracyanoquinodimethanide salt at room temperature, 1996, pp. 304 - 308, vol. 249, with kind permission from Elsevier Science -NL, Sara Burgerhartstraat 25, 1055 KV Amsterdam, The Netherlands (reference 29).

is estimated to be 5.72 emu/mol, which corresponds to 1.03 x 10^{-3} μ_B/molecule. An accurate T_C value is not yet determined, but might be close to the decomposition point of this radical anion salt (ca. 500 K). In addition to the ferromagnetic behavior, it should be noted that this salt exhibits a large χ_{dia} (-6.30 x 10^{-4} emu/mol).

It should be quite astonishing that this purely organic radical anion salt indeed exhibited ferromagnetism even at room temperature in view of extremely low T_C for the other organic ferromagnets as obtained so far (*8-12*, *25*, *26*). Although the cause of a remarkable increase in T_C is at present unknown, the perpendicular arrangement of $TCNQF_4^{-\cdot}$ (**A** donor molecules) and $TCNQF_4$ (**B** acceptor molecules) to each other as well as the CT between the donor and acceptor molecules (*30*) might play a crucial role. The two different contacts between the $TCNQF_4$ and $TCNQF_4^{-\cdot}$ molecules are present, in which one occurs between the ring carbons of a $TCNQF_4^{-\cdot}$ moleclue and the F atoms of a $TCNQF_4$ molecule, and the other between the N atoms of a $TCNQF_4^{-\cdot}$ molecule and the ring carbons of a $TCNQF_4$ molecule. Because of a weak n-π or σ-π type of CT (*31*) between the $TCNQF_4$ and $TCNQF_4^{-\cdot}$ molecules, the electron transferred from the donor to the acceptor molecule is expected very small in amount. However, it is possible that the electrons appeared on each $TCNQF_4$ (**B**) molecule interact ferromagnetically with each other as a result of π orbital orthogonality between the $TCNQF_4$ (**B**) and $TCNQF_4^{-\cdot}$ (**A**) molecules, accomplishing a two-dimensional spin ordering in each $TCNQF_4/TCNQF_4^{-\cdot}$ layer. The intervention of an $N(CH_3)_4^+$ layer between the neighboring $TCNQF_4/TCNQF_4^{-\cdot}$ layers makes spin interaction between the layers weak, so that this ferromagnet is very soft in character.

Acknowledgments

This work was supported by a Grant-in-Aid for Scientific Research on Priority Area "Molecular Magnetism" (Area No. 228/04242104) from the Ministry of Education, Science and Culture, Japan.

References

(1) Acker, D. S.; Hertler, W. R. *J. Am. Chem. Soc.* **1962**, *84*, 3370.
(2) Wheland, R. C.; Martin, E. L. *J. Org. Chem.* **1975**, *40*, 3101.
(3) Ferraris, J. P.; Cowan, D. O.; Walatka, V. V.; Perlstein, J. H. *J. Am. Chem.Soc.* **1973**, *95*, 948.
(4) Coleman, L. B.; Cohen, M. J.; Sandman, D. J.; Yamagishi, F. G.; Garito, A. F.; Heeger, A. J. *Solid State Commun.* **1973**, *12*, 1125.
(5) Jerome, D.; Mazaud, A.; Ribault, M.; Bechgaard, K. *J. Phys. Lett.* **1980**, *41*, L95.
(6) Mukai, K. *Bull. Chem. Soc. Jpn.* **1969**, *42*, 40.
(7) Itoh, K., Report on *Molecular Magnetism*, a Scientific Research on Priority Area supported by a Grant-in-Aid from the Ministry of Education, Science and Culture, 1993.
(8) Tamura, M.; Nakazawa, Y.; Shiomi, D.; Nozawa, K.; Hosokoshi, Y.; Ishikawa, M.; Takahashi, M.; Kinoshita, M. *Chem. Phys. Lett.* **1991**,*186*, 401.
(9) Chiarelli, R.; Novak, M. A.; Rassat, A.; Tholence, J. L. *Nature* **1993**, *363*, 147.
(10) Nogami, T.; Tomioka, K.; Ishida, T.; Yoshikawa, H.; Yasui, M.; Iwasaki, F.; Iwamura, H.; Takeda, N.; Ishikawa, M. *Chem. Lett.* **1994**, 29.
(11) Ishida, T.; Tuboi, H.; Nogami, T.; Yoshikawa, H.; Yasui, M.; Iwasaki, F.; Iwamura, H.; Takeda, N.; Ishikawa, M. *Chem. Lett.* **1994**, 919.

(12) Sugawara, T.; Matsushita, M. M.; Izuoka, A.; Wada, N.; Takeda, N.; Ishikawa, M. *J. Chem. Soc., Chem. Commun.*, **1994**, 1723.
(13) Sugimoto, T.; Tsujii, M.; Matsuura, H.; Hosoito, N. *Chem. Phys. Lett.* **1995**, *235*, 183.
(14) Jones, M. T.; Maruo, T.; Jansen, S.; Roble, J.; Rataiczak, R. D. *Mol. Cryst. Liq. Cryst.* **1986**, *134*, 21.
(15) Shull, C. G. ; Strauser, W. A.; Wollan, E. O. *Phys. Rev.* **1951**, *83*, 333.
(16) Erickson, R. A. *Phys. Rev.* **1953**, *90*, 779.
(17) Matarrese, L. M.; Stout, J. W. *Phys. Rev.* **1954**, *94*, 1792.
(18) Stout, J. W.; Catalano, E. *J. Chem. Phys.* **1955**, *23*, 2013.
(19) Tomiyoshi, S.; Yano, T.; Azuma, N.; Shoga, M.; Yamada, K.; Yamauchi, J. *Phys. Rev.* **1994**, *B49*, 16031.
(20) Kremer, R. K.; Kanellakopulos, B.; Bele, P.; Neugebauer, F. A. *Chem. Phys. Lett.* **1994**, *230*, 255.
(21) Bleaney, B.; Bowers, K. D. *Proc. R. Soc. London* **1952**, *A214*, 451.
(22) Dzialoshinski, I. E. *J. Phys. Chem. Solid* **1958**, *4*, 241.
(23) Moriya, T. *Phys. Rev.* **1960**, *117*, 635.
(24) Hoekstra, A.; Spoelder, T.; Vos, A. *Acta Cryst.* **1972**, *B 28*, 14.
(25) Sugimoto, T.; Tsujii, M.; Suga, T.; Matsuura, H.; Hosoito, N.; Takeda, N.; Ishikawa, M.; Shiro, M *Technical Report of ISSP, Ser. A* **1994**, *No. 2854*, 1.
(26) Sugimoto, T.; Tsujii, M.; Suga, T.; Hosoito, N.; Ishikawa, M.; Takeda, N.; Shiro, M. *Mol. Cryst. Liq. Cryst.* **1995**, *272*, 183.
(27) Sugimoto, T.; Tsujii, M.; Murahashi, E.; Nakatsuji, H.; Yamauchi, J.; Fujita, H.; Kai, Y.; Hosoito, N. *Mol. Cryst. Liq. Cryst.* 1993, **232**, 117.
(28) Allemand, P.-M.; Khemani, K. C.; Koch, A.; Wudl, F.; Holczer, K.; Donovan, S.; Gruner, G.; Thompson, J. D. *Science* **1991**, *253*, 301.
(29) Sugimoto, T,; Ueda, K.; Tsujii, M.; Fujita, H.; Hosoito, N.; Kanehisa, N.; Shibamoto, Y.; Kai, Y. *Chem. Phys. Lett.* **1996**, *249*, 304.
(30) Mulliken, R. S. *J. Am. Chem. Soc.* **1952**, *74*, 811.
(31) Foster, R. *Organic Charge-Transfer Complexes* ; Academic Press: London & New York, 1969.

EXTENDED METAL-BASED SYSTEMS

Chapter 20

Spin-Transition Molecular Materials for Display and Data Processing

Olivier Kahn, Epiphane Codjovi, Yann Garcia, Petra J. van Koningsbruggen, René Lapouyade, and Line Sommier

Laboratoire des Sciences Moléculaires, Institut de Chimie de la Matière, Condensée de Bordeaux, Laboratoire Propre du Centre National de la Recherche Scientifique 9484, 33608 Pessac, France

Some Fe(II)-1,2,4-triazole derivatives exhibit very abrupt thermally induced spin transitions around room temperature, accompanied by a well pronounced change of color, between purple in the low-spin (LS) state and white in the high-spin (HS) state. Our objective is to obtain such compounds for which the ambiant temperature exactly falls within the thermal hysteresis loop. If it is so, at room temperature, the state of the system depends on its history, which confers a memory effect on this system. Such compounds could be used as active elements of memory devices or displays.

In some transition metal compounds, in particular in those containing $3d^4$-$3d^7$ metal ions in octahedral surroundings, a crossover between a low-spin (LS) and a high-spin (HS) state may occur. In a first approximation, such a situation is achieved when the energy gap Δ between the high-lying e_g and low-lying t_{2g} orbitals is of the same order of magnitude as the mean spin pairing energy P. The phenomenon is schematized in Figure 1 in the case of a $3d^6$ ion, for instance Fe(II). At the molecular scale the spin transition corresponds to an intraionic electron transfer with spin flip of the transferred electrons. The word «intraionic» means that the electrons are transferred between the e_g and t_{2g} orbitals but remain in the immediate vicinity of the metal ion. Such a spin crossover may be induced by a variation of temperature, of pressure or by a light irradiation (*1-9*). This phenomenon probably represents the most spectacular example of molecular bistability (*10,11*). Our main objective in the field of spin-crossover compounds is to design and synthesize compounds exhibiting a large thermal hysteresis, and to explore to what extent these compounds could be used as active elements in memory devices. Our first step along this line deals with display devices.

0097–6156/96/0644–0298$15.00/0

Cooperativity, Thermal Hysteresis, and Polymeric Structures

In order to utilize spin transition materials in memory devices, some requirements must be fulfilled, which can be summed up as follows : (i) the transition must be abrupt both in the warming mode at $T_C\uparrow$ and in the cooling mode at $T_C\downarrow$; (ii) it must occur with a large thermal hysteresis effect. The bistability is achieved in the temperature range between $T_C\uparrow$ and $T_C\downarrow$; (iii) for most of the applications, the transition temperatures must be as close as possible to room temperature. The ideal situation is that where the ambiant temperature falls in the middle of the thermal hysteresis loop; (iv) the transition between LS and HS states must be accompanied by an easily detectable response, such as a change of color; (v) finally, the system must be chemically stable, and present no fatiguability over successive thermal cycles (*12*). The first two requirements are related to the cooperativity, i.e. to the magnitude of the interactions between active sites within the crystal lattice. In purely molecular crystals containing spin-crossover mononuclear entities the intermolecular interactions are very weak, of the van der Waals type. Such mononuclear compounds seem not to be appropriate for our purpose. In many cases, the spin-transition mononuclear entities are hydrogen bonded within the crystal lattice. These hydrogen bonds may lead to stronger intermolecular interactions. This situation is most often achieved when the crystal lattice contains some non-coordinated solvent molecules. Even larger interactions between active sites may be anticipated if these active sites are covalently linked by chemical bridges, in particular if these bridges are conjugated. Our starting idea, illustrated in Figure 2, is that in polymeric spin-transition compounds the cooperativity should be magnified. This cooperativity might be increased further by hydrogen bondings connecting polymeric units, anions, and solvent molecules.

The first experimental result supporting this idea was provided by the compound $[Fe(btrz)_2(NCS)_2]\cdot H_2O$ with btrz = 4,4'-bis-1,2,4-triazole whose structure is depicted in Figure 3 (*13*). Each iron atom is surrounded by two NCS groups in *trans* position and four btrz ligands coordinated through the nitrogen 1- and 1'-positions. The btrz ligands bridge the iron atoms, which results in a two-dimensional structure. The non-coordinated water molecules are hydrogen bonded to the nitrogen atoms of btrz occupying the 2- and 2'-positions. This two-dimensional compound shows extremely abrupt spin transitions at $T_c\uparrow$ = 144.5 K and $T_c\downarrow$ = 123.5 K (see Figure 3). The thermal hysteresis width is of 21 K. Between $T_c\uparrow$ and $T_c\downarrow$ the state of the system depends on its history. Moreover, this state can be read through the color of the sample. Indeed, the compound is purple in the LS state and white in the HS state (vide infra). However, the ambiant temperature does not fall within the thermal hysteresis loop.

Spin Transition Materials with a large thermal hysteresis

It was in 1977 that for the first time a Fe(II) - triazole compound showing a spin transition above room temperature accompanied by a purple - white change of color was mentioned (*14*). Later on, a Russian group reported on a series of iron(II) - 1,2,4-triazole compounds, several of them exhibiting spin transitions around room temperature with thermal hysteresis and thermochromic effects (*15,16*). More recently Goodwin and coworkers also investigated these

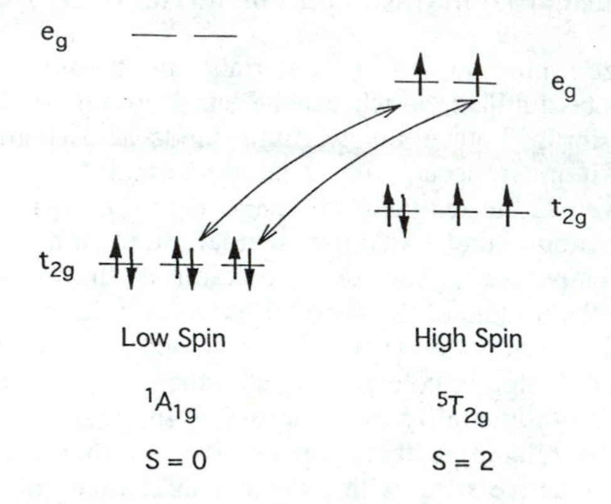

Figure 1 : Spin crossover in a $3d^6$ octahedral compound.

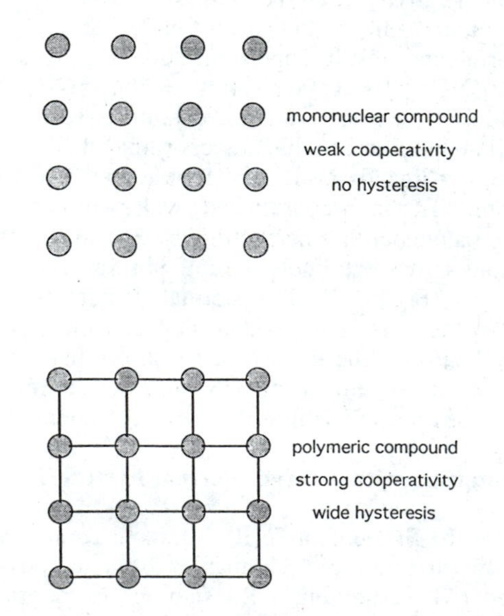

Figure 2 : Cooperativity in spin transition mononuclear and polymeric species (reproduced with the permission from reference 19).

compounds (*17*). A few years ago, it occurred to us that the compounds of this family could fulfill the criteria mentioned above, and we decided to investigate them in a thorough manner.

Let us focus on one of these compounds. The reaction of $[Fe(H_2O)_6](BF_4)_2$ with 1,2,4-1*H*-triazole affords two compounds depending on the experimental conditions. The former compound is $[Fe(Htrz)_2(trz)](BF_4)$; a spontaneous deprotonation of one out of three Htrz ligands occurs. The latter compound is $[Fe(Htrz)_3](BF_4)_2 \cdot H_2O$; no deprotonation occurs, and the three ligands are neutral. We will restrict the discussion here to one of the modifications of the former compound, obtained by adding a solution of 1,2,4-1*H*-trz dissolved in ethanol to a solution of the iron(II) salt dissolved in water (*18*).

A large variety of physical techniques have been utilized to characterize the spin transition regime of this compound. Both in the warming and cooling mode the spin transition is exceptionally abrupt. It is also accompanied by a well pronounced thermochromism between the purple color in the LS state and the white color in the HS state. The purple color in the LS state arises from the $^1A_{1g} \rightarrow {}^1T_{1g}$ d - d absorption at 520 nm. In the HS, the spin-allowed absorption of lowest energy, $^5T_{2g} \rightarrow {}^5E_g$, appears at 800 nm, at the limit of the infra-red range. The transition temperatures are found to be very slightly dependent on both the sample and of the technique used to detect the transition. They may be given as $T_c\uparrow = 383\pm3$ K and $T_c\downarrow = 345\pm2$ K. The thermal hysteresis width is then about 40 K.

Let us briefly comment on the physical data. As already pointed out, the spin transition in all the Fe(II) - 1,2,4-trz compounds is accompanied by a dramatic change of color, so that the transitions can be detected optically, using the set up schematized in Figure 4 (*19*). This set up records changes in optical properties of spin transition compounds in a remarkably simple and reliable way. The sample holder containing about 50 mg of microcrystalline compound is fitted to the common end of a double optical fiber, and plugged into a cryogenic system. The temperature rises and decreases at the rate of 1 Kelvin per minute without thermal oscillations. A photomultiplier provided with a 520 nm interferential filter converts the light reflected by the sample into an electric current whose intensity depends on the color change. As the spin transition compound absorbs at 520 nm in the LS state and not in the HS state, the intensity recorded is null for the purple color (LS state), and maximum for the white color (HS state). A personal computer is used to control the process and record the data. One can then convert the arbitrary intensities into normalized HS molar fractions, identical to what is obtained with the magnetic technique. The optical detection of the bistability for $[Fe(Htrz)_2(trz)](BF_4)$ is shown in Figure 5. In the present case, successive thermal cycles do not modify significantly the thermal hysteresis loop.

The Mössbauer spectroscopy confirms the magnetic data. Typical spectra are represented in Figure 6. The spectrum is characteristic of a HS iron(II) compound above 400 K, and of a LS iron(II) compound below 340 K. The two spectra coexist within a few Kelvin both in the warming and cooling mode. The spectra do not show any dynamical broadening of the Mössbauer lines, thus indicating that the conversion rates remain lower than the Mössbauer hyperfine frequency (around 10^7 s^{-1}), despite the fact that the transitions take place above room temperature. The slowing down of the conversion rates as compared to

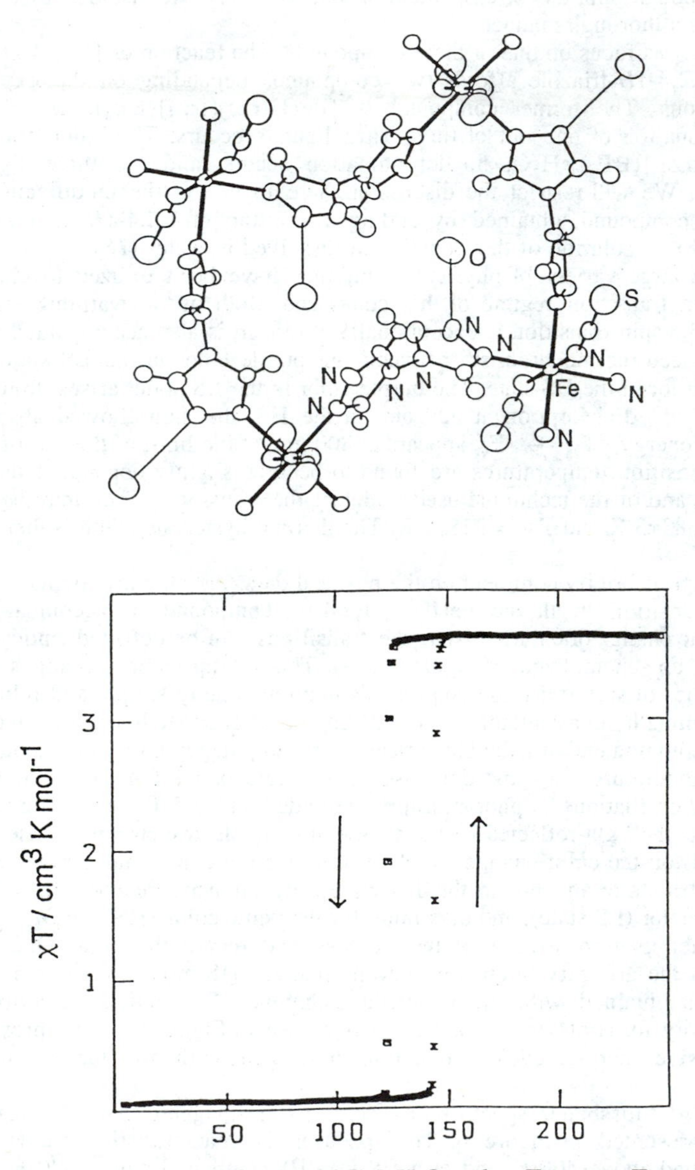

Figure 3 : Structure of $[Fe(btrz)_2(NCS)_2]{\cdot}H_2O$ and $\chi_M T$ versus T curves in the warming and cooling modes for this compound (reproduced with the permission from reference 13).

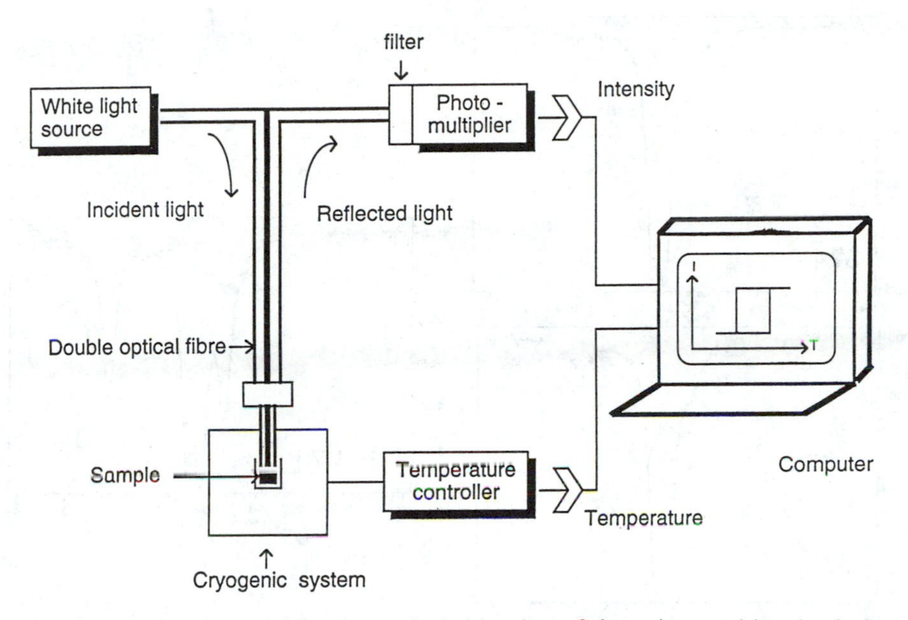

Figure 4 : Device used for the optical detection of the spin transition in the Fe(II) - 1,2,4-triazole compounds (reproduced with the permission from reference 19).

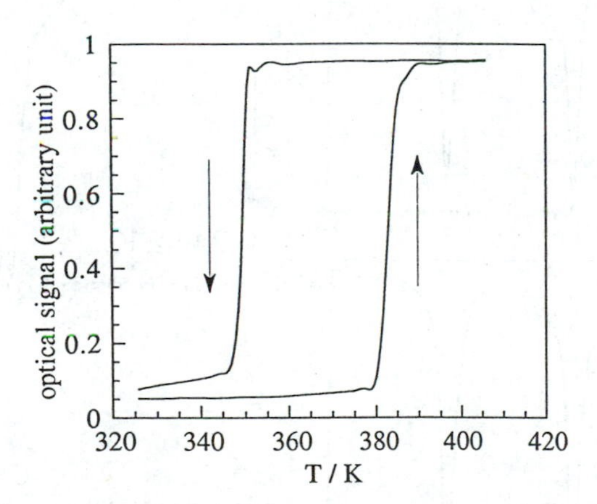

Figure 5 : Optical detection of the spin transition for $[Fe(Htrz)_2(trz)](BF_4)$ (reproduced with the permission from reference 18).

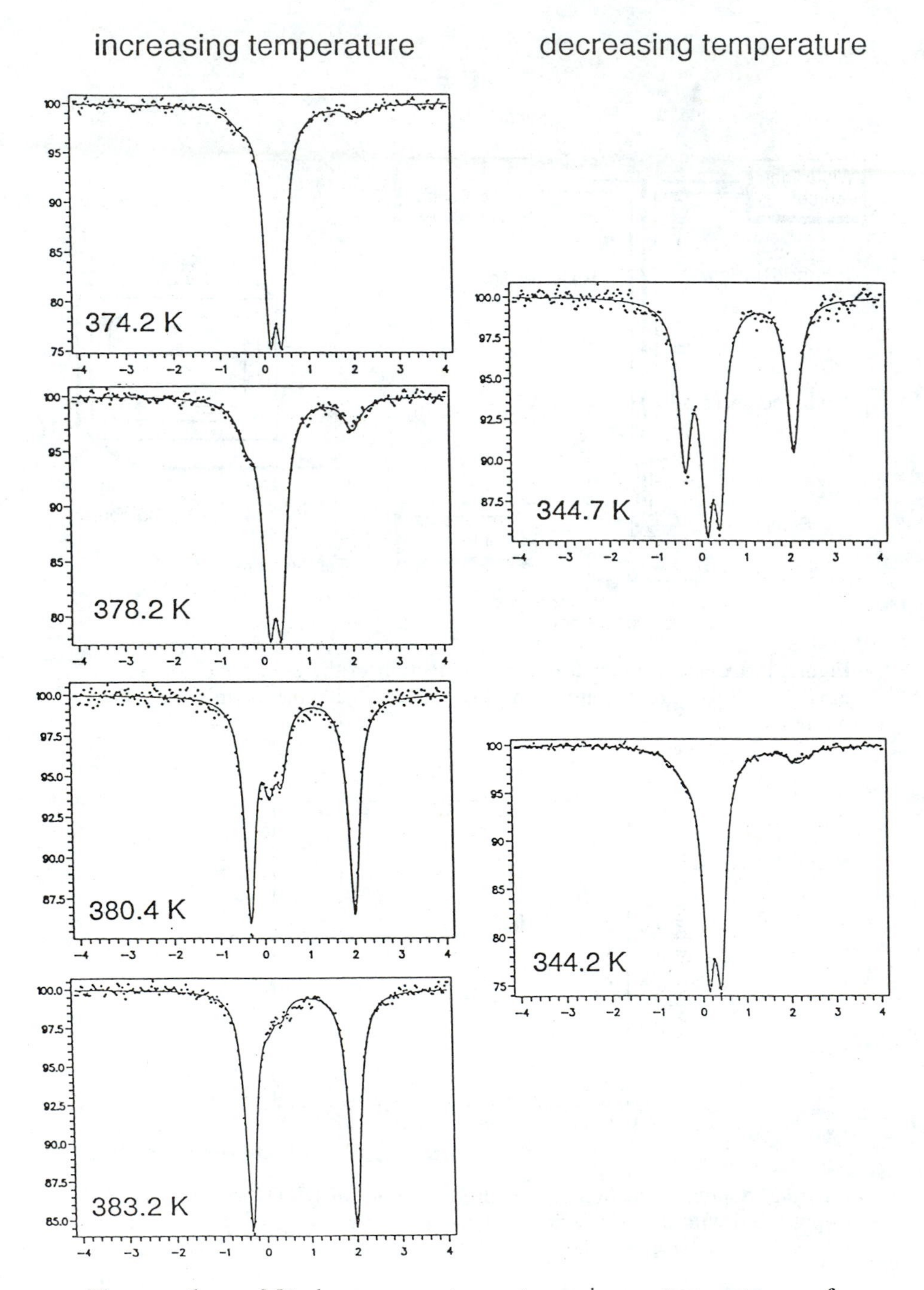

Figure 6 : Mössbauer spectra at various temperatures for $[Fe(Htrz)_2(trz)](BF_4)$ (reproduced with the permission from reference 18).

other iron(II) spin transition compounds might be due to the polymeric nature of the compound.

The differential scanning calorimetry (DSC) curves are represented in Figure 7. They show an abrupt endothermic process at $T_c\uparrow$ in the warming mode, and an even more abrupt exothermic process at $T_c\downarrow$ in the cooling mode. The enthalpy variations associated with these processes are found to be equal to 27.8 and 28.6 kJ mol^{-1}, respectively. These values are much higher than those found for the mononuclear iron(II) spin-transition compounds, usually in the 5 - 15 kJ mol^{-1} range. For instance, the enthalpy variation for $Fe(phen)_2(NCS)_2$ with phen = *ortho*-phenanthroline is equal to 8.59 kJ mol^{-1} (*3*).

In spite of many efforts it has not been possible yet to grow single crystals of $[Fe(Htrz)_2(trz)](BF_4)$, and structural information has been obtained from alternative techniques such as EXAFS at the iron edge and X-ray powder diffraction (20).

The EXAFS data led to a rather accurate description of the basic structure. This structure consists of linear chains in which the Fe(II) ions are triply bridged by triazole ligands through the nitrogen atoms occupying the 1- and 2-positions, as represented in Figure 8. One out of three triazole rings is deprotonated, and probably a proton hole in $[Fe(Htrz)_2(trz)](BF_4)$ is disordered over each $Fe(Htrz)_2(trz)Fe$ bridging network, so that the three-fold symmetry of the chain compound depicted in Figure 8 is retained. Actually, all the Fe - 1,2,4-triazole derivatives with abrupt spin transitions reported so far seem to possess the same chain structure. It is for instance the case for $[Fe(NH_2trz)_3](NO_3)$, with NH_2trz = 4-NH_2-1,2,4-triazole (*21*). For this compound, the transition temperatures are found as $T_c\uparrow$ = 348 K and $T_c\downarrow$ = 313 K. The transitions occur closer to the ambiant temperature than for $[Fe(Htrz)_2(trz)](BF_4)$.

Synergy Between Spin Transition and Dehydration-Rehydration Process for $[Fe(NH_2trz)_3](tos)_2{\cdot}2H_2O$, tos = tosylate

We have seen that one of our main targets is to synthesize spin-transition compounds for which room temperature exactly falls in the middle of the thermal hysteresis loop. Using the trial and error method, such compounds can be obtained. We synthesized, for instance, $[Fe(NH_2trz)_3]Br_2{\cdot}H_2O$ which shows a spin transition regime with $T_c\uparrow$ = 307 K and $T_c\downarrow$ = 279 K. The center of the hysteresis loop is 293 K, i.e. exactly at room temperature. However, the transition in the cooling mode is rather gradual, so that this compound cannot be used in a display device (12). We are presently exploring the iron(II) - triazole system as a whole. We will not enter into the details of our findings here. Rather, we would like to mention an interesting material showing an apparent hysteresis of about 80 K.

The reaction of Fe(II) tosylate with NH_2trz = 4-NH_2-1,2,4-triazole in methanol affords a compound of formula $[Fe(NH_2trz)_3](tos)_2{\cdot}2H_2O$ (22). Warming a sample of this compound reveals a very abrupt LS → HS transition at $T_c\uparrow$ = 361 K, accompanied by the removal of the two non-coordinated water molecules. This dehydration process was confirmed by thermogravimetric analysis. Cooling this dehydrated compound $[Fe(NH_2trz)_3](tos)_2$ reveals a slightly smoother HS → LS transition transition at $T_c\downarrow$ = 279 K, as shown in Figure 9. A subsequent thermal cycle allows us to determine the transition

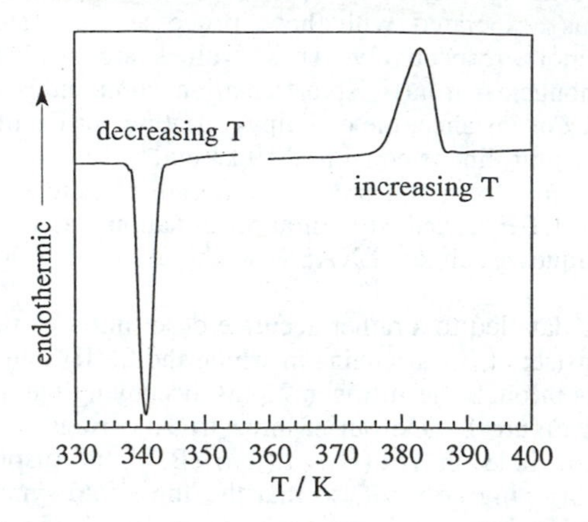

Figure 7 : Differential scanning calorimetry curves $[Fe(Htrz)_2(trz)](BF_4)$ (reproduced with the permission from reference. 18).

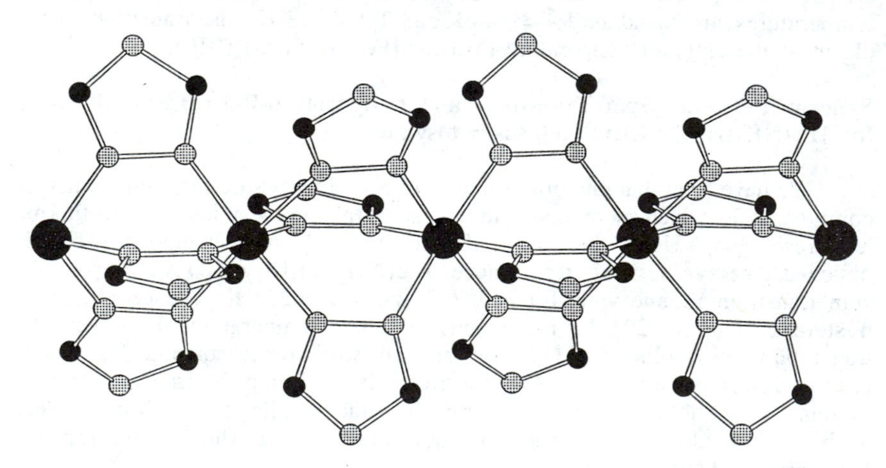

Figure 8 : Structure of the polymeric Fe(II) - 1,2,4-triazole compounds, as deduced from the EXAFS investigation (reproduced with the permission from reference 20).

temperature in the warming regime for $[Fe(NH_2trz)_3](tos)_2$ as $T_c\uparrow = 296$ K. In the presence of humidity, and at room temperature, $[Fe(NH_2trz)_3](tos)_2$ transforms back into the violet hydrated phase. The transition temperature in the cooling regime for $[Fe(NH_2trz)_3](tos)_2{\bullet}2H_2O$ was determined by embedding the compound in a PVC polymer matrix, which prevented the removal of the non-coordinated water molecules. This $T_c\downarrow$ value was found as 340 K. What is quite interesting concerning this material, it is that the synergy between spin transition and dehydration - rehydration process affords a system presenting an exceptionally large thermal hysteresis loop; the apparent hysteresis width is about 80 K. Moreover, even though the room temperature is not exactly in the middle of the hysteresis loop, it falls within this loop. Its implementation in display devices still requires to ensure the stability of its performances, and thus to achieve the dehydration - rehydration process.

Spin Transition Molecular Alloys

In this section, we focus on another approach to obtain the ideal compound, based on the concept of spin-transition molecular alloys.

Let us be explicit what we mean by "molecular alloy". We know that the compounds with the stoechiometry $[Fe(Rtrz)_3]A_2$ or $[Fe(Rtrz)_2(trz)]A$ where A is a monovalent anion have a one-dimensional polymeric structure. Let us assume now that we can synthesize two compounds of this sort, of formula $[Fe(R_1trz)_3]A_2$ and $[Fe(R_2trz)_3]A_2$, respectively, R_1trz and R_2trz being two different 4-substituted-1,2,4-triazole ligands. We assume further that both compounds exhibit a spin transition, the former around the temperature T_1 and the latter around T_2. T_1 and T_2 may be defined as the middles of the thermal hysteresis loops. If we mix 50% of $[Fe(R_1trz)_3]A_2$ and $[Fe(R_2trz)_3]A_2$ in the solid state, we obviously end up with a material containing 50% of chains $[Fe(R_1trz)_3]_\infty$ and 50% of chains $[Fe(R_2trz)3]_\infty$. This material will display a two-step spin transition regime, one step occurring around T_1 and the other step occurring around T_2. On the other hand, if we react the $[Fe(H_2O)_6]A_2$ salt with a mixture of 50% of R_1trz and 50% of R_2trz dissolved in a common solvent, we end up with another material in which now all the chains have a mixed composition with 50% of each of the two ligands, as emphasized in Figure 10. Such a material is defined as a molecular alloy, and owing to the interactions between the Fe(II) ions along the chain, a mean effect may be expected. In such a case, rather abrupt transitions around $T = (T_1+T_2)/2$ might be observed. In other words, the molecular alloy of formula $[Fe(R_1trz)_{1.5}(R_2trz)_{1.5}]A_2$ could exhibit a spin transition regime that neither of the two pure compounds exhibits. More generally, the spin transition regime of the molecular alloys $[Fe(R_1trz)_{3-3x}(R_2trz)_{3x}]A_2$ might be fine-tuned through the x value specifying the chemical composition of the alloy.

We are engaged in a thorough investigation of this completely new phenomenon. Here, we restrict ourselves to one example, concerning the system $[Fe(Htrz)_{3-3x}(NH_2trz)_{3x}](ClO_4)_2$. The variations of the transition temperatures, $T_c\uparrow$ and $T_c\downarrow$, as a function of the composition of the alloy defined by x are shown in Figure 11. Both $T_c\uparrow$ and $T_c\downarrow$ decrease almost linearly with x according to :

$T_c\uparrow/K = 319.7 - 129.7x$
$T_c\downarrow/K = 303.0 - 116.4x$

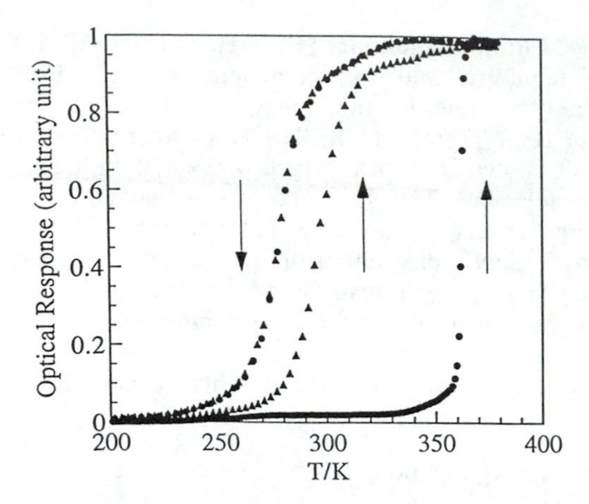

Figure 9 : Optical detection of the spin transition for the couple $[Fe(NH_2trz)_3](tos)_2 \cdot 2H_2O$ - $[Fe(NH_2trz)_3](tos)_2$; (●) first thermal cycle; (Δ) subsequent thermal cycle in the absence of water (reproduced with the permission from reference 22).

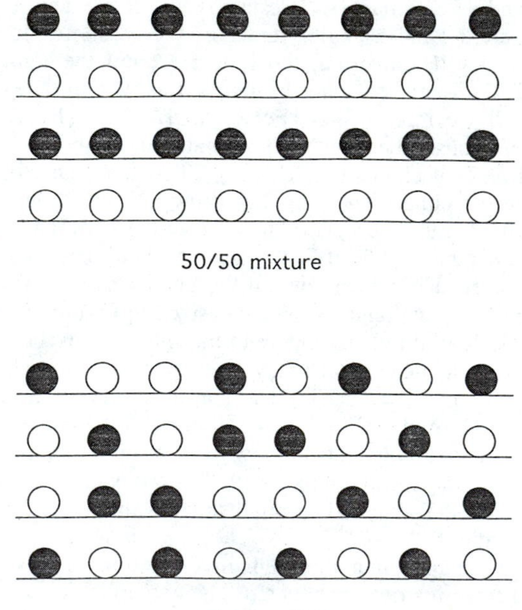

Figure 10 : Schematic representation of the structural difference between a 50/50 mixture and a molecular alloy with the same composition. The white and dark balls symbolize the R_1trz and R_2trz ligands, respectively (reproduced with the permission from reference 19).

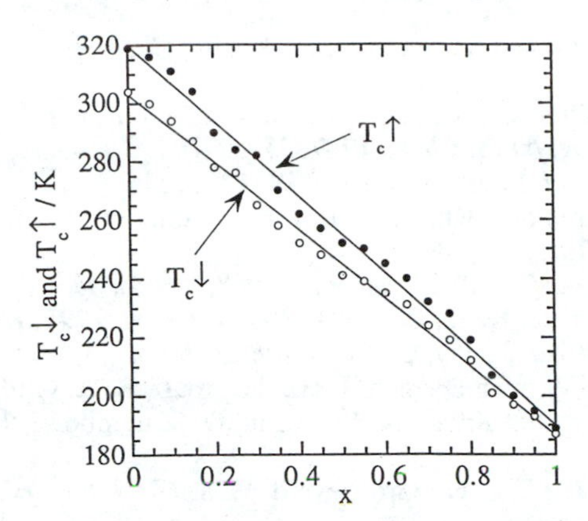

Figure 11 : T_c↑ and T_c↓ versus x curves for the spin transition molecular alloys $[Fe(Htrz)_{3-3x}(NH_2trz)_{3x}](ClO_4)_2$.

For x = 0.15, the transition occurs on either side of room temperature, with T_c↑ = 300.3 K and T_c↓ = 285.5 K. At 293 K, one may observe a purple compound in its S = 0 state or a white compound in a S = 2 state, depending on the history of the sample (*23*). Both states are stable at room temperature. A gentle cooling (for instance by dipping the tube containing the alloy in acetone and evaporating acetone) or a gentle warming of the sample is sufficient to induce the transition. Let us also mention that the width of the hysteresis loop is slightly increased if the compound contains non-coordinated water molecules.

Concluding Remarks

Spin transition compounds exhibiting abrupt transitions associated with a change of color are quite interesting candidates as information display and storage medium, and prototypes have already been described (*12*). Their ease of implementation and their compatibility with flexible substrates are the practical arguments which make them appealing in comparison with liquid crystal-based technology.

Fundamentally, the compounds have to fulfill some key requirements for being completely adapted to genuine display applications. Two of these requirements are particularly important and difficult to control, namely a working temperature (the center of the thermal hysteresis loop) close to room temperature, and a hysteresis width covering the temperature range of utilization. Ideally, a hysteresis width of at least 50 K seems to be necessary. Our work suggests that the ideal compound might not be far from being found, and that it is worth pursuing our efforts in this direction.

References

1. Kahn, O. *Molecular Magnetism*, VCH, New York, 1993.
2 Goodwin, H. A. *Coord. Chem. Rev.* **1976**, *18*, 293.
3. Gütlich, P. *Struct. Bonding (Berlin)* **1981**, *44*, 83.
4 Gütlich, P.; Hauser, A. *Coord. Chem. Rev.* **1990**, *97*, 1.

5 Gütlich, P.; Hauser, A.; Spiering, H. *Angew. Chem. Int. Ed. Engl.* **1994**, *33*, 2024.
6. König, E.; Ritter, G.; Kulshreshtha, S. K. *Chem. Rev.* **1985**, *85*, 219.
7 König, E. *Prog. Inorg. Chem.* **1987**, *35*, 527.
8 König, E. *Struct. Bonding (Berlin)* **1991**, *76*, 51.
9 Roux, C.; Zarembowitch, J.; Gallois, B.; Granier, T.; Claude, R. *Inorg. Chem.* **1994**, *33*, 2273.
10 Kahn, O.; Launay, J. P. *Chemtronics* **1988**, *3*, 140.
11 Zarembowitch, J.; Kahn, O. (1991) *New J. Chem.* **1991**, *15*, 181.
12 Kahn, O.; Kröber, J.; Jay, C. *Adv. Mater.* **1992**, *4*, 718.
13 Vreugdenhil, W.; van Diemen, J. H.; de Graaff, R. A. G.; Haasnoot, J. G.; Reedijk, J.; van der Kraan, A. M.; Kahn, O., Zarembowitch, J. *Polyhedron* **1990**, *9*, 2971.
14 Haasnoot, J. G.; Vos, G.; Groeneveld, W. L. Z. *Naturforsch.* **1977**, *B 32*, 421.
15 Lavrenova, L. G.; Ikorskii, V. N.; Varnek, V. A.; Oglezneva, I. M., Larionov S. V. *Koord. Khim.* **1986**, *12*, 207.
16 Lavrenova, L. G.; Ikorskii, V. N.; Varnek, V. A.; Oglezneva, I. M.; Larionov S. V. *Koord. Khim.* **1991**, *16*, 654.
17 Sugiyarto, K. H.; Goodwin, H. A. *Aust. J. Chem.* **1994**, *47*, 263.
18 Kröber, J.; Audière, J. P.; Claude, R.; Codjovi, E.; Kahn, O., Haasnoot, J. G.; Grolière, F.; Jay C.; Bousseksou, A.; Linarès, J.; Varret, F.; Gonthier-Vassal, A. *Chem. Mater.* **1994**, *6*, 1404.
19 Kahn, O.; Codjovi, E. *Phil. Trans. R. Soc. London A* **1996**, *354*, 359.
20 Michalowicz, A.; Moscovici, J.; Ducourant, B.; Cracco, D.; Kahn, O. *Chem. Mater.* **1995**, *7*, 1833.
21 Lavrenova, L. G.; Ikorskii, V. N.; Varnek, V. A.; Oglezneva, L. M.; Larionov, S. V. *Zhurnal Struk. Khim.* **1993**, *34*, 145.
22 Codjovi, E.; Sommier, L.; Kahn, O.; Jay, C. *New. J. Chem.* in press.
23 Kröber, J.; Codjovi, E.; Kahn, O.; Groslière, F.; Jay, F. *J. Am. Chem. Soc.* **1993**, *115*, 9810.

Chapter 21

Improved Synthesis of the V(tetracyanoethylene)$_x$•y(solvent) Room-Temperature Magnet: Doubling of the Magnetization at Room Temperature

Jie Zhang[1], Joel S. Miller[1,4], Carlos Vazquez[2], Ping Zhou[3], William B. Brinckerhoff[3], and Arthur J. Epstein[3]

[1]Department of Chemistry, University of Utah, Salt Lake City, UT 84112
[2]DuPont Experimental Station E–328, Wilmington, DE 19880–0328
[3]Department of Physics, Ohio State University, 174 West 18th Street, Columbus, OH 43210–1106

The reaction of $Na[V^{-I}(CO)_6]\cdot 2diglyme$, $V^{I}(C_5H_5)(CO)_4$, $[Et_4N][V^{-I}(CO)_6]$, $V^{o}(C_5H_5)(C_7H_7)$, $V^{o}(CO)_6$, $[V^{I}(C_6H_6)_2][PF_6]$, $[V^{I}\text{-}(C_6H_6)_2][V^{-I}(CO)_6]$, $V^{II}(C_5H_5)_2$, $[V^{II}(NCMe)_6][BPh_4]_2$, $[V^{II}\text{-}(NCMe)_6]\text{-}[V^{-I}(CO)_6]_2$, and $[V^{II}(THF)_6][V^{I}(CO)_6]_2$, with tetracyanoethylene, TCNE, at room-temperature in CH_2Cl_2 to form a room-temperature magnet was studied. Reaction with $V^{o}(CO)_6$, $[V^{I}(C_6H_6)_2][V^{-I}(CO)_6]$, $[V^{II}(NCMe)_6][V^{-I}(CO)_6]_2$, and $[V^{II}\text{-}(THF)_6][V^{-I}(CO)_6]_2$ led to the formation of room-temperature magnets and the reaction with $V^{o}(CO)_6$ was studied in detail. All six CO's were lost upon the reaction with TCNE and the material was nominally compositonally similar to that prepared from $V(C_6H_6)_2$ and TCNE except that the route is more facile. In contrast, the magnetization of the magnet prepared from $V^{o}(CO)_6$ is enhanced by more than a factor of two at room-temperature with respect to the magnet prepared from $V(C_6H_6)_2$.

The quest for molecule-based magnets possessing spin-bearing organic species is a subject of increasing research activity worldwide.[1,2] $V(TCNE)_x\cdot y$(solvent) prepared via the reaction of

[4]Corresponding author

0097–6156/96/0644–0311$15.00/0

$V^{o}(C_6H_6)_2$ and TCNE at room temperature, has been characterized to be a magnet below a critical temperature estimated to be ~400 K.[1a,3] It is a known disordered material, hence an improved preparative route avoiding the difficult-to-obtain $V(C_6H_6)_2$ was sought.

Experimental Section

All reactions and electrochemistry were carried out under an inert atmosphere in an Vacuum Atrmosphere's DriLab. The electrochemistry was carried out under nitrogen at room temperature in dry CH_2Cl_2 with 0.1 *M* $[Bu_4N][ClO_4]$ supporting electrolyte and Pt working and counter electrodes vs. a Ag/AgCl reference electrode and are reported vs. SCE. $V(C_6H_6)_2$ is oxidatively unstable and its reversible cyclic voltamagram was obtained at -55 °C in glyme and a scan rate of 100 Vs^{-1} employing the same electrodes and electrolyte. For $V(C_6H_6)_2$ at -55 °C a second reversible one-electron oxidation was observed at -2.67 V. IR mulls were prepared in the Vacuum Atrmosphere's DriLab by placing Nujol or fluorolube mulls between NaCl plates. The IR spectra were obtained on Perkin Elmer 783 IR spectrometer. The magnetic susceptibility was determined on a either a Faraday[4] or Quantum Design MPMS SQUID magnetometers. Powder diffraction was obtained on a Rigaku Miniflexpowder diffractometer using a home-built air-tight sample holder with a mylar window and loaded under nitrogen in the DriLab.

$Na[V^{-I}(CO)_6]\cdot 2$diglyme,[5b] and $V^{I}(C_5H_5)(CO)_4$,[5f] were used as purchased from Strem Chemicals while $[Et_4N][V^{-I}(CO)_6]$,[5a] $V^{o}(C_5H_5)(C_7H_7)$,[5c] $V^{o}(CO)_6$,[5a] $[V^{I}(C_6H_6)_2][PF_6]$,[5d] $[V^{I}(C_6H_6)_2][V^{-I}(CO)_6]$[5e] $V^{II}(C_5H_5)_2$,[5g] $[V^{II}(NCMe)_6][BPh_4]_2$,[5h] $[V^{II}(NCMe)_6][V^{-I}(CO)_6]_2$,[5i] and $[V^{II}(THF)_6][V^{I}(CO)_6]_2$,[5j] were synthesized according to the literature methods.

In a typical preparation of $V(TCNE)_x\cdot yCH_2Cl_2$, TCNE (25 mg; 0.2 mmol) dissolved in CH_2Cl_2 (5 mL) was added to a stirred 5 mL solution of $V(CO)_6$ (22 mg; 0.1 mmol) dissolved in CH_2Cl_2. Upon mixing bubbles of CO were vigorously liberated and a dark precipitate immediately formed. After 30 min of additional stirring at ambient temperature, the precipitate was filtered with a medium frit, washed with CH_2Cl_2 (2 mL x 3), and dried under vacuum (yield 21 mg) . Reactions of other vanadium compounds with TCNE were carried out as described above.

Results and Discussion

The reactions of TCNE with $[Et_4N][V^{-I}(CO)_6]$,[5a] $Na[V^{-I}(CO)_6]\cdot 2diglyme$,[5b] $V^o(C_5H_5)(C_7H_7)$,[5c] $V^o(CO)_6$, $[V^I(C_6H_6)_2][PF_6]$,[5d] $[V^I(C_6H_6)_2][V^{-I}(CO)_6]$[5e] $V^I(C_5H_5)(CO)_4$,[5f] $V^{II}(C_5H_5)_2$,[5g] $[V^{II}(NCMe)_6][BPh_4]_2$,[5h] $[V^{II}(NCMe)_6][V^{-I}(CO)_6]_2$,[5i] and $[V^{II}(THF)_6]$-$[V^I(CO)_6]_2$,[5j] were studied with the aim of preparing a room-temperature magnet. In all cases the reaction of the V-source and TCNE in a solvent (CH_2Cl_2, MeCN, or THF) lead to the isolation of an amorphous, insoluble black solid. Unlike $V^o(C_6H_6)_2$, the reaction of isoelectronic $V^o(C_5H_5)(C_7H_7)$ with TCNE did not afford a room-temperature magnet. This is consistent with the greater oxidation potential of $V^o(C_5H_5)(C_7H_7)$ (0.34 V vs. SCE) vs. $V^o(C_6H_6)_2$ (-0.28 V). Similarly, the reactions of $[TCNE]^{\cdot -}$ and $[V^I(C_6H_6)_2]^+$ or TCNE and $V^{II}(C_5H_5)_2$ did not afford a room-temperature magnet. Since the reaction of $V^o(C_6H_6)_2$ and TCNE leads to a strongly magnetic material and, but the reaction of $[V^I(C_6H_6)_2]^+$ and $[TCNE]^{\cdot -}$ does not, the mechanism of the reaction is crucial in the formation of the room-temperature magnet. In contrast, the room temperature reactions of $V^o(CO)_6$, $[V^I(C_6H_6)_2][V^{-I}(CO)_6]$, $[V^{II}(NCMe)_6][V^{-I}(CO)_6]_2$, $[V^{II}(NCMe)_6]^{2+}$, and $[V^{II}(THF)_6][V^{-I}(CO)_6]_2$ with TCNE lead to room-temperature magnets; however, the reactions of the more stable $[Et_4N][V^{-I}(CO)_6]$ and $Na[V^{-I}(CO)_6]\cdot 2diglyme$ did not. Therefore, the reaction of $V^o(CO)_6$ with TCNE was selected for further study.

The elemental composition of the new magnets, like those reported earlier,[3] could not determined without ambiguity due to their extreme sensitivity toward both oxygen and moisture as all analyses reveal some oxygen. However, the data suggest that the magnet prepared from both $V(C_6H_6)_2$ and $V(CO)_6$ have similar compositions. Since the magnets were prepared and handled under an oxygen- and moisture-free environment, the possibility of the oxygen arising from the CO was addressed. The similarity of the IR spectra with respect to the magnet prepared from $V(C_6H_6)_2$ and the absence of $\upsilon_{C\equiv O}$ stretches in the 1800 to 2000 cm^{-1} region, Figure 1, strongly suggest that all the carbonyls are expelled from the vanadium coordination sphere upon the reaction of $V(CO)_6$ with TCNE.[7] Nonetheless, since $\upsilon_{C\equiv N}$ absorptions (2000 to 2100 cm^{-1}) are very broad, these bands could obscure overlapping C≡O absorptions. Hence, the CO-loss was quantified via Toepler pump experiments[8] for:

$$V(CO)_6 + y\,TCNE \rightarrow V(TCNE)_y(CO)_{6-x} + x\,CO$$

and

$$[NEt_4][V(CO)_6] + y\,TCNE \rightarrow V(TCNE)_y(CO)_{6-x} + x\,CO.$$

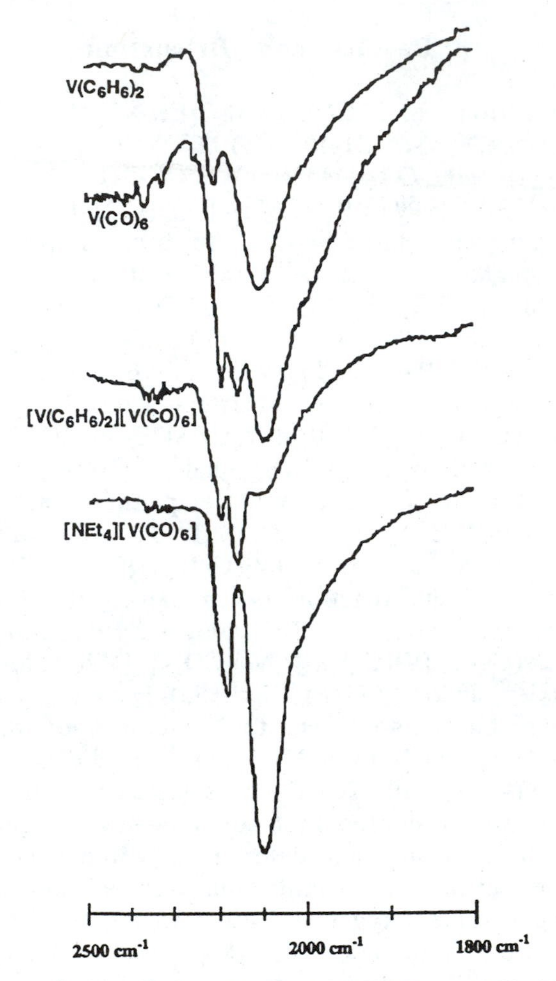

Figure 1 υ_{CN} regions of the IR for $V(TCNE)_x{\cdot}y(CH_2Cl_2)$ prepared from $V(C_6H_6)_2$, $V(CO)_6$, $[V(C_6H_6)_2][V(CO)_6]$, and $[NEt_4][V(CO)_6]$.

For both reactions, x was determined to be 5.9 ± 0.1; thus, within experimental error, all the six carbonyls are substituted.

The vanadium-based magnets are disordered cross-linked polymers; hence the structural characterization has been difficult to elucidate. At present the key method to elucidate aspects of the structure is the IR spectra in the 1800 - 2200 cm^{-1} region where the magnets have strong, multiple, broad and sometimes overlapping bands assigned to υ_{CN} absorptions. Analysis of the IR spectra may reveal information about CN-V binding modes as well as oxidation states from the position, number, and relative intensities of the υ_{CN} stretches. For example, the differences between the υ_{CN} absorptions of the magnet prepared from $V(CO)_6$ and that prepared from $V(C_6H_6)_2$ (Figure 1) clearly indicate that the two magnets are structurally inequivalent. This is in agreement with the fact that the magnetic behavior of the two materials are not identical, even though they both have T_c above room temperature. The information available from the IR spectra, however, is very limited and inadequate to characterize fully this class of materials. First, crucial structural information regarding the binding mode(s) associated with specific IR absorptions has yet to be unambiguously established. Second, preliminary experiments show that the magnetic properties of the magnets are not directly correlated to the positions of the CN absorptions. For example, materials from the reaction of TCNE and both $V(C_6H_6)_2$ and $[Et_4N][V(CO)_6]$ have essentially identical υ_{CN} absorptions, Figure 1, but only the former is a magnet at room temperature. Furthermore, although the magnetic phases may have certain structural features resulting in common IR absorptions, it is not necessary that all the magnetic phases have identical

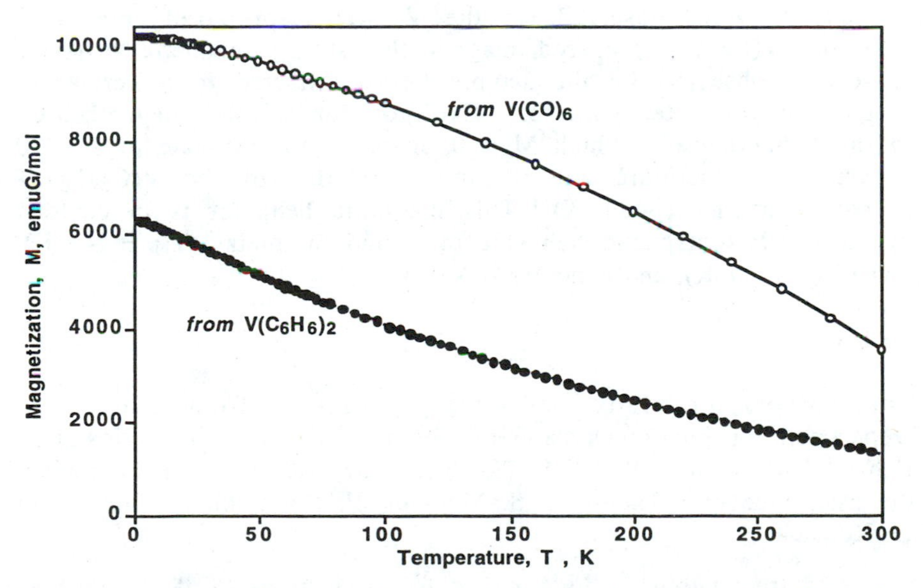

Figure 2 Magnetization, M, as a function of temperature, T, at 1 kG for $V(TCNE)_x \cdot y(CH_2Cl_2)$ prepared from $V(CO)_6$ (o) and $V(C_6H_6)_2$ (•).

υ_{CN} absorptions, nor it is true that materials having similar IR absorptions have similar magnetic behavior. This discrepancy is the consequence of the IR absorptions being primarily sensitive to the local chemical bonding environment, while magnetism is a bulk cooperative property that involves long range structure in addition to local bonding.

The dc SQUID magnetic susceptibility and magnetization, M, of $V(TCNE)_x \cdot y(CH_2Cl_2)$ prepared from $V(CO)_6$ was measured between 1.4 and 300 K (Figure 2). The magnetization for this sample was 10,300 emuG/mol at 4.2 K and 3600 emuG/mol at room temperature; representing increases of 67% and 233% at these temperatures, respectively.

Whereas M decreases monotonically with increasing T for the $V(C_6H_6)_2$-prepared magnet, this is less evident at low temperature for the $V(CO)_6$-prepared magnet. The unusual linear decrease of M with increasing T is consistent with extensive disorder in the sample, and suggests that the $V(CO)_6$-prepared magnet is less disordered. Hysteresis with a coercive field of 15 Oe is observed at room temperature and at 4.2 K for the $V(CO)_6$-prepared magnet. Minor loops (maximum applied field of 100 Oe) have a coercive field of less than 1 Oe in the vicinity of room temperature, increasing to several Oe at 4.2 Oe. This is significantly lower than the 60 Oe value observed for the $V(C_6H_6)_2$-prepared magnet.[3] Like the $V(C_6H_6)_2$-prepared magnet the strong magnetic behavior is readily observed by the sample being attracted to a permanent magnet at room temperature. Extrapolation of the magnetization to the temperature at which M = 0, leads to the estimate of a ~370 K critical temperature and is similar to that of the $V(C_6H_6)_2$-prepared magnet (~400 K).[3] This magnetic behavior is in contrast to the antiferromagnetic behavior of vanadium metal (T_N = 245 K), VO (T_N = 117 K), and nonordered V_2O_3.[9]

Acknowledgment

The authors gratefully acknowledge support from the U. S. Department of Energy Division of Materials Science Grant Nos. DE-FG03-93ER45504 and DE-FG02-86ER45271.A000. We appreciate the assistance provided by R. S. McLean, E. Delawski, X. Wang, and G. Kodama.

⊗ Abstract published in *Advance ACS Abstracts*, XXX X, 1995.

Literature Cited

1. Recent reviews: (a) Miller, J. S.; Epstein, A. J., *Angew. Chem.*, **1994**, *106*, 399; *Angew. Chem. internat. edit.*, **1994**, *33*, 385. (b) Buchachenko, A. L., *Russ. Chem. Rev.*, **1990**, *59*, 307; *Usp. Khim.* **1990**, *59*, 529. (c) Gatteschi, D., *Adv. Mat.*, **1994**, *6*, 635. (d) Kahn, O., *Molecular Magnetism*, VCH Publishers, Inc. 1993.
2. Conference Proceedings in the conference on *Ferromagnetic and High Spin Molecular Based Materials* (Eds.: Miller, J. S.; Dougherty, D. A.), *Mol. Cryst., Liq. Cryst.* **1989**, *176*. Conference Proceedings in the conference on *Molecular Magnetic Materials: NATO ARW Molecular Magnetic Materials* (Eds.: Kahn, O.; Gatteschi, D.; Miller, J. S.; Palacio, F.), Kluwer Acad. Pub., London, **1991**, E198.Conference Proceedings in the conference on *Chemistry and Physics of Molecular Based Magnetic Materials* (Eds.; Iwamura, H.; Miller, J. S.), *Mol. Cryst., Liq. Cryst.* **1993**, *232/233*. Conference Proceedings in the conference on *Molecule-based Magnets* (Eds.: Miller, J. S. and Epstein, A. J.), *Mol. Cryst., Liq. Cryst.* **1995**, *271-274*.
3. Manriquez, J. M.; Yee, G. T.; McLean, R. S.; Epstein, A. J.; Miller, J. S., *Science* **1991**, *252*, 1415. Miller, J. S.; Yee, G. T.; Manriquez, J. M.; Epstein, A. J., in the Proceedings of Nobel Symposium #NS-81 *Conjugated Polymers and Related Materials: The Interconnection of Chemical and Electronic Structure*, Oxford University Press, **1993**, 461. *La Chim. & La Ind.* **1992**, *74*, 845. Epstein, A. J.; Miller, J. S., in the Proceedings of Nobel Symposium #NS-81 *Conjugated Polymers and Related Materials: The Interconnection of Chemical and Electronic Structure*, Oxford University Press, **1993**, 475. *La Chim. & La Ind.* **1993**, *75*, 185.
4. Miller, J. S.; Dixon, D. A.; Calabrese, J. C.; Vazquez, C.; Krusic, P. J.; Ward, M. D.; Wasserman, E.; Harlow, R. L. *J. Am. Chem. Soc.* **1990**, *112*, 381.
5. (a) Ellis, J. E.; Faltynek, R. A.; Rochfort, L.; Stevens, R. E.; Zank, G. A., *Inorg. Chem.* **1980**, *19*, 1082. (b) Werner, R. P. M.; Podall, H. E., *Chem & Ind. (London)* **1961**, 144. (c) King, R. B.; Stone, F. G. A., *J. Am. Chem. Soc.* **1959**, *81*, 5363. (d). Palm, C.; Fischer, E. O.; Reckziegel, A., *Tet. Lett.* **1962**, 253. (e) Calderazzo, F. *Inorg. Chem.* **1963**, *3*, 810. (f) King, R. B.; Stone, F. G. A., *Inorg. Syn.* **1963**, *7*, 99 (g) Fischer, E. O.; Vigoureux, S., *Chem. Ber.* **1958**, 91, 2205. (h) Anderson, S. J.; Wells, F. J.; Wilkinson, G.; Hussain, H.; Hursthouse, M. B., *Polyhed.* **1988**, *7*, 2615. (i) W. Hieber, W.; Peterhans, J.; Winter, E., *Chem. Ber.* **1961**, *94*, 2572. (j) Hieber, W.; Winter, E.; Schubert, E., *Chem. Ber.* **1962**, *95*, 3070.

6. (a) Calderazzo, F., *Inorg. Chem.* **1964**, *3*, 1207. (b) Gambarotta, S.; Floriani, C.; Chiesi-Villa, A.; Guastini, C., *Inorg. Chem.* **1984**, *23*, 1739.
7. Selected IR data (Fluorolube): $Na[V^{-I}(CO)_6]\cdot$2diglyme: 1840 vs br cm^{-1}; $V^{o}(CO)_6$: 1973vs cm^{-1}; $[V^{I}(C_6H_6)_2]^+[V^{-I}(CO)_6]^-$: 1844vs br cm^{-1}; $[Et_4N][V^{-I}(CO)_6]$: 1845 vs br cm^{-1}; $V^{I}(CO)_4Cp$ (Nujol): 2070w, 2017vw, 1988m cm^{-1};[6a] $Cp^*{}_2V^{II}(CO)$ (Nujol): 1845 cm^{-1}; $Cp^*{}_2V^{III}(CO)(CN)$ (Nujol): 1910 cm^{-1} (υCN: 2070 cm^{-1});[6b] $[V^{II}(NCMe)_6][V^{-I}(CO)_6]_2$: 1856vs br cm^{-1} (υCN: 2321m and 2291m cm^{-1}); $[V^{II}(THF)_6][V^{-I}(CO)_6]_2$: 1848vs br cm^{-1}. In contrast, if difficult-to-reduce nitriles are reacted with $V(CO)_6$, prominent CO stretches are observed at around 1850 cm^{-1}, consistent with incomplete loss of CO.
8. Shriver, D. F.; Drezdzon, M. A., *The Manipulation of Air-Sensitive Compounds*, John Wiley & Sons, New York, 1986. pp 113-115.
9. Goodenough, J. B., *Magnetism and the Chemical Bond*, Wiley-Interscience: New York, 1963. p98, 100, and 104.

Chapter 22

Design and Magnetism of New Bimetallic Assemblies with Fe(III)–CN–Ni(II) Linkages

Hisashi Okawa and Masaaki Ohba

Department of Chemistry, Faculty of Science, Kyushu University, Hakozaki 6–10–1, Higashi-ku, Fukuoka 812, Japan

Three types of bimetallic FeNi assemblies comprised of $[Fe(CN)_6]^{3-}$ and $[Ni(diamine)_2]^{2+}$ (diamine = en, pn) are described: (1) $PPh_4[Ni(pn)_2][Fe(CN)_6]\cdot H_2O$, (2) $[Ni(en)_2]_3[Fe(CN)_6]_2\cdot 2H_2O$ and (3) $[Ni(pn)_2]_2[Fe(CN)_6]X\cdot 2H_2O$ (X = ClO_4^-, BF_4^-, PF_6^-). $PPh_4[Ni(pn)_2][Fe(CN)_6]$ forms a 1-D chain of the alternate array of *trans*-$[Ni(pn)_2]^{2+}$ and $[Fe(CN)_6]^{3-}$. $[Ni(en)_2]_3[Fe(CN)_6]_2$ has a rope-ladder structure where two 1-D chains, formed by the alternate array of *cis*-$[Ni(en)_2]^{2+}$ and $[Fe(CN)_6]^{3-}$, are combined by *trans*-$[Ni(en)_2]^{2+}$. $[Ni(pn)_2]_2[Fe(CN)_6]ClO_4\cdot 2H_2O$ have a 2-D sheet of a square structure. Type 1 and type 2 of 1-D structures show no magnetic ordering over balk whereas type 3 of a 2-D structure shows a long-range magnetic ordering.

The design of highly ordered molecular systems using paramagnetic constituents is a current subject with the aim to obtain magnetic materials exhibiting spontaneous magnetization (*1-2*). A family of bimetallic assemblies comprised of hexacyanometallate $[M(CN)_6]^{n-}$ and a simple metal ion have been extensively studied (*3-6*) and significantly high magnetic phase-transition temperatures are reported for some of them. Those bimetallic assemblies are presumed to have a face-centered cubic structure extended by M-CN-M' linkage, but structurally characterized assemblies are very few. To understand magnetostructural correlations of those high T_C ferromagnets, it is of great value to study analogous bimetallic assemblies that contain a metal complex instead of the simple metal ion. In this study three types of bimetallic assemblies comprised of $[Fe(CN)_6]^{3-}$ and $[Ni(diamine)_2]^{2+}$ are described: $PPh_4[Ni(pn)_2][Fe(CN)_6]\cdot H_2O$, $[Ni(en)_2]_3[Fe(CN)_6]_2\cdot 2H_2O$ and $[Ni(pn)_2]_2[Fe(CN)_6]$-$X\cdot 2H_2O$ (X=ClO_4^-, BF_4^-, PF_6^-). X-ray crystallography for those bimetallic assemblies has proved diverse bridging modes of $[Fe(CN)_6]^{3-}$ to give different network structures. Magnetic properties of the assemblies are discussed together with their network structures.

$PPh_4[Ni(pn)_2][Fe(CN)_6]\cdot H_2O$

In our preliminary study this assembly was obtained as microcrystals by the 1 : 1 reaction of $[Ni(pn)_2]Cl_2$ and $K_3[Fe(CN)_6]$ in aqueous solution in the presence of

0097–6156/96/0644–0319$15.00/0

PPh_4Cl, but eventually we found that the use of $[Ni(pn)_3]Cl_2$ gave the same compound as large crystals. In the latter reaction, the slow dissociation of $[Ni(pn)_3]^{2+}$ into $[Ni(pn)_2]^{2+}$ in aqueous solution may lead to the growing of large crystals of the assembly. The assembly shows two ν(CN) modes at 2130 and 2110 cm^{-1}. In comparison in IR with $K_3[Fe(CN)_6]$ the weaker band at 2130 cm^{-1} is attributed to bridging CN group.

Crystal structure. Crystal data: Formula = $C_{36}H_{42}N_{10}OPFeNi$, formula weight = 388.16, crystal system = monoclinic, space group = P2/c(#13), a = 12.958(3), b = 8.437(3), c = 17.250(2) Å, β = 99.96(1)°, V = 1857.4(7) $Å^3$, Z = 2, Dc = 1.388 g cm^{-3}, μ(Mo-Kα) = 9.84 cm^{-1}, R = 0.081, R_w = 0.076.

Figure 1a shows the ORTEP view of the asymmetric unit that consists of each one-half of $[Fe(CN)_6]^{3-}$ anion, *trans*-$[Ni(pn)_2]^{2+}$ cation, tetraphenylphosphate ion, and water molecule. The Ni resides at the inversion center (0, 0, 0) and Fe (0, 0.3262, 0.25) and P(1) (0.5, 0.0691, 0.25) ions reside on the same mirror plane. The $[Fe(CN)_6]^{3-}$ coordinates to two adjacent *trans*-$[Ni(pn)_2]^{2+}$ molecules with its two CN nitrogens in cis, providing an 1-D zigzag chain with the alternate array of $[Fe(CN)_6]^{3-}$ and *trans*-$[Ni(pn)_2]^{2+}$ ions, running along c axis (Figure 1b). The Fe ion retains the octahedral surrounding with the Fe-C bond lengths ranging from 1.995(10) to 1.976(8) Å. The Ni-N(1) bond distance is 2.146(6) Å and the C(1)-N(1)-Ni angle is 158.3(8)°. The Ni has a trans six-coordinate geometry with the Ni-N bonds of 2.067(8) - 2.146(6) Å.

An projection of the chains on a plane is shown in Figure 1b. The nearest interchain Fe--Fe(x, y+1, z), Ni--Ni(x, y+1, z) and Fe--Ni(-x, -y+1, -z) separations along b axis are 8.437, 8.437 and 7.135 Å, respectively. The corresponding interchain separations along a axis are 12.958, 12.958 and 13.220 Å, respectively. The PPh_4^+ cation and lattice water reside among the chains.

Magnetism. Cryomagnetic property of this assembly, determined on powdered sample, is shown in Figure 2 in the form of $\chi_M T$ *vs*. T plot. The $\chi_M T$ value at room temperature is 1.94 cm^3 K mol^{-1} (3.94 μ_B) per FeNi that is larger than the value expected for non-coupled Fe(III)(S=1/2)-Ni(II)(S=1). The $\chi_M T$ gradually decreases with decreasing temperature to the round minimum of 1.75 cm^3 K mol^{-1} (3.74 μ_B) at 50 K and then increases up to the maximum of 2.34 cm^3 K mol^{-1} (4.33 μ_B) at 4.3 K. The decrease in $\chi_M T$ from 290 to 50 K can be attributed to the temperature-dependence of the $^2T_{2g}$ term of the low-spin Fe(III) under pseudo Oh symmetry. The magnetic behavior in the temperature range below 50 K indicates a ferromagnetic interaction between the adjacent Fe(III) and Ni(II) ions through cyanide bridge. The ferromagnetic coupling between the metal ions can be rationalized in terms of 'strict orthogonality of magnetic orbitals' (t_{2g}^1(Fe) ⊥ e_g^2(Ni)) (*9-11*). The maximal $\chi_M T$ is slightly larger than the value for S_T=3/2 of ferromagnetically coupled Fe(III)(S=1/2)-Ni(II)(S=1). But no long-range magnetic ordering occurs in this assembly, as anticipated from the large interchain separation.

$[Ni(en)_2]_3[Fe(CN)_6]_2 \cdot 2H_2O$

This assembly was obtained as black crystals by the 3:2 reaction of $[Ni(en)_3]Cl_2$ and $K_3[Fe(CN)_6]$ in an aqueous solution (*7*). A similar reaction using *trans*-$[NiCl_2(en)_2]$ gave the same product but in a polycrystalline form. Three ν(CN) modes appear at 2150, 2130 and 2110 cm^{-1}, indicating a lowered symmetry about $[Fe(CN)_6]^{3-}$ in the lattice.

Crystal Structure. The asymmetric unit consists of each one $[Fe(CN)_6]^{3-}$ and *cis*-$[Ni(en)_2]^{2+}$, one half of *trans*-$[Ni(en)_2]^{2+}$ and one water molecule, with the inversion

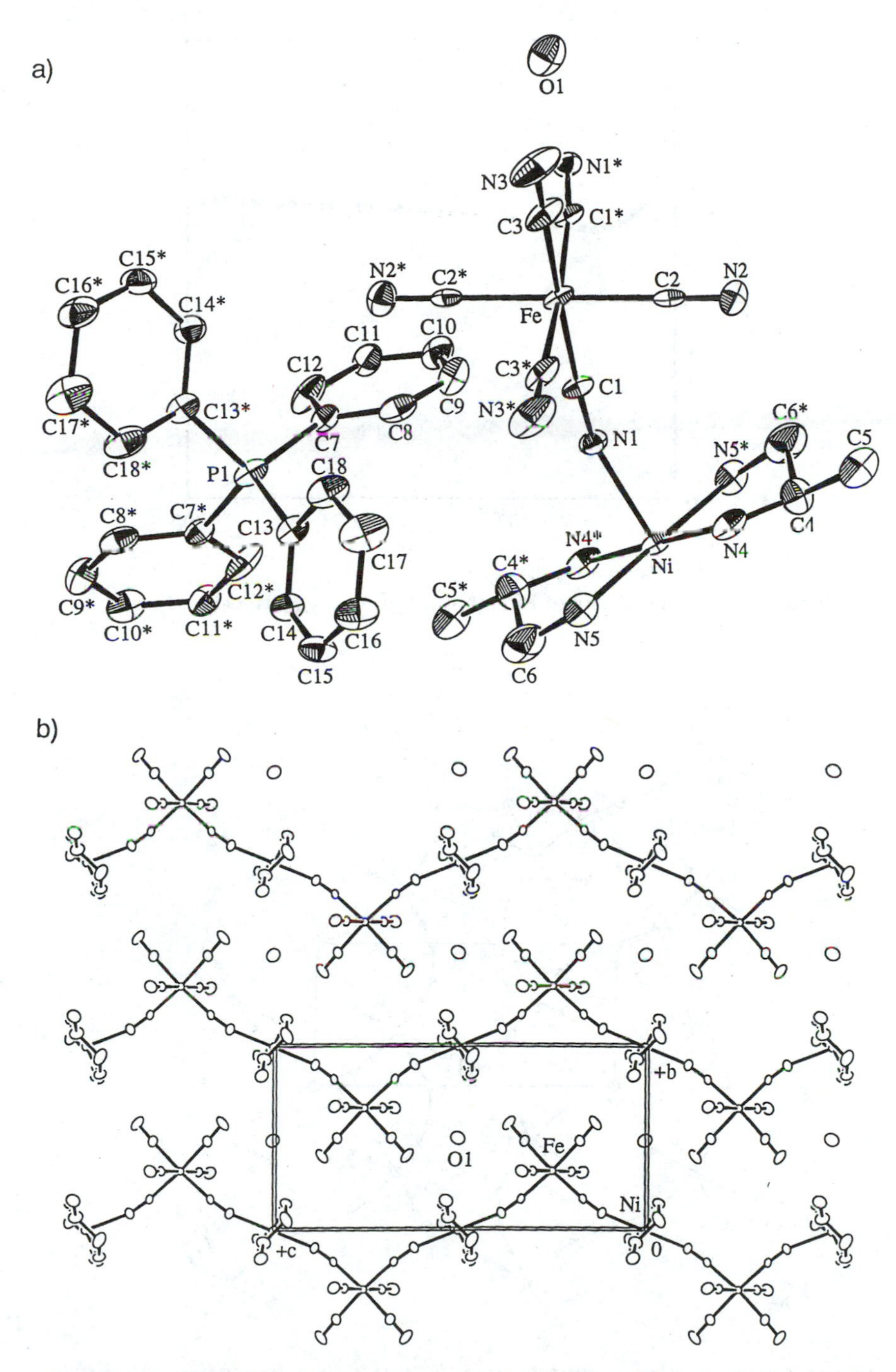

Figure 1. (a) ORTEP drawing of the asymmetric unit and (b) a projection of 1-D zigzag chains of $PPh_4[Ni(pn)_2][Fe(CN)_6]\cdot H_2O$ (PPh_4^+ cations are omitted for clarity).

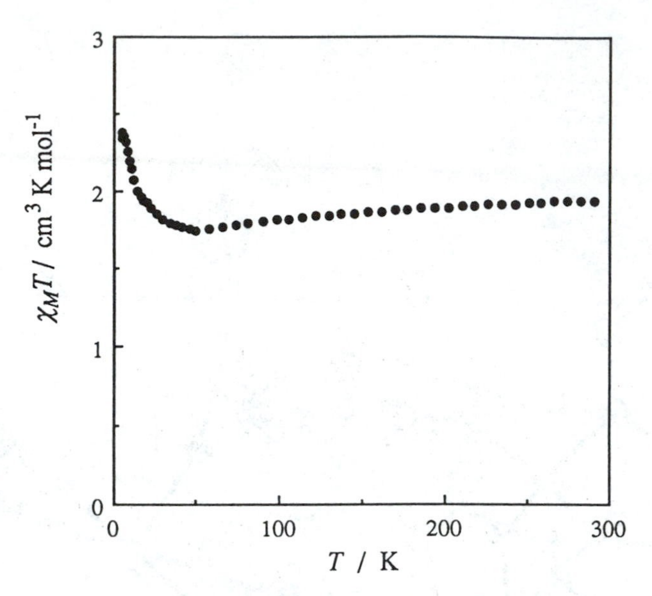

Figure 2. $\chi_M T$ *vs.* T plot of $PPh_4[Ni(pn)_2][Fe(CN)_6]\cdot H_2O$ (per FeNi).

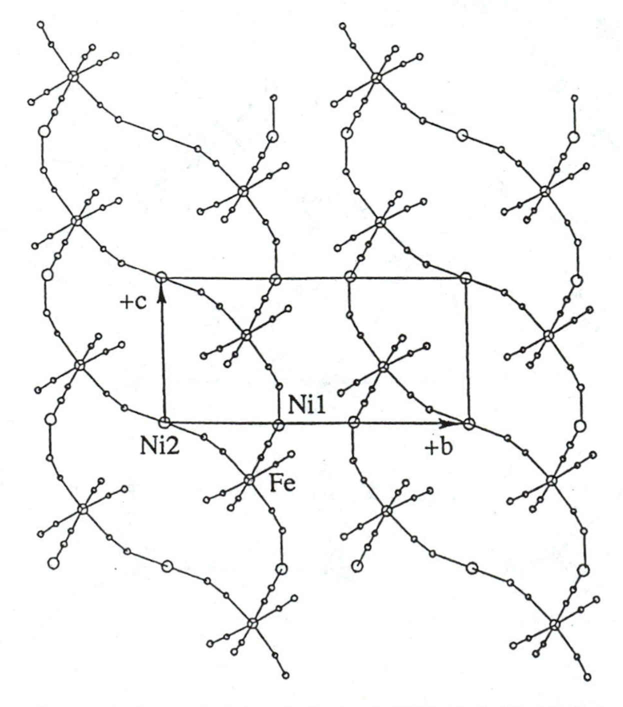

Figure 3. A projection of 1-D chains of $[Ni(en)_2]_3[Fe(CN)_6]_2\cdot 2H_2O$ (en molecules are omitted for clarity).

center at the Ni of *trans*-$[Ni(en)_2]^{2+}$ (*7*). A 1-D zigzag chain is formed by the alternate array of $[Fe(CN)_6]^{3-}$ and *cis*-$[Ni(en)_2]^{2+}$ ions and two zigzag chains are combined by *trans*-$[Ni(en)_2]^{2+}$, providing a novel rope-ladder structure running along *c* axis (Figure 3). The $[Fe(CN)_6]^{3-}$ coordinates to three Ni(II) ions with its CN nitrogens in the meridional mode. The Fe--Ni(1) and Fe--Ni(2) distances are 5.145(2) and 4.993(2) Å, respectively. In the crystal, the chains align along the diagonal line of the *ab* plane providing a pseudo 2-D sheet. The lattice water molecules are captured between the sheets. The nearest interchain Ni1--Ni1(-x-1, -y+1, -z) and Ni1--Fe(-x-1, -y+1, -z) separations in the sheet are 5.375 and 6.295 Å, respectively. The nearest intersheet Ni1--Fe(x-1, y, z), Ni1--Ni2(x-1, y, z), and Ni1--Ni1(x+1, y, z) separations are 6.494, 7.713, and 9.709 Å, respectively. A disorder is seen in *trans*-$[Ni(en)_2]^{2+}$ with respect to the en configuration in the lattice.

Magnetism. In our preliminary study magnetic measurements were made on polycrystalline sample obtained by the reaction with *trans*-$[NiCl_2(en)_2]$ (*7*). The $\chi_M T$ value at room temperature was 4.95 cm^3 K mol^{-1} (6.29 μ_B) that increased with decreasing temperature, very abruptly below 20 K, up to a large maximum value of 86.1 cm^3 K mol^{-1} at 14 K, and then decreased below this temperature (see Figure 4). The intrachain interaction between the adjacent Fe(III) and Ni(II) ions is evidently ferromagnetic. The abrupt increase in $\chi_M T$ around 20 K suggested the onset of long-range magnetic ordering and this was proved by our magnetization studies (T_C = 18.6 K) (*7*). In Figure 5 is given the field-dependence of the magnetization that shows an inflection at *ca.* 1 T and a tendency of saturation to M_S =7.6 Nμ_B at 5 T. The saturation magnetization well corresponds to 8 Nμ_B expected for ferromagnetically coupled $Ni^{II}_3Fe^{III}_2$. Based on the magnetization study this assembly is metamagnetic in nature.

It is noticed that the crystalline sample with the rope-ladder structure differs in magnetic behavior from the polycrystalline sample described above. In this context we have to retract the previous conclusion based on the assumption that both the crystalline and polycrystalline samples have the same 1-D rope-ladder structure (*7*). The $\chi_M T$ *vs.* *T* curve of the crystalline sample shows a maximum at 14 K (see Figure 4) and the maximal value (10.22 cm^3 K mol^{-1}) corresponds to the value for S_T=4 of ferromagnetically coupled $Ni^{II}_3Fe^{III}_2$. The drop in $\chi_M T$ below 14 K suggests the operation of antiferromagnetic intersheet interaction. Non-zero magnetic moment exists in this compound that can be attributed to the residual $\chi_\perp$ contribution (*12*). Evidently no long-range magnetic ordering occurs in the crystalline sample. It is likely that a disorder in the network structure occurs in the preparation of the polycrystalline sample by rapid precipitation. A candidate is the intercrossing of *trans*-$[Ni(en)_2]^{2+}$ between the rope-ladders, providing quasi 2-D (or 3-D) domains in the lattice.

$[Ni(pn)_2]_2[Fe(CN)_6]X\cdot 2H_2O$ ($X=ClO_4^-$, BF_4^-, PF_6^-)

The bimetallic assemblies were obtained as black crystals by the 1:2 reaction of $K_3[Fe(CN)_6]$ and $[Ni(pn)_3]X_2$ ($X=ClO_4^-$, BF_4^-, PF_6^-) in an aqueous solution (*8*). Analogous complexes with en ligand have been obtained which, however, are not included here because our spectroscopic studies have suggested a differing network structure for the en series. Two ν(CN) modes are observed for this type at 2140 and 2110 cm^{-1} of which the former band, attributable to the bridging CN group, is more intense in this case.

Crystal Structure of $[Ni(pn)_2]_2[Fe(CN)_6]ClO_4\cdot 2H_2O$. The asymmetric unit consists of each one-half of $[Fe(CN)_6]^{3-}$ and ClO_4^- anions, one $[Ni(pn)_2]^{2+}$ and water molecule (*8*). All the metal ions and the Cl atom are at the special equivalent

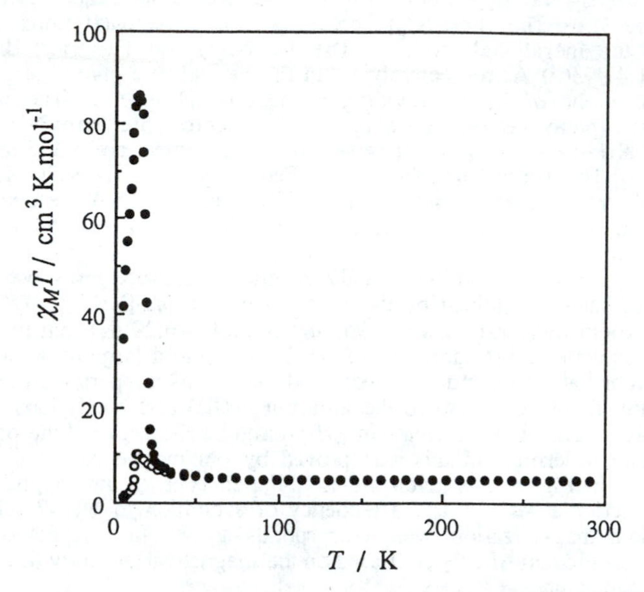

Figure 4. $\chi_M T$ *vs.* T plots (per Fe_2Ni_3) of $[Ni(en)_2]_3[Fe(CN)_6]_2 \cdot 2H_2O$: (●) polycrystalline sample and (○) crystalline sample.

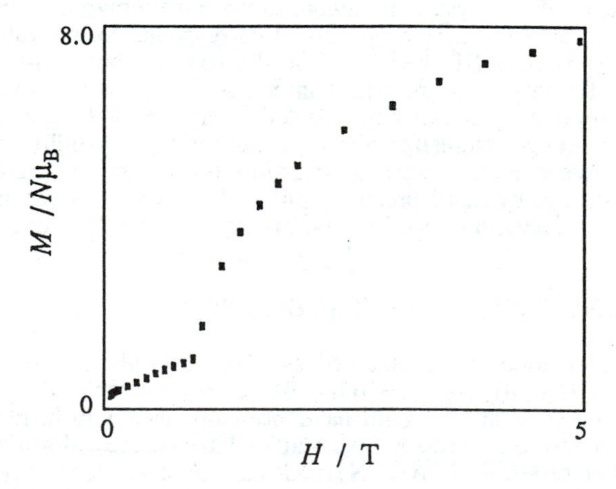

Figure 5. The field-dependence of magnetization of $[Ni(en)_2]_3[Fe(CN)_6]_2$-$\cdot 2H_2O$ obtained at 7 K.

positions: Fe (0, 1/2, 0), Ni(1) (1/2, 1/2, 0), Ni(2) (0, 0, 0), Cl (1/2, 0, 0). The $[Fe(CN)_6]^{3-}$ ion coordinates to four adjacent $[Ni(pn)_2]^{2+}$ cations through its four cyano nitrogens on a plane, providing a 2-D network of square-shaped structure (Figure 6). The mean Fe-C, C-N, and N-Ni bond distances are 1.949, 1.142, and 2.105 Å, respectively, and the C(1)-N(1)-Ni(1) and C(2)-N(2)-Ni(2) bridge angles are 157.2(6) and 164.1(7) Å, respectively. The Fe--Ni(1) and Fe--Ni(2) separations are 5.062(2) and 5.126(3) Å, respectively. It is seen that *trans*-$[Ni(pn)_2]^{2+}$ align along *a* and *b* axes, giving a slight distortion of the quadrangle formed by four Fe ions to a rhombus. Perchlorate ion resides in each quadrangle and assumes two arrangements. In the bulk the 2-D sheets align along *c* axis with the intersheet Fe--Fe(0, 1/2, 1) separation of 8.978(5) Å. The nearest intersheet Fe--Ni(1) (x, y, z+1), Fe--Ni(2) (x, y+1, z+1) and Ni(1)--Ni(2) (x+1, y+1, z+1) separations are 9.852, 8.613 and 9.754 Å, respectively.

Magnetism. The three salts of this type are similar in magnetic behavior. Magnetic property of the perchlorate salt is described. The $\chi_M T$ *vs.* T curve is shown in Figure 7 and the χ_M *vs.* T plots in the low temperature region is given in the insert. The $\chi_M T$ *vs.* T curve has a maximum of 11.31 cm^3 K mol^{-1} (9.51 μ_B) at 10 K that is very large for S_T=5/2 of ferromagnetically coupled $Ni^{II}_2Fe^{III}$ (4.38 cm^3 K mol^{-1} and 5.91 μ_B based on spin-only formalism). The χ_M *vs.* T plots show a round maximum at 10 K and a minimum at 6 K. The above results mean a ferromagnetic ordering occurs within the 2-D sheet and an antiferromagnetic interaction operates between the sheets. Near liquid helium temperature a non-zero magnetic susceptibility is observed which can be attributed to a residual $\chi_\perp$ contribution (*12*).

Magnetization studies have been made for this assembly (Figure 8). The FCM curve obtained under a weak applied field of 3 G shows a peak at 10 K and a break near 8 K. Below 8 K a remnant magnetization is observed. The ZFCM curve shows two maxima at 8.0 and 9.9 K. The magnetic behavior in the temperature range of 8.0-9.9 K is typical of metamagnetic ordering and the peak at 8.0 K implies that the applied field is too weak to move domain walls below this temperature. Magnetic hysteresis studies at 4.2 K support the metamagnetic nature of this assembly (see Figure 9). An abrupt increase in magnetization occurs around 3800 G, probably due to a spin flipping based on the intersheet antiferromagnetic interaction. The magnetization is not saturated at 20 T probably due to canted local spins. Such a canting of spin may arise from the magnetic anisotropy of the constituting metal ions. The remnant magnetization is 400 cm^3 G mol^{-1} and the coercive field is 290 G. The observed small remnant magnetization adds a support to the non-zero magnetism at zero field. Both the tetrafluoroborate and hexafluorophosphate salts showed similar metamagnetic ordering at 10.1 and 9.4 K, respectively. The remnant magnetization and the coercive field for the hexafluorophosphate salt are 260 cm^3 G mol^{-1} and 200 G, respectively. The field-dependence of the magnetization shows an inflection at *ca.* 4.5 kG and a gradual increase to M_S =3.9 Nμ_B at 80 kG. The maximum M_S value is smaller than 5 Nμ_B expected for ferromagnetically coupled $Ni^{II}_2Fe^{III}$, suggesting that each local spin is canted and a very large magnetic field is necessary for overcoming this canting (Figure 10).

Conclusion

From above discussion it is shown that $[Fe(CN)_6]^{3-}$ can adopt various bridging modes providing bimetallic assemblies of different network structures. The Ni/Fe ratio primarily relates to the dimensionality of the network structures. That is, type 1 of Ni/Fe=1 has a 1-D zigzag chain structure, type 2 of Ni/Fe=1.5 has a 1-D rope-ladder structure and type 3 of Ni/Fe=2 has a 2-D square sheet structure. All the three types show a ferromagnetic interaction operating within each chain or sheet through CN bridge. No magnetic ordering occurs in type 1 and type 2 of 1-D structure whereas a magnetic ordering occurs in type 3 of 2-D structure. The results demonstrate that

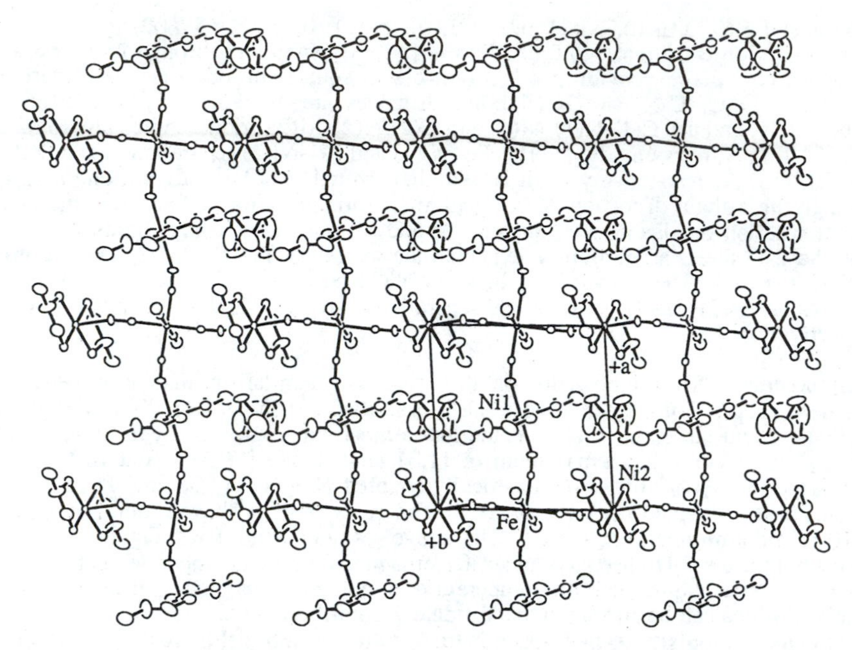

Figure 6. A projection of 2-D sheet of $[Ni(pn)_2]_2[Fe(CN)_6]ClO_4 \cdot 2H_2O$.

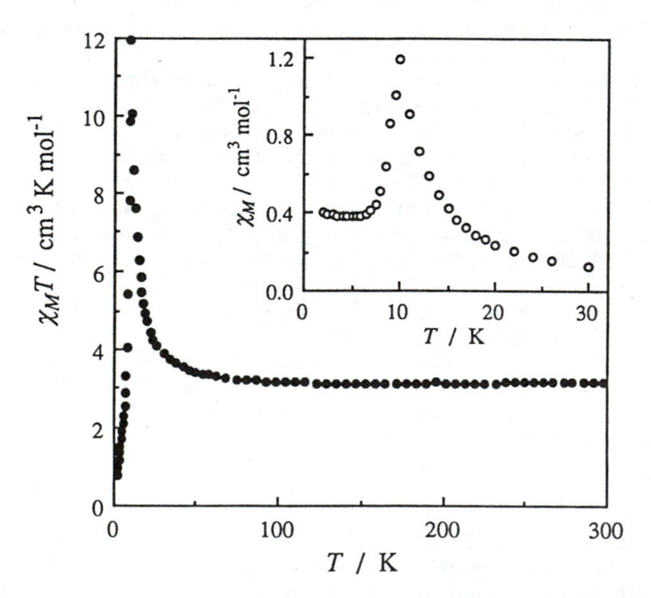

Figure 7. The $\chi_M T$ *vs.* T curve of $[Ni(pn)_2]_2[Fe(CN)_6]ClO_4 \cdot 2H_2O$. The insert is χ_M *vs.* T plot in the low temperature region.

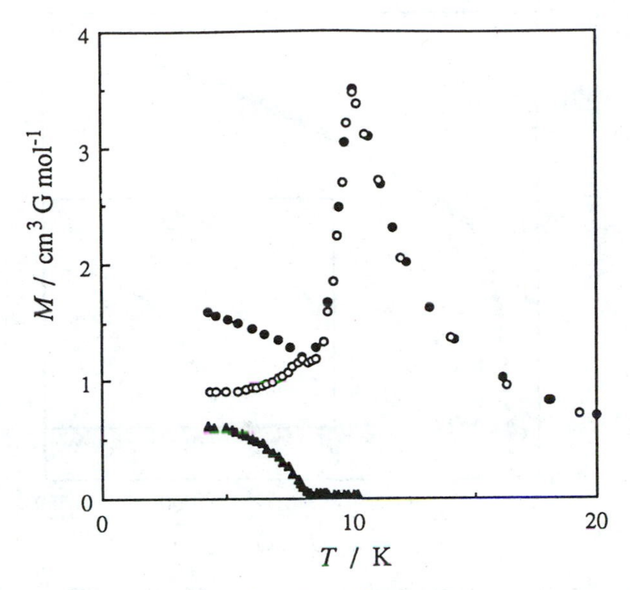

Figure 8. Temperature-dependences of magnetization (under 3 G) of $[Ni(pn)_2]_2[Fe(CN)_6]ClO_4\cdot 2H_2O$; (●) FCM, (▲) RM, (○) ZFCM.

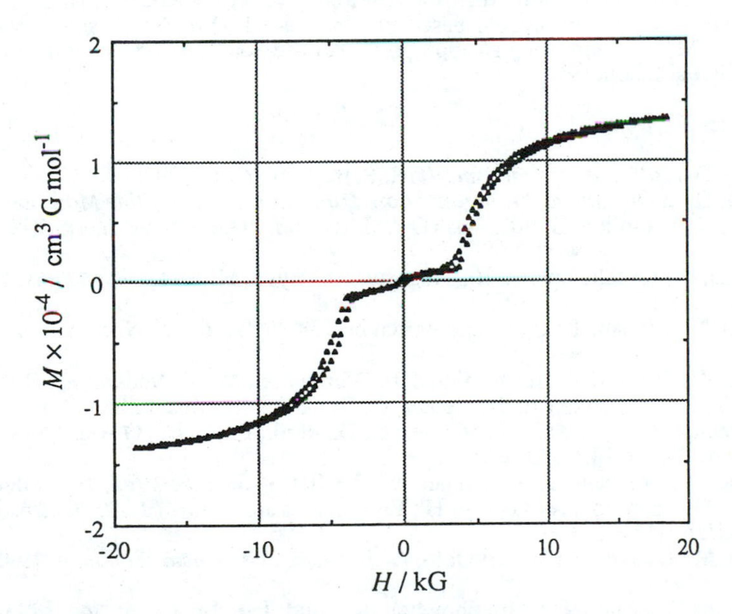

Figure 9. Magnetic hysteresis of $[Ni(pn)_2]_2[Fe(CN)_6]ClO_4\cdot 2H_2O$ obtained at 4.2 K.

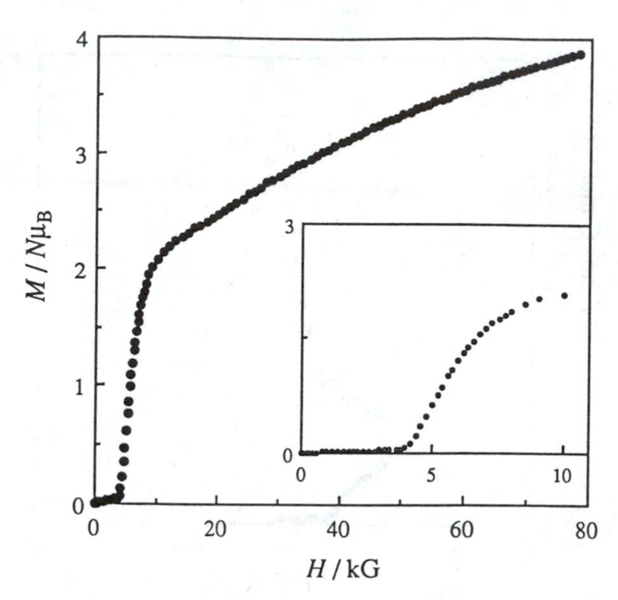

Figure 10. The field-dependence of magnetization of $[Ni(pn)_2]_2[Fe(CN)_6]$-$ClO_4 \cdot 2H_2O$ obtained at 4.2 K.

3-D bimetallic network is important for developing ferromagnetic materials and such a 3-D network can be constructed based on, in general, $[M(CN)_6]^{n-}$ and a planar M' complex in M'/M=3 and using an appropriate counter ion. Studies in this line are in progress in our laboratory.

References

1. Kahn, O. *Molecular Magnetism*, VCH. Pub.: New York, 1993.
2. Kahn, O. In *Organic and Inorganic Low-Dimensional Crystalline Materials*, Delhaes, P., Drillon, B, Eds.; NATO ASI Ser.168; Plenum: New York, 1987; pp. 93.
3. Klenze, H.; Kanellalopulos, B.; Tragester, G.; Eysel, H. H. *J. Chem. Phys.* **1980**, *72*, 5819.
4. Gadet, V.; Mallah, T.; Castro, I.; Verdaguer, M. *J. Am. Chem. Soc.* **1992**, *114*, 9213.
5. Gadet, V.; Bujoli-Doeuff, M.; Force, L.; Verdaguer, M.; El Malkhi, K.; Deroy, A.; Besse, J. P.; Chappert, C.; Veillet, P.; Renard, J. P.; Beauvillain, P. In *Molecular Magnetic Material*; Gatteschi, D., et al., Eds.; NATO ASI Series 198; Kluwer: Dordrecht, 1990; p 281.
6. Mallah, S.; Thiebaut, S.; Verdaguer, M.; Veillet, P. *Science* **1993**, *262*, 1554.
7. Ohba, M.; Maruono, N.; Okawa, H.; Enoki, T.; Latour, J. M. *J. Am. Chem. Soc.* **1994**, *116*, 11566.
8. Ohba, M.; Okawa, H.; Ito, T.; Ohto, A. *J. Chem. Soc., Chem. Commun.* **1995**, 1545.
9. Journaux, Y.; Kahn, O.; Zarembowitch, L.; Jaud, J. *J. Am. Chem. Soc.* **1983**, 105, 7585.
10. Pei, Y.; Journaux, Y.; Kahn, O. *Inorg. Chem.* **1989**, *28*, 100.
11. Ohba, M.; Tamaki, H.; Matsumoto, N.; Okawa, H.; Kida, S. *Chem. Lett.* **1991**, 1157.
12. Carlin, R. L. *Magnetochemistry*, Springer: Berlin, **1986**.

INDEXES

Author Index

Affiliation Index

Subject Index

N